Toward Sustainability: Bike-Sharing Systems Design, Simulation and Management

Toward Sustainability: Bike-Sharing Systems Design, Simulation and Management

Editors

Leonardo Caggiani

Rosalia Camporeale

MDPI • Basel • Beijing • Wuhan • Barcelona • Belgrade • Manchester • Tokyo • Cluj • Tianjin

Editors

Leonardo Caggiani
Department of Civil,
Environmental, Land, Building
Engineering and Chemistry
Polytechnic University of Bari
Bari
Italy

Rosalia Camporeale
Department of Technology and
Society – Division of Transport
and Roads
Lund University
Lund
Sweden

Editorial Office
MDPI
St. Alban-Anlage 66
4052 Basel, Switzerland

This is a reprint of articles from the Special Issue published online in the open access journal *Sustainability* (ISSN 2071-1050) (available at: www.mdpi.com/journal/sustainability/special_issues/bike_sharing_systems).

For citation purposes, cite each article independently as indicated on the article page online and as indicated below:

LastName, A.A.; LastName, B.B.; LastName, C.C. Article Title. *Journal Name* **Year**, *Volume Number*, Page Range.

ISBN 978-3-0365-1706-3 (Hbk)
ISBN 978-3-0365-1705-6 (PDF)

Contents

Contents

About the Editors

Leonardo Caggiani

Leonardo Caggiani is a research fellow at the Department of Civil, Environmental, Land, Building Engineering, and Chemistry of the Polytechnic University of Bari, Italy. He was awarded a Ph.D. in Road and Transportation Systems, Land and Technological Innovation at Polytechnic University of Bari, Italy, in 2009. He received the best Ph.D. thesis award from the Italian Society of Transportation Academics in 2012 and has been rewarded, together with co-authors, with the paper award at the mobil.TUM 2016 conference. He participated in an Italian research project (PRIN2007) on container ports competitiveness in the Mediterranean; he also took part in the Green Intermodal Freight Transport European project (GIFT) of the SEE program. His research interests concern quantitative methods for sustainable mobility, transportation networks design, and equity issues in transportation. Among others, bike-sharing systems are the subject of the most important studies he has carried out.

Rosalia Camporeale

Rosalia Camporeale is an Associate Senior Lecturer at the Department of Technology and Society, Division of Transport and Roads, Lund University, Sweden. She has expertise in transportation network design problems, sustainable transport, accessibility, and equity issues. During her Ph.D., granted at the Polytechnic University of Bari in March 2017, she developed a methodology to plan and design public transport routes to meet the needs of communities fostering equitable accessibility. Rosalia is currently involved in two JPI Urban Europe projects with a focus on competing and complementary mobility in urban contexts, and she is Project Manager of two national research projects: one dealing with accessibility analysis and distributional effects connected to Swedish railway investments, and the other one concerning the vulnerability of public transport in the COVID-19 era.

sustainability

MDPI

Editorial

Toward Sustainability: Bike-Sharing Systems Design, Simulation and Management

Leonardo Caggiani [1,*] and Rosalia Camporeale [2]

[1] Department of Civil, Environmental, Land, Building Engineering and Chemistry, Polytechnic University of Bari, Via Edoardo Orabona, 4, 70125 Bari, Italy

[2] Division of Transport and Roads, Department of Technology and Society, Lund University. P.O. Box 118, 22100 Lund, Sweden; rosalia.camporeale@tfh.lth.se

* Correspondence: leonardo.caggiani@poliba.it

Citation: Caggiani, L.; Camporeale, R. Toward Sustainability: Bike-Sharing Systems Design, Simulation and Management. *Sustainability* **2021**, *13*, 7519. https://doi.org/10.3390/su13147519

Received: 1 July 2021
Accepted: 3 July 2021
Published: 6 July 2021

1. Introduction

Bike-sharing systems (BSSs) are a mobility service of public bicycles available for shared use that is becoming increasingly popular in urban contexts. These shared systems provide city users with an alternative, low-carbon, and ecologically sustainable transportation mode (especially suited for short-distance trips) that can significantly reduce traffic congestion, air pollution and noise in city centers, and supports the growth of greener urban environments.

Different issues and challenges have been discussed in previous studies with regard to these systems [1]. Among them are BSS planning and design problems, especially concerning station locations, system simulation and operation problems, such as user demand forecasting and bicycle relocation [2,3]. In this framework, new possible solutions are constantly suggested, each one with its own strengths and weaknesses. Dockless systems (also known as free-floating BSSs) have started to become popular alongside station-based ones, both in big cities and smaller urban environments [4]. At the same time, together with regular bicycles, electric/pedal-assisted bicycles are also being used [5]: in the Vélib' BSS in Paris, for example, a mixed system with both traditional and electric bicycles has recently been implemented [6].

The goal of this Special Issue is to discuss new challenges in the simulation and management problems of both traditional and innovative BSSs, to ultimately encourage the competitiveness and attractiveness of BSSs and contribute to the further promotion of sustainable mobility. We have selected thirteen papers for publication in this Special Issue. Their contributions are summarized and discussed in the following section.

2. Synopsis of the Contributions

One of the common challenges facing all BSS operators is managing the practical problem of mismatch of bike supply and user BSS demand. To maintain the quality of service to a certain level, these systems need bicycle relocation operations to compensate for imbalances in the network.

Jia et al. (2021) (contribution 1) contribute in this sense by suggesting a new bike-sharing rebalancing problem that considers multi-energy mixed fleets and traffic restrictions (aspects mostly neglected in previous studies), using a mixed-integer programming model with the objective of minimizing the total rebalancing cost of the fleet. Their results and sensitivity analysis seem to confirm the efficacy of the algorithm to reduce the total cost associated with BSS rebalancing operations.

A different approach is proposed by Lahoorpoor et al. (2019) (contribution 2), with their bottom-up cluster-based model. They start from an investigation of spatial and temporal patterns of bike-sharing trips, aiming at discovering groups of correlated stations with an agglomerative clustering method. Intra-cluster and inter-cluster rebalancing levels are considered, and relocation tours are optimized using a single objective genetic algorithm that minimizes the tour length significantly, which ultimately corresponds to a direct cost to the operator and indirect cost to the sustainability of BSSs.

Another crucial issue not often discussed concerns the disorderly parking of free-floating shared bikes. In their study, Jiang et al. (2019) (contribution 3) try to collect, as comprehensively as possible, the causes of such parking behavior through a two-phase questionnaire survey followed by factor analysis. Their investigation, carried out in China (where the problem is particularly acute), aims at facilitating decision-making by governments and enterprises for reference.

Several studies have already attempted to explore the factors that may affect the willingness to use BSS: individual socio-demographic characteristics (gender, age, occupation, education level, monthly income, household bicycle ownership, etc.), individual travel patterns (trip mode, travel time, trip purpose, etc.), transportation infrastructure, land use and built environment characteristics, bike-sharing facilities, and environmental conditions. Different angles and perspectives are presented in the papers collected within this Special Issue.

For instance, the stated preference survey designed and conducted by Politis et al. (2020) (contribution 4) targets car and bus users as well as pedestrians. The results highlight a distinctive set of factors and patterns: the choice of preferred transport mode is most sensitive to travel time and cost of the competitive travel options. According to their findings from Thessaloniki, Greece, BSS seems to be a more attractive option for certain socio-demographic groups and seems to mainly attract bus users and pedestrians rather than car users.

When looking at the infrastructure, the provision of a connected bikeway network has been proven one of the main measures to motivate cycling, since it is directly connected to cyclists' safety. In this regard, Shui and Chan (2019) (contribution 5) propose a novel bikeway design problem that combines a genetic algorithm and an elimination heuristic, and that aims at covering all demand sources and minimizing the total travel time of all cyclists under budget constraints. Their model, tested in two Hong Kong new towns, is not only applicable to new system designs but can also capture the existence of built bikeways and bike stations for system expansion.

Wu and Chen (2019) (contribution 6) support improvement of the night visibility of cyclists by evaluating the differences between shared and private bikes with five types of visibility aids. Their goal is to help policymakers incorporate suitable visibility aids within bike-sharing programs, enhancing the overall traffic safety conditions.

It is also of great importance to understand the motivations and barriers underlying the usage of shared bicycles. The study by Xu et al. (2020) (contribution 7) focuses on free-floating BSSs, adopting an extended theory of planned behavior (TPB) to examine psychological determinants of intention and actual behavior of users. The results, based on an online survey in Beijing, show important implications for planners and lead to several suggestions proposed to support the policymaking of the system.

More specifically, Xiao and Wang (2020) (contribution 8) target as research object of their study the brand choice of bike-sharing in China (namely Hellobike, Mobike, and Ofo). Using a conditional Logic model calibrated on data from an online questionnaire survey, they explore the influence of socio-economic attributes of cyclists and their subjective evaluations, providing a basis for traffic management departments to quantitatively evaluate performances of bike-sharing companies, and assessing the distribution of the total volume among them.

Bardi et al. (2019) (contribution 9) focus on e-bike sharing programs for cruise tourists, an additional niche of operation for bike-sharing systems. They try to understand the major driving forces that lead to the development of these programs, and the major motivating factors for cruise tourists to participate in e-bike sharing services. An ordered probit model is specified to identify the relationship among the variables influencing e-bike sharing

usage and satisfaction of cruise tourists, and interesting interpretations are provided in terms of the relative importance of significant variables.

Most existing studies mainly discuss the relationship between BSSs and external environments, while studies from the perspective of the relationship between internal stations of BSSs are insufficient. Yao et al. (2019) (contribution 10) try to fill this gap with their research. They construct the public bicycle networks of different urban areas (based on real-time data of the Nanjing public BSS) using Gephi software. Using complex network theory and a geographic visualization method, they aim to analyze internal correlation characteristics of BSSs and better understand the station usage.

Considering the large spread of BSSs, it is crucial to gain a better comprehension of the differences between these systems, hence the search for strategies to classify and compare them. An interesting possible approach for clustering different bike-sharing systems around the world can be found in the article by Mátrai and Tóth (2020) (contribution 11). They have gathered data about existing BSSs, grouping them into four categories (public, private, mixed, and other) as the first step for further identification of their common features, which can help to find similar systems and identify problems and best practices in early stages.

Moreover, Caggiani et al. (2021) (contribution 12) suggest a method to evaluate the efficiency of BSSs based on data envelopment analysis (DEA) in order to assist the decisions regarding the performance evaluation of BSS stations. A pool of input and output variables supported by literature, reports, and BSS planning guides is considered, and application to the Malmöbybike system, in Sweden, shows how this approach can provide a reliable evaluation of BSS efficiency.

We conclude the synopsis of this Special Issue's contributions with the study by Nikitas (2019) (contribution 13), which has the ambition to reinvent the formula of long-term success for bike-sharing operations by developing policy and business lessons that will help policymakers and transport providers in establishing and managing these (and other micro-mobility) schemes more effectively. Their findings are supported by primary data from two survey-based studies in Sweden and Greece.

3. What the Future of BSS Holds

The future of bike-sharing systems is of course unknown, but some speculations are possible based on what has been observed, and what trends seem to be arising.

Technological innovations are definitely contributing to a considerable change in the way of using and owning all kinds of vehicles and having an impact on all transport systems. The idea of geofencing—that is, a virtual boundary around a predetermined area or building [7]—might represent a compromise between traditional station-based and free-floating BSSs, facilitating the benefits and alleviating the challenges associated with these systems [4,8,9]. Designated operating areas to pick up and drop off vehicles could help in overcoming some docked BSS limitations (i.e., insufficient racks or station malfunctions), retaining to a certain extent the parking flexibility provided by free-floating BSSs without hindering pedestrians and/or blocking cycle paths or traffic flows.

A larger differentiation among vehicles can be foreseen. Alongside traditional bike sharing, BSSs with alternative vehicles (mixed-fleet) can attract more users and help satisfy more necessities. One possible option is represented by BSSs using e-bikes, which are superior to conventional bicycles in the ability to traverse longer distances and reach higher speeds, and in greater ease of use, especially over hilly terrains [10]. Another option can be BSSs using traditional and cargo, or e-cargo, bikes. This type of bicycle has recently gained attention as a possible urban mode of transport, particularly for families with children, or to carry heavy shopping or goods [11,12].

The most critical key usage barrier to the future development of BSSs concerns the lack of adequate cycling infrastructure (e.g., bike lanes, cycle paths, parking racks) that, in turn, is directly related to better road safety for cyclists. Poor traffic safety and insufficient bike-friendly infrastructure are the main reasons that cause reluctance to use BSSs (contribution 13, [13,14]).

Also quite relevant is the need for strategic solutions and infrastructure investments that could help in reducing pollution on urban cycle paths. Cycling in downtown areas, especially during the commute, may expose cyclists to air pollutants harmful to human health in large quantities [15,16]. Moreover, because of their physical activity, cyclists often have much higher respiration rates than people who travel by car, and consequently inhale more air pollutants over the same time [17]. The choice of paths is very important to reduce cyclists' exposure to air pollution [18]. Longer cycling routes toward the preferred destination could sometimes significantly lower this exposure. For instance, a recent study done in Coimbra [19] has shown that a 6% increase in distance and time can reduce the exposure to particulate matter and carbon monoxide related to traffic emissions by almost one-third, without requiring any additional physical effort. Hence, it is essential to acquire proper knowledge of the parameters influencing air pollution and noise along cycling facilities to better inform the planning and design of urban bicycle networks [20].

Finally yet equally importantly, there is a need to resolve challenges related to the "new normal" after the COVID-19 pandemic. The role of sustainable transport has been, in a way, redefined, and although cycling per se seems to have had a positive surge [21], the shared use of equipment in BSSs may cause concerns. Micro-mobility systems (such as bike-sharing) can definitely provide a safe alternative transport mode, but it is important that operators expand their efforts in performing and communicating precautionary actions and policies to support community health, to maintain and promote BSSs' roles during the last stages of the pandemic and afterwards [22].

4. List of Contributions

1. Jia, Y., Zeng, W., Xing, Y., Yang, D., Li, J. The Bike-Sharing Rebalancing Problem Considering Multi-Energy Mixed Fleets and Traffic Restrictions. *Sustainability* **2021**, *13*, 270.
2. Lahoorpoor, B., Faroqi, H., Sadeghi-Niaraki, A., Choi S.-M. Spatial Cluster-Based Model for Static Rebalancing Bike Sharing Problem. *Sustainability* **2019**, *11*, 3205.
3. Jiang, Q., Ou, S.-J., Wei, W. Why Shared Bikes of Free-Floating Systems Were Parked Out of Order? A Preliminary Study based on Factor Analysis. *Sustainability* **2019**, *11*, 3287.
4. Politis, I., Fyrogenis, I., Papadopoulos, E., Nikolaidou, A., Verani, E. Shifting to Shared Wheels: Factors Affecting Dockless Bike-Sharing Choice for Short and Long Trips. *Sustainability* **2020**, *12*, 8205.
5. Shui, C.S., Chan, W.L. Optimization of a Bikeway Network with Selective Nodes. *Sustainability* **2019**, *11*, 6531.
6. Wu, C., Chen, D. The Difference in Night Visibility between Shared Bikes and Private Bikes during Night Cycling with Different Visibility Aids. *Sustainability* **2019**, *11*, 7035.
7. Xu, D., Bian, Y., Shu, S. Research on the Psychological Model of Free-floating Bike-Sharing Using Behavior: A Case Study of Beijing. *Sustainability* **2020**, *12*, 2977.
8. Xiao, G., Wang, Z. Empirical Study on Bikesharing Brand Selection in China in the Post-Sharing Era. *Sustainability* **2020**, *12*, 3125.
9. Bardi, A., Mantecchini, L., Grasso, D., Paganelli, F., Malandri, C. Flexible Mobile Hub for E-Bike Sharing and Cruise Tourism: A Case Study. *Sustainability* **2019**, *11*, 5462.
10. Yao, Y., Zhang, Y., Tian, L., Zhou, N., Li, Z., Wang, M. Analysis of Network Structure of Urban Bike-Sharing System: A Case Study Based on Real-Time Data of a Public Bicycle System. *Sustainability* **2019**, *11*, 5425.
11. Mátrai, T., Tóth, J. Cluster Analysis of Public Bike Sharing Systems for Categorization. *Sustainability* **2020**, *12*, 5501.
12. Caggiani, L., Camporeale, R., Hamidi, Z., Zhao, C. Evaluating the Efficiency of Bike-Sharing Stations with Data Envelopment Analysis. *Sustainability* **2021**, *13*, 881.
13. Nikitas, A. How to Save Bike-Sharing: An Evidence-Based Survival Toolkit for Policy-Makers and Mobility Providers. *Sustainability* **2019**, *11*, 3206.

Author Contributions: The authors' contribution is equal. All authors have read and agreed to the published version of the manuscript.

Funding: This research received no external funding.

Institutional Review Board Statement: Not applicable.

Informed Consent Statement: Not applicable.

Data Availability Statement: Not applicable.

Acknowledgments: We would like to thank all the authors who have contributed to this Special Issue for their professional work. We are grateful to the reviewers whose critical support and valuable comments have contributed to improving significantly the quality of the collection. Special thanks go to the editorial assistance office of MDPI for their support throughout the review and publication process of this Special Issue.

Conflicts of Interest: The authors declare no conflict of interest.

References

1. Ricci, M. Bike sharing: A review of evidence on impacts and processes of implementation and operation. *Res. Transp. Bus. Manag.* **2015**, *15*, 28–38. [CrossRef]
2. Caggiani, L.; Camporeale, R.; Ottomanelli, M.; Szeto, W.Y. A modeling framework for the dynamic management of free-floating bike-sharing systems. *Transp. Res. Part C Emerg. Technol.* **2018**, *87*, 159–182. [CrossRef]
3. Eren, E.; Uz, V.E. A review on bike-sharing: The factors affecting bike-sharing demand. *Sustain. Cities Soc.* **2020**, *54*, 101882. [CrossRef]
4. Chen, Z.; van Lierop, D.; Ettema, D. Dockless bike-sharing systems: What are the implications? *Transp. Rev.* **2020**, *40*, 333–353. [CrossRef]
5. Zhu, S. Optimal fleet deployment strategy: Model the effect of shared e-bikes on bike-sharing system. *J. Adv. Transp.* **2021**, *2021*, 6678637. [CrossRef]
6. Vélib'. Available online: https://en.wikipedia.org/wiki/V%C3%A9lib%27 (accessed on 26 June 2021).
7. AlFayad, F.S. Geofencing in the GCC and China: A Marketing Trend That's Not Going Away. *J. Mark. Consum. Res.* **2018**, *46*, 84–98.
8. Zhang, Y.; Lin, D.; Mi, Z. Electric fence planning for dockless bike-sharing services. *J. Clean. Prod.* **2019**, *206*, 383–393. [CrossRef]
9. Mahmoodian, V.; Zhang, Y.; Charkhgard, H. *Design of a Hybrid Rebalancing Strategy to Improve Level of Service of FreeFloating Bike Sharing Systems*; CTECH Final Report; eCommons: Cornell's Digital Repository-Cornell University Library: Ithaca, NY, USA, 2019.
10. Ghamami, M.; Shojaei, M. Introducing a Design Framework for a Multi-Modal Public Transportation System, Focusing on Mixed-Fleet Bike-Sharing Systems. *Transp. Res. Rec.* **2018**, *2672*, 103–115. [CrossRef]
11. Becker, S.; Rudolf, C. Exploring the potential of free cargo-bikesharing for sustainable mobility. *Gaia-Ecol. Perspect. Sci. Soc.* **2018**, *27*, 156–164. [CrossRef]
12. Hess, A.K.; Schubert, I. Functional perceptions, barriers, and demographics concerning e-cargo bike sharing in Switzerland. *Transp. Res. Part D Transp. Environ.* **2019**, *71*, 153–168. [CrossRef]
13. Ge, Y.; Qu, W.; Qi, H.; Cui, X.; Sun, X. Why people like using bikesharing: Factors influencing bikeshare use in a Chinese sample. *Transp. Res. Part D Transp. Environ.* **2020**, *87*, 102520. [CrossRef]
14. Podgórniak-Krzykacz, A.; Trippner-Hrabi, J. Motives and factors that determine city residents' use of public bicycles. The case of Lodz, Poland. *Case Stud. Transp. Policy* **2021**, *9*, 651–662. [CrossRef]
15. De Hartog, J.J.; Boogaard, H.; Nijland, H.; Hoek, G. Do the health benefits of cycling outweigh the risks? *Environ. Health Perspect.* **2010**, *118*, 1109–1116. [CrossRef] [PubMed]
16. Apparicio, P.; Carrier, M.; Gelb, J.; Séguin, A.M.; Kingham, S. Cyclists' exposure to air pollution and road traffic noise in central city neighbourhoods of Montreal. *J. Transp. Geogr.* **2016**, *57*, 63–69. [CrossRef]
17. Panis, L.I.; de Geus, B.; Vandenbulcke, G.; Willems, H.; Degraeuwe, B.; Bleux, N.; Mishra, V.; Thomas, I.; Meeusen, R. Exposure to particulate matter in traffic: A comparison of cyclists and car passengers. *Atmos. Environ.* **2010**, *44*, 2263–2270. [CrossRef]
18. Carreras, H.; Ehrnsperger, L.; Klemm, O.; Paas, B. Cyclists' exposure to air pollution: In situ evaluation with a cargo bike platform. *Environ. Monit. Assess.* **2020**, *192*, 470. [CrossRef]
19. Giménez-Gaydou, D.A.; dos Santos, A.C.; Mendes, G.; Frade, I.; Ribeiro, A.S. Energy consumption and pollutant exposure estimation for cyclist routes in urban areas. *Transp. Res. Part D Transp. Environ.* **2019**, *72*, 1–16. [CrossRef]
20. Minet, L.; Stokes, J.; Scott, J.; Xu, J.; Weichenthal, S.; Hatzopoulou, M. Should traffic-related air pollution and noise be considered when designing urban bicycle networks? *Transp. Res. Part D Transp. Environ.* **2018**, *65*, 736–749. [CrossRef]
21. Nikitas, A.; Tsigdinos, S.; Karolemeas, C.; Kourmpa, E.; Bakogiannis, E. Cycling in the Era of COVID-19: Lessons Learnt and Best Practice Policy Recommendations for a More Bike-Centric Future. *Sustainability* **2021**, *13*, 4620. [CrossRef]
22. Jobe, J.; Griffin, G.P. Bike share responses to COVID-19. *Transp. Res. Interdiscip. Perspect.* **2021**, *10*, 100353.

sustainability

MDPI

Article

The Bike-Sharing Rebalancing Problem Considering Multi-Energy Mixed Fleets and Traffic Restrictions

Yongji Jia *, Wang Zeng, Yanting Xing, Dong Yang and Jia Li *

Glorious Sun School of Business & Management, Donghua University, Shanghai 200051, China; zeng_wish@163.com (W.Z.); yt__xing@163.com (Y.X.); yangdong@dhu.edu.cn (D.Y.)
* Correspondence: yjjia@dhu.edu.cn (Y.J.); jia.li@mail.dhu.edu.cn (J.L.)

Abstract: Nowadays, as a low-carbon and sustainable transport mode bike-sharing systems are increasingly popular all over the world, as they can reduce road congestion and decrease greenhouse gas emissions. Aiming at the problem of the mismatch of bike supply and user demand, the operators have to transfer bikes from surplus stations to deficiency stations to redistribute them among stations by vehicles. In this paper, we consider a mixed fleet of electric vehicles and internal combustion vehicles as well as the traffic restrictions to the traditional vehicles in some metropolises. The mixed integer programming model is firstly established with the objective of minimizing the total rebalancing cost of the mixed fleet. Then, a simulated annealing algorithm enhanced with variable neighborhood structures is designed and applied to a set of randomly generated test instances. The computational results and sensitivity analysis indicate that the proposed algorithm can effectively reduce the total cost of rebalancing.

Keywords: sustainable transport; bike-sharing rebalancing problem; multi-energy mixed fleets; traffic restrictions; simulated annealing; variable neighborhood structures

Citation: Jia, Y.; Zeng, W.; Xing, Y.; Yang, D.; Li, J. The Bike-Sharing Rebalancing Problem Considering Multi-Energy Mixed Fleets and Traffic Restrictions. *Sustainability* **2021**, *13*, 270. https://doi.org/10.3390/su13010270

Received: 16 December 2020
Accepted: 27 December 2020
Published: 30 December 2020

1. Introduction

Nowadays, bike-sharing systems (BSSs), as a low-carbon and sustainable transport mode, are becoming more and more popular across the global, as they can reduce road congestion and decrease greenhouse gas emissions caused by motorized transportation [1,2]. The first BSS was introduced in Amsterdam in 1965 [3] and there are now more than 1500 active BSSs [4] and this number is growing at an increasing rate [5,6]. Recently, some scholars begin to pay attention to the practical problem of mismatch of bike supply and user demands in the BSS, which is a common challenge to all BSS operators [7]. Some operators are trying to meet user demand by placing bikes in cities in large numbers, but this creates congestion on city streets and is not sustainable. In China, the government has introduced policies to restrict operators from placing too many bikes. Thus, these operators have to transfer bikes from the surplus stations to the deficiency stations by means of vehicles so that the BSS can be rebalanced. This problem is known as the Bike-sharing Rebalancing Problem (BRP) [8].

Originally, the BSS operators employ internal combustion vehicles (ICVs) to do the rebalancing operations. However, the development of electric vehicles (EVs) has made great progress over the past few years and many BSS operators are discovering that EVs bring more advantages than ICVs beyond environmental benefits, such as less maintenance, less noise pollution and reduced driving cost [9]. The Chinese government has strongly supported companies to develop sustainable EVs in recent years, providing many policy bonuses for the purchase and use of EVs. Furthermore, more and more cities have implemented traffic restrictions on ICVs to reduce carbon emissions and alleviate traffic congestions [10]. In these cities, ICVs cannot provide services for certain BSS stations in the restricted areas, while EVs can. Therefore, in many practical situations, a multi-energy mixed fleet, consisting of both EVs and ICVs, are used to perform the rebalancing operations.

In this study, the original version of the BRP is extended by assuming the fleet of vehicles to be multi-energy types. The BRP considering Multi-energy Mixed Fleets and Traffic Restrictions (BRP-MMFTR) is considered to be an NP-hard problem, for it originates from the classical Vehicle Routing Problem (VRP). We formulate the BRP-MMFTR as a mixed integer programming model based on one commodity pickup and delivery problem [11], with the objective of minimizing the total rebalancing cost of the mixed fleet. Our model has more complicate constraints than general BRP, such as battery capacity limits for EVs and traffic restrictions for ICVs. It is therefore more complex and more difficult to solve. As the size of bike-sharing stations increases, the calculation time will increase exponentially. Hence, the real-life BRP-MMFTR instances cannot be solved exactly within acceptable computation time.

To handle BRP-MMFTR in a runtime acceptable to BSS operators, we propose the Simulated Annealing algorithm with Variable Neighborhood structures (SAVN) to obtain the optimal solution. To avoid getting into the trap of local optima and enhance the exploratory capability, several variable neighborhood structures are incorporated in our algorithm.

The contribution of this paper lies on the following:

- To introduce a new and practical bike-sharing rebalancing problem considering multi-energy mixed fleets and traffic restrictions;
- To present a mixed integer programming model to formulate the problem above;
- To propose a simulated annealing algorithm with several variable neighborhood structures to solve it.

The remainder of this paper is organized as follows. Section 2 presents the literature review on the related problems. Section 3 formally presents a mixed integer programming formulation for BRP-MMFTR. Section 4 describes the procedure and key components of SAVN. Section 5 contains the computational experiments and a sensitivity analysis. Finally, Section 6 presents some concluding remarks about this work.

2. Literature Review

The BRP has been receiving considerable attention from the literatures during the past decade. Most of studies have focused on optimization models that maximize profits or minimize costs. Dell' Amico et al. provide four mixed integer programming models with the objective of minimizing total costs, where a fleet of capacitated vehicles is employed to relocate the bikes [12]. Erdoğan et al. and Cruz et al. solve the same model proposed by Dell' Amico with a single vehicle, where multi-visit strategy is considered at each station [11,13]. Duan et al. focus on multi-vehicle BRP with the objective of minimizing the total travel distance, and a greedy algorithm is proposed [14]. Casazza et al. and Bulhões et al. incorporate time into the constraints such that each route does not exceed service time limitation [5,15].

All the problems above need to fully meet the rebalancing demands in the BSS, which is difficult to achieve in reality. Hence, some researchers are trying to relax these constraints, and setting the objective to maximize the satisfied rebalancing demand. Papazek et al. set the primary goal to minimize the absolute deviation between target and final fill levels for all stations [16]. Gaspero et al. solve the problem with the aim of minimizing the weighted sum of the total travel time and the total absolute deviation from the target number of bikes [17], while Raviv and Kolka take it as a penalty cost [18]. Faulty bikes are considered by Wang and Szeto with the objective of minimizing the total carbon emission of all vehicles [19]. Usama et al. also consider replacing faulty bikes in the system with the following objectives: User dissatisfaction and vehicle routing costs [20].

Actually, BRP is a large-scale uncertainty problem, due to the existence of uncertainty of user demand [21]. Hence, some scholars have turned their attention to the BRP with uncertain user demand. There are roughly three types of methods for dealing with these uncertainty problems. First, prediction is a common method to resolve uncertainty. Alvarez-Valdes et al. estimate the unsatisfied demand to guide rebalancing algorithms [3]. Zhang et al. propose a dynamic bike rebalancing method that considers both bike rebalancing, vehicle routing and the prediction of inventory level and user arrivals [22]. Second, some scholars divide the uncertainty into multiple independent stochastic scenarios for research. Dell'Amico et al. develop the stochastic programming model with the objective of minimizing the travel cost and the penalty costs for unfulfilled demands [23]. Maggioni et al. propose the two-stage and multi-stage stochastic optimization models to determine the optimal number of bikes to assign to each station [24]. Yuan et al. present a mixed integer programming model with the objective of minimizing the daily costs including the capital cost and the expected operating cost [25]. Third, some studies deal uncertainty directly with dynamic factors. Legros focuses on dynamic rebalancing strategy in the BSS with the objective of minimizing the rate of arrival of unsatisfied users, and solve it by a Markov decision process approach [26]. You develops a constrained mathematical model to deal with a multi-vehicle BRP, aiming to develop dynamic decisions to minimize the sum of the travel costs and unmet costs under service level constraints over a planning horizon [27].

However, none of these literature works considers multi-energy mixed fleets. Goncalves et al. consider the vehicle routing problem with pickup and delivery, and propose a heterogeneous vehicle routing model based on ICVs and EVs [28]. Sassi et al. formulate the heterogeneous electric vehicle routing problem with time dependent charging costs, in which a set of customers have to be served by a mixed fleet of vehicles composed of ICVs and EVs [29]. A mixed fleet of ICVs and EVs are also considered in Goeke and Schneider [30]. The authors formulate the electric vehicle routing problem with time windows and mixed fleet, which is solved through adaptive large neighborhood search algorithm. Charging times vary according to the battery level when the EV arrives at the charging station, and charging is always done up to maximum battery capacity. Macrina et al. present a new variant of the green vehicle routing problem with time windows and propose an iterative local search heuristic to optimize the routing of a mixed vehicle fleet, composed of EVs and ICVs [31].

In summary, there are abundant studies about BSS, but no model considers multi-energy mixed fleets and traffic restrictions. For real-life BSSs in big cities, it is more realistic and more useful to consider a fleet of EVs and ICVs. Therefore, in this paper, we focus on a new and practical variant of BRP under the background of multi-energy mixed fleets (composed of EVs and ICVs) and traffic restrictions to ICVs. Our contribution to the development and application of the BSS model is twofold. On the one hand, we formulate BRP-MMFTR as a mixed integer programming model with the objective of minimizing the total rebalancing cost. On the other hand, because this model is too complex to be solved accurately, we develop SAVN to solve it. Our work will expand the existing knowledge on modeling BSS.

3. Model Formulation

3.1. Model Description and Notations

As shown in Figure 1, BRP-MMFTR studied in this paper can be described as follows: There are two types of vehicles available, namely EVs and ICVs. Vehicles start from the depot with no inventory of bikes, then serve all stations by sequentially loading excess bikes or unloading insufficient bikes, and finally return to the depot after serving all stations. Each station can be served by a vehicle only once. During the rebalancing process, the number of bikes carried by a vehicle cannot exceed its maximal capacity. EVs need to visit the charging stations during the service process. To reduce the complexity of our model, the EV battery must be fully charged at any charging station and the depot. In addition, ICVs are not allowed to enter the traffic restricted area. The objective of BRP-MMFTR is to minimize the total rebalancing cost including the vehicles' fixed costs and traveling energy costs, recharging costs for EVs, and carbon emissions for ICVs.

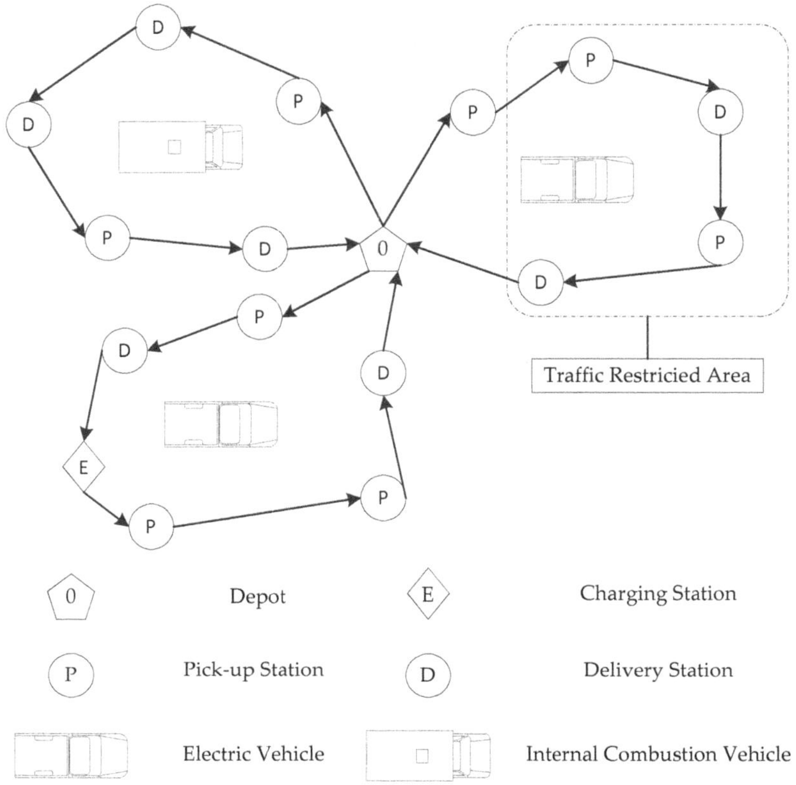

Figure 1. A schematic example of BRP-MMFTR.

The assumptions related to BRP-MMFTR are given as follows:

- All stations' demands are known and fixed;
- The traveling energy costs and carbon emissions are only related to the travel distances;
- The loading and unloading time are neglected;
- The residual charge level of the EV grows linearly with the charging time;
- All homogeneous vehicles run at a uniform speed;
- The number of vehicles in the depot and the number of charging piles in each charging stations are sufficient;

The notations of sets, parameters, and decision variables used in our model are listed in Table 1.

Table 1. Description of notations.

Sets	Description
0	The depot.
N	The set of stations, $N = \{1, 2, \ldots, n\}$.
A	The set of arcs, $A = \{(i, j)/i, j \in N \cup \{0\}, i \neq j\}$.
E	The set of charging stations, $E = \{1, 2, \ldots, e\}$.
K_1	The set of ICVs, $K_1 = \{1, 2, \ldots, k_1\}$.
K_2	The set of EVs, $K_2 = \{k_1 + 1, k_1 + 2, \ldots, k_2\}$.
K	The set of vehicles, $K = K_1 \cup K_2$.
S	The set of auxiliary variables avoiding sub-loops, $S \subseteq N$.
D	The set of stations in restricted area, $D \subseteq N$.
Parameters	**Description**
Q^k	The maximal capacity of vehicle k.
F^k	The fixed cost of vehicle k.
C_{ij}^k	The unit energy cost from vertex i to vertex j of vehicle k.
d_{ij}	The distance from vertex i to vertex j.
R	The maximum battery level of EVs.
u_{ik}^1	The residual charge of EV k when reaching vertex i.
u_{ik}^2	The residual charge of EV k when leaving vertex i.
λ	The unit charging rate of EV.
r	The unit battery energy consumption of EV.
μ	The coefficient of safe-residual-charge level of EV.
t_e^k	The charging time of EV k at charging station e.
C_w	The unit cost of charging time.
L	The unit carbon emission of ICV.
C_L	The unit carbon trade cost.
G_i	The bike demand at station i.
g_{ij}^k	The number of bikes loaded when vehicle k is on arc (i, j).
Decision variables	**Description**
x_{ij}^k	Binary variable, 1 if vehicle k traverses arc (i, j), and 0, otherwise.
y_e^k	Binary variable, 1 if vehicle k is charged at charging station e, and 0, otherwise.

3.2. The Total Rebalancing Cost

The total rebalancing cost of BRP-MMFTR include four parts: Vehicles' fixed costs, vehicles' traveling energy costs, EV recharging costs, and ICVs' carbon emission costs. The vehicles' fixed costs and traveling energy costs are the basic components of objectives in most traditional VRP models. In the objective of our model, the recharging costs reflect the additional charging time during the operation of EVs, while the carbon emission costs are related to the greenhouse gas emission of ICVs.

(1) Fixed costs (C_1)

The fixed costs of vehicles are the costs of using vehicles for rebalancing operations. They are different for EVs and ICVs, and the equation of vehicles' fixed costs in our model is shown as follows:

$$C_1 = \sum_{k \in K} \sum_{j \in N,\, j \neq 0} x_{0j}^k \cdot F^k \tag{1}$$

(2) Traveling energy costs (C_2)

The traveling energy costs are composed of two types of costs, that is, fuel costs of ICVs and electricity costs of EVs. Both of them are only associated with the travel distance. And the equation of vehicles' traveling energy costs is shown as follows:

$$C_2 = \sum_{k \in K} \sum_{(i,j) \in A, i \neq j} C_{ij}^k \cdot d_{ij} \cdot x_{ij}^k \tag{2}$$

(3) Recharging costs (C_3)

The travel distance of an EV is limited by its maximum battery level. Hence, it needs to be charged when its residual charge level lower than the safe residual charge level. The recharging cost is only related to the charging time. The equation of recharging costs is shown as follows:

$$C_3 = \sum_{k\in K_2}\sum_{e\in E} C_w \cdot t_e^k \cdot y_e^k \tag{3}$$

(4) Carbon emission costs (C_4)

In the process of rebalancing, a large amount of CO_2 is generated by ICVs from their fuel consumptions, resulting in greenhouse effect. By reducing the costs of carbon emissions, the total rebalancing cost is reduced. The equation of carbon emission costs is shown as follows:

$$C_4 = \sum_{k\in K_1}\sum_{(i,j)\in A, i\neq j} L \cdot C_L \cdot d_{ij} \cdot x_{ij}^k \tag{4}$$

3.3. Model Establishment

The mixed integer programming model for BRP-MMFTR can be written as follows:

$$\begin{aligned} min\ Z = \ & C_1 + C_2 + C_3 + C_4 = \sum_{k\in K}\sum_{j\in N, j\neq 0} x_{0j}^k \cdot F^k + \sum_{k\in K}\sum_{(i,j)\in A, i\neq j} C_{ij}^k \cdot d_{ij} \cdot x_{ij}^k \\ & + \sum_{k\in K_2}\sum_{e\in E} C_w \cdot t_e^k \cdot y_e^k + \sum_{k\in K_1}\sum_{(i,j)\in A, i\neq j} L \cdot C_L \cdot d_{ij} \cdot x_{ij}^k \end{aligned} \tag{5}$$

Subject to:

$$\sum_{j\in N, j\neq 0} x_{0j}^k = \sum_{i\in N, i\neq 0} x_{i0}^k = 1 \ \forall k \in K \tag{6}$$

$$\sum_{k\in K}\sum_{j\in N, j\neq i} x_{ij}^k = 1 \ \forall i \in N \tag{7}$$

$$\sum_{j\in N, j\neq i} x_{ij}^k - \sum_{j\in N, j\neq i} x_{ji}^k = 0 \ \forall i \in N, \forall k \in K \tag{8}$$

$$\sum_{i\in S}\sum_{j\in S} x_{ij}^k \leq |S| - 1 \ \forall S \subseteq N, |S| \geq 2 \tag{9}$$

$$\sum_{j\in D, j\neq i} x_{ij}^k = 0 \ \forall i \in N, \forall j \in D, \forall k \in K_1 \tag{10}$$

$$max\{0, G_i, G_j\} \cdot x_{ij}^k \leq g_{ij}^k \leq min\{Q^k, Q^k + G_i, Q^k - G_j\} \cdot x_{ij}^k \ \forall i,j \in N,\ i \neq j \tag{11}$$

$$\sum_{k\in K}\sum_{j\in N, j\neq i} g_{ij}^k - \sum_{k\in K}\sum_{j\in N, j\neq i} g_{ji}^k = G_i \ \forall i \in N \tag{12}$$

$$0 \leq g_{ij}^k \leq Q^k \cdot x_{ij}^k \quad \forall i,j \in N,\ i \neq j, \forall k \in K \tag{13}$$

$$t_e^k = y_e^k \cdot \left[\left(R - u_{ek}^1\right)/\lambda\right] \quad \forall k \in K_2, \forall e \in E \tag{14}$$

$$0 \leq u_{jk}^1 \leq u_{ik}^2 - r \cdot d_{ij} \cdot x_{ij}^k + R \cdot \left(1 - x_{ij}^k\right) \quad \forall i,j \in N,\ i \neq j, \forall k \in K_2 \tag{15}$$

$$u_{ik}^1 \geq \mu \cdot R \quad \forall i \in N, \forall k \in K_2 \tag{16}$$

$$u_{0k}^2 = R \quad \forall k \in K_2 \tag{17}$$

$$u_{ek}^2 = y_e^k \cdot R \quad \forall k \in K_2, \forall e \in E \tag{18}$$

$$u_{ik}^1 = u_{ik}^2 \quad \forall i \in N, \forall k \in K_2 \tag{19}$$

$$x_{ij}^k,\ y_e^k = [0,1] \quad \forall i \in N, \forall j \in N, \forall k \in K, \forall e \in E \tag{20}$$

Objective function (5) minimizes the total rebalancing cost, including vehicles' fixed costs, vehicles' traveling energy costs, EVs' recharging costs, and ICVs' carbon emission costs. Constraints (6) ensure that each vehicle starts at the depot and returns to the depot at the end of its route. Constraints (7) guarantee that each station is served exactly once. Constraints (8) refer to the usual flow conservation. Constraints (9) can avoid subtours and thus guarantee route-connectivity. Constraints (10) indicate that ICVs are restricted from entering the restricted area. Constraints (11) give the upper and lower bounds of the number of bikes loaded by vehicles. Constraints (12) and (13) ensure that the vehicle's maximum capacity is not exceeded. Constraints (14) and (15) are the EVs' charging functions and power consumption functions, respectively. Constraints (16) indicate the safe residual charge constraints of EVs. Constraints (17) and (18) show that EVs are fully charged when leaving the depot or a charging station. Constraints (19) guarantee the residual charges of EVs are the same after serving a BSS station. Constraints (20) are the binary decision variables.

4. Simulated Annealing Algorithm with Variable Neighborhood Structures

Simulated Annealing (SA) is a heuristic method to solve various NP-hard optimization problems. It can expand the exploration capability by accepting worsen solutions with some probability. This has benefit to reduce the probability of getting trapped in local optima. In order to improve the search efficiency and get solutions with higher quality, we introduce the variable neighborhood structures [32] into the framework of SA, and propose SAVN to deal with the real-life BRP-MMFTR instances. The procedure and key components of our algorithm are described in detail below.

4.1. The Procedure of Our Algorithm

Given an initial solution, SAVN starts from the initial temperature T_0. During the search process, the algorithm randomly selects a neighborhood structure to transform the current solution S into a randomly generated feasible neighbor S'. Note that the cost of a feasible solution S, namely $Z(S)$, is evaluated with the Equation (5). If the cost of S' is less than the cost of S, S' will definitely be accepted. Otherwise, the acceptance probability of S' is $p = exp\left(-\frac{Z(S')-Z(S)}{T}\right)$, where T is the current temperature. For each T, this process is performed *Len* times. Then, T decreases by multiplying the cooling rate α.

Repeating the above processes until the stop criterion is met, that is, the unimproved number of the best solution so far reaches the pre-specified MaxUN. Furthermore, when T is less than 0.01, T will increase to help the algorithm escape from the local optima [33]. We double T_b first (but no more than T_0) and then set $T = T_b$. The pseudocode of SAVN is shown in Algorithm 1.

4.2. Initial Solution Generation

Considering the particularity of BRP-MMFTR, we firstly classify all BSS stations into two types by region: stations in restricted areas and stations in non-restricted areas. Then a three-step algorithm is proposed to generate the high-quality initial solution.

Step 1: For stations in the restricted areas, EVs must be employed. We use the insert algorithm to construct the initial routes for these stations. First, the station with the largest distance from the depot is selected as the first customer of an EV. Then, if the capacity constraints are satisfied, the other stations are inserted into the current EV route in turn. Otherwise, a new EV route will be generated and serve the remainder stations. Repeat this insertion process until all stations in the restricted areas are serviced.

Step 2: For the stations in non-restricted areas, we first judge whether a station can be inserted into the existing EV routes. For the stations that can be inserted, we insert them to the positions with the lowest incremental cost in the existing EV routes. For the other stations, we employ the same insert algorithm to generate some new ICV routes.

Step 3: Charging stations are allocated to EV routes. If the residual charge level of an EV is enough, this step can be skipped. Otherwise, when the residual charge level of an EV is lower than the designated safe-residual-charge level, the nearest charging station will be inserted into the EV route to increase its mileage.

Algorithm 1 The procedure of SAVN

Parameters: S^* (the best solution so far), T_0 (the initial temperature), T_b (the temperature with which S^* is found), MaxUN (the maximum unimproved number of the best solution), *Len* (maximum number of iterations at the current temperature), N_k (the neighborhood structures, $k = 1, \cdots, k_{max}$), α (the cooling rate)
Construct S as an initial solution
$S^* = S, T = T_0, T_b = T_0$, count = 0
while count < MaxUN
  for $n = 1$ to *Len*
    Select randomly a neighborhood structure N_k
    Generate a feasible solution S' from S with N_k
    if $Z(S') < Z(S)$
      $S = S'$
    else
      Set $S = S'$ with probability $p = exp\left(-\frac{Z(S')-Z(S)}{T}\right)$
    end
    if $Z(S) < Z(S^*)$
      $S^* = S, T_b = T$, count = 0
    else
  count = count + 1
  end
  end
  $T = \alpha * T$
  if $T < 0.01$
  $T_b = \min\{2 * T_b, T_0\}, T = T_b$
  end
end
Return S^*

4.3. Neighborhood Structures

Four neighborhood structures are employed in our algorithm. In the procedure of SAVN, when a new solution is needed, one of these four structures is randomly selected with equal probability. As shown in Figure 2, all structures are described by an instance with 5 stations.

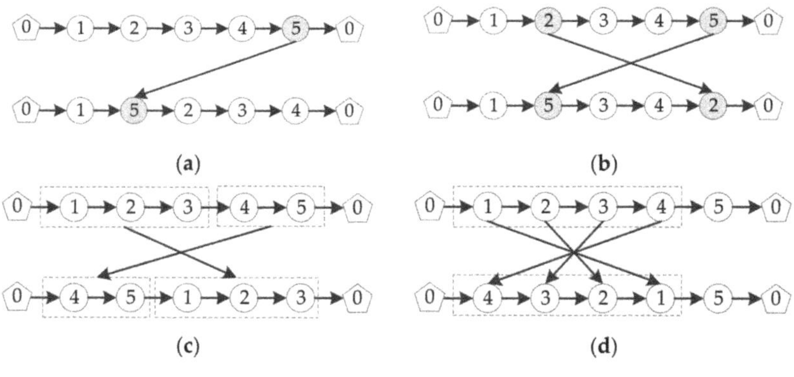

Figure 2. The neighborhood structures.

Neighborhood structure (a): A randomly selected station is relocated to another position in the current solution.

Neighborhood structure (b): Two randomly selected stations exchange their positions.

Neighborhood structure (c): Two randomly selected segments exchange their positions, and their lengths are at most 3.

Neighborhood structure (d): A pair of stations is randomly selected and all the stations between them, including themselves, are reversed.

5. Computational Results

5.1. Parameters and Experimental Data

We have implemented SAVN using MATLAB R2018a, and all computational experiments have been carried on an Intel Core i5-8259U with 2.3 GHz CPU and 8 GB RAM running the macOS Catalina operating system.

The test instances of BRP-MMFTR are generated randomly. The BSS stations' coordinates are randomly generated in the range of abscissa [0,200] and ordinate [0,150]. The coordinate range of the traffic restricted area is the abscissa [80,160] and the ordinate [30,60]. The rebalancing demand of each BSS stations is randomly generated between [−20,20]. The depot coordinate is (100,75).

Table 2 shows an example instance with one depot, 18 BSS stations and five charging stations. And Table 3 displays the parameters of EVs and ICVs. The other parameters used in BRP-MMFTR instances are as follows: R = 75 kW·h, r = 0.5 kW·h/km, μ = 0.3, λ = 0.5 kW·h/min, C_w = 0.4 CNY/min, C_L = 0.06 CNY/kg [34], and L = 5 kg/km. Through preliminary tests, the parameters of SAVN are set as follows: T_0 = 50, α = 0.96, MaxUN = 50, and $Len = n^2$ (where, n is the number of BSS stations).

5.2. Comparisons of Computational Results

To verify the effectiveness of our SAVN, we compare it with SA and VNS (Variable Neighborhood Search) on the instance depicted above. Note that both SA and VNS use the same parameters of our SAVN. All the three algorithms were executed 10 times to weaken the randomness of the heuristic algorithm.

Figure 3 illustrates the vehicle routes obtained by SA, VNS and SAVN, respectively. The red lines represent the route of the ICV, while the blue lines represent the route of the EV. It can be observed that the solution obtained by SAVN has fewer detours, which make the vehicle's travel distance smaller than those obtained by SA and VNS. Furthermore, its traveling energy costs, recharging costs and carbon emission costs can be also decreased.

Table 2. Information of the depot, BSS stations and charging stations.

	X-Axis	Y-Axis	Demand	Index
Depot	100	75	0	0
BSS stations	86	143	−6	1
	75	35	3	2
	86	57	6	3
	99	36	11	4
	132	25	6	5
	138	44	−7	6
	108	18	−9	7
	100	98	−15	8
	172	60	6	9
	123	55	3	10
	110	128	16	11
	34	35	8	12
	161	112	−14	13
	140	103	7	14
	191	71	−12	15
	76	101	7	16
	48	118	5	17
	19	65	−4	18
Charging stations	30	66		C1
	48	35		C2
	126	73		C3
	52	101		C4
	145	27		C5

Table 3. Parameters of EVs and ICVs.

Vehicle Type	Maximum Mileage/km	Capacity	Fixed Cost/CNY	Traveling Energy Cost/(CNY·km^{-1})
EV	150	15	150	0.8
ICV	500	25	200	1.5

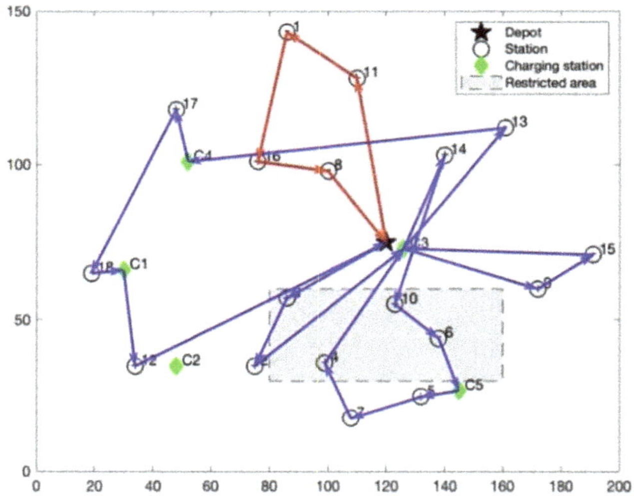

(**a**)

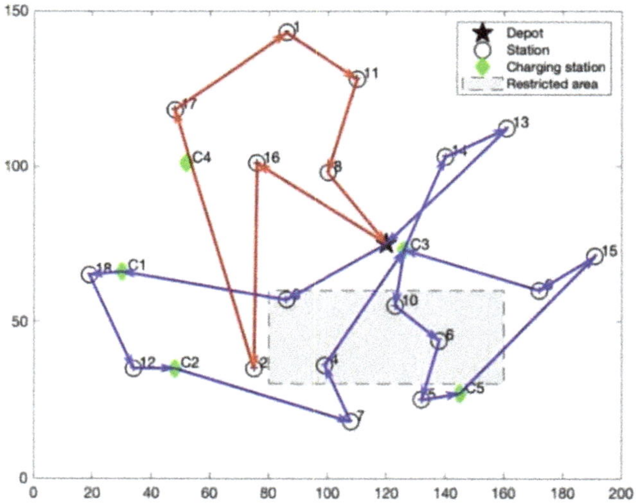

(**b**)

Figure 3. *Cont.*

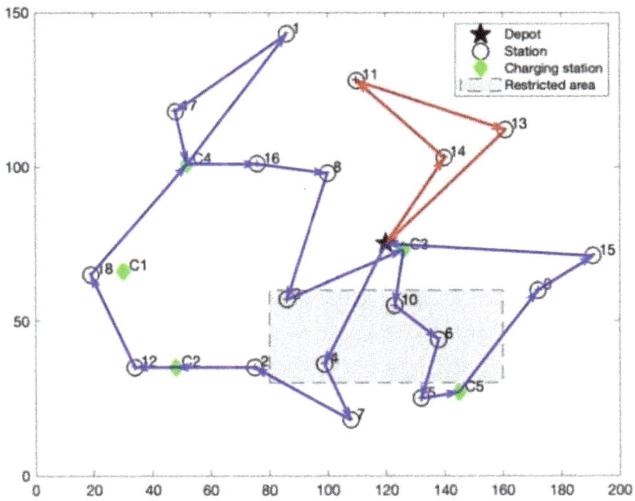

(c)

Figure 3. Optimal vehicle routes obtained by three algorithms. (**a**) SA (**b**) VNS (**c**) SAVN.

Figure 4 displays the comparison results of these algorithms. The X-coordinate is the number of iterations and the Y-coordinate is the total rebalancing cost. It can be seen that the final solution obtained by SAVN is better than SA and VNS, although in the early stage of the search process, SAVN may be inferior to VNS. This is due to the better optimization performance of SAVN, which can better escape from the trap of local optima. Hence, our SAVN can be regarded as an effective algorithm to solve BRP-MMFTR.

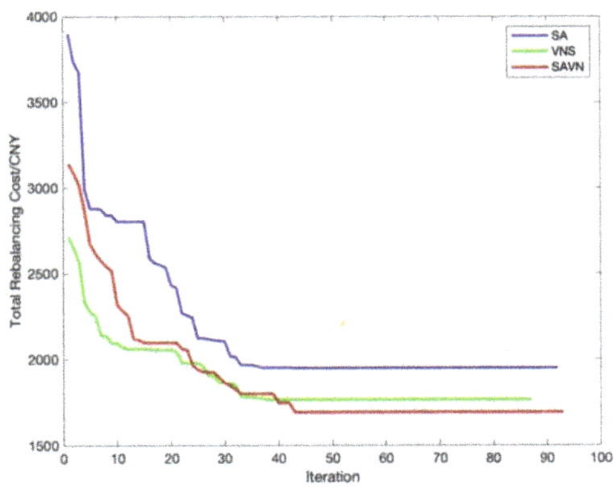

Figure 4. Comparison of total rebalancing cost.

Table 4 lists the components of the total rebalancing cost in the BRP-MMFTR optimal solution obtained by SA, VNS, and SAVN. Obviously, for these three algorithms, SAVN is the best and SA is the worst for the total rebalancing cost. Although the electricity costs of EV obtained by SAVN is slightly worse than that of VNS, the fuel cost of ICV has been drastically reduced. In this way, SAVN's traveling energy cost is less than VNS, because it is the sum of the electricity cost of EV and the fuel cost of ICV. Similarly, although SAVN's recharging costs is higher than VNS for its EV has a longer travel distance, its carbon emission costs is lower. For practical BSS operators, SAVN's solution is significantly better than VNS, because it employs EVs to perform more rebalancing operations. As we know, EVs are environmentally friendly and they will replace ICVs in the near future.

Table 4. Components of the total rebalancing cost in the optimal solution.

Costs/CNY	SA	VNS	SAVN
Fixed cost	350	350	350
Electricity cost of EV[1]	691	478	546
Fuel cost of ICV[2]	281	437	227
Recharging cost	571	414	524
Carbon emission cost	56	87	45
Total rebalancing cost	1949	1766	1692

Traveling energy cost is the sum of Electricity cost of EV[1] and Fuel cost of ICV[2].

5.3. Computational Results of More Instances

To further verify the effectiveness and universality of SAVN, we tested the three algorithms in further instances. Theywere evaluated by nine test instances (three types, three instances per type), displayed in Table 5. The three types are small-size (n = 20), medium-size (n = 50) and large-size (n = 100). The instances are also generated randomly according to the method introduced in Section 5.1.

Table 5. Nine instances of three types.

Instances	Small-Size			Medium-Size			Large-Size		
	1	2	3	4	5	6	7	8	9
n	20	20	20	50	50	50	100	100	100
Depot					(100,75)				
Abscissa range					[0,200]				
Ordinate range					[0,150]				
Demand range					[−20,20]				

To provide reliable statistics, each algorithm is executed 10 times for each instance. And the average CPU time (t/s), the average total rebalancing cost (TRC) and the standard deviation (Sd) are displayed in Table 6. The best values are marked with bold fonts.

Table 6. Comparative results on nine instances.

Instances	Stations	SA			VNS			SAVN		
		t/s	TRC	Sd	t/s	TRC	Sd	t/s	TRC	Sd
1	20	6.8	3071	29	8	2721	41	**5.5**	**2669**	**13**
2	20	**8.4**	2992	18	9.6	2917	33	9.4	**2848**	**6**
3	20	**4.4**	2402	30	7.4	2301	38	6.7	**2162**	**16**
4	50	**44.6**	5625	66	89.5	5220	58	115.6	**5093**	**32**
5	50	**39.3**	5452	58	88.3	4861	35	97.8	**4720**	**31**
6	50	**32.4**	6684	109	90.6	6214	60	66.4	**6045**	**39**
7	100	**129.2**	10733	144	508	9532	154	378.6	**9072**	**63**
8	100	**185.6**	10361	111	540.4	9510	80	394	**9299**	**56**
9	100	**174.3**	9920	105	448.5	8177	72	572.9	**8032**	**13**
Small	20	**6.5**	2821.7	25.6	8.3	2646.3	37.3	7.2	**2559.7**	**11.7**
Medium	50	**38.8**	5920.3	77.7	89.5	5431.7	51	93.3	**5286**	**34**
Large	100	**163**	10351.3	120	499	9073	102	448.5	**8801**	**44**

The best values are marked with bold.

From Table 6, we can draw some conclusions as follows: For the average CPU time, SA has obvious advantages, but its solution quality is the worst. In addition, the calculation time of SAVN and VNS is difficult to distinguish the pros and cons. For the average total rebalancing cost, the solution obtained by SAVN can reduce 9.2% compared to SA or 3.4% compared to VNS in the small-size instances, 10.7% or 2.7% in the medium-size instances, and 15% or 3% in the large-size instances. With the number of stations increases, the solution quality of SAVN is getting better and better. Hence, from the perspective of solution quality and CPU time, SAVN can obtain a better solution than SA and VNS.

Furthermore, the standard deviation can reflect the stability of the algorithm. And the smaller the standard deviation is, the better the stability of the algorithm is. From Table 6, it is observed that the standard deviation of SAVN is the smallest in all the instances. Therefore, SAVN is more stable than the other two algorithms.

5.4. Discussion on the Value of μ

Parameter μ (the coefficient of safe-residual-charge level of EV) directly affects the charging time and then the recharging costs and the total travel distances. Hence, it is important to set a suitable value for μ. If the value of μ is set too low, there will be a probability of failing to drive to the nearest charging station. If the value of μ is set too high, the EV may frequently go to the charging station, which will cause a waste of energy and charging time. In this section, the sensitivity analysis of μ is analyzed using the 18-station instance in Section 5.2. The different values of μ are from 0.15 to 0.5 in increment of 0.05.

The SAVN is run 10 times for each μ and the trends of change with the increase of μ are illustrated in Figure 5. Obviously, the total rebalancing cost shows a clear trend of decreasing and then increasing. $\mu = 0.3$ is the point with the lowest total rebalancing cost. In summary, parameter μ plays an important role in BRP-MMFTR and its value should be reasonably determined according to the layout of charging station system. If the quantity of charging stations is sufficient, the value of μ can be appropriately set lower. Otherwise, it should be set higher. Therefore, setting an appropriate value for μ is helpful to reduce the total rebalancing cost.

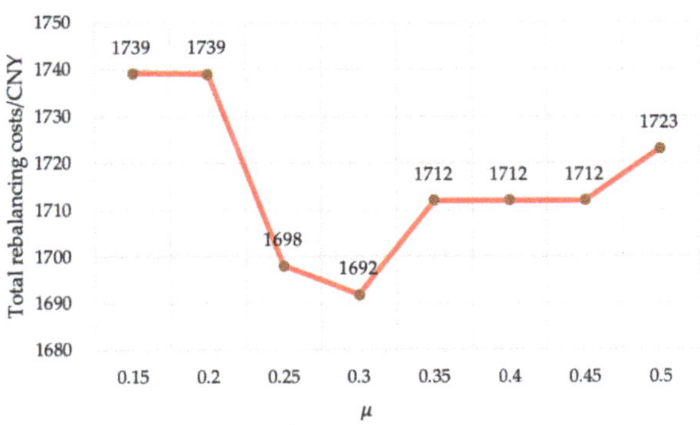

Figure 5. The trend of TRC changes with the increase of μ.

6. Conclusions

As a low-carbon and ecologically sustainable transportation mode, BSS has become a way to deal with the growing menace of global warming. In this study, we have proposed, modeled and solved BRP-MMFTR, which is a variant of BRP considering multi-energy mixed fleets and traffic restrictions. We first formulate BRP-MMFTR as a mixed integer programming model with the objective of minimizing the total rebalancing cost composed of vehicles' fixed costs, vehicles' traveling energy costs, EVs' recharging costs, and ICVs' carbon emission costs. Then SAVN is designed to solve this model, which is the simulated annealing algorithm enhanced with variable neighborhood structures. To illustrate the efficiency and efficacy of our algorithm, some test instances of BRP-MMFTR are generated randomly. The computational results reveal the huge advantage of SAVN, compared with SA and VNS. SAVN can achieve better solution in terms of solution quality and CPU time, outperforming those obtained by SA and VNS. In addition, SAVN is more stable than SA and VNS. Finally, the sensitivity analysis results of parameter μ indicate that as μ increases, the total rebalancing cost shows a clear trend of decreasing and then increasing. Therefore, setting an appropriate value for μ is helpful to reduce the total rebalancing cost. In addition, the value of μ is not necessarily constant in the practical rebalancing operations, and it can be dynamically adjusted according to the real-time conditions.

For BSS operators, we provide the optimal vehicle scheduling suggestions for multi-energy mixed fleets to minimize the total rebalancing cost when bike supply and user demand are not matched. In addition, we also focus on carbon emissions during the rebalancing process. BSS is originally a sustainable transportation mode, but if their operators use ICVs during rebalancing operations, the low-carbon benefits of BSS will be partially offset by the carbon emissions of ICVs. Therefore, it is obviously beneficial to create a green and low-carbon sustainable transportation system by using EVs instead of ICVs. The government should vigorously promote the sustainable development of EVs. Some policies, such as traffic restrictions on ICVs, can be introduced to encourage the BSS operators to purchase more EVs to minimize the carbon emissions.

Future research can extend our model and algorithm to solve more complex variants of BRP-MMFTR, such as considering the impact of average speed and speed variations to the traveling energy costs. Furthermore, dynamic or stochastic BRP-MMFTR is also a good research direction.

Author Contributions: Conceptualization, Y.J.; methodology, Y.J. and J.L.; software, W.Z. and Y.X.; validation, Y.J. and J.L.; writing—original draft preparation, Y.J. and W.Z.; writing—review and editing, D.Y. and J.L.; visualization, W.Z. and Y.X. All authors have read and agreed to the published version of the manuscript.

Funding: This research was funded by a grant from Shanghai Planning Office of Philosophy and Social Science, grant No. 2018BGL018.

Institutional Review Board Statement: Not applicable.

Informed Con sent Statement: Not applicable.

Data Availability Statement: Data is contained within the article.

Conflicts of Interest: The authors declare no conflict of interest.

References

1. Macioszek, E.; Świerk, P.; Kurek, A. The Bike-Sharing System as an Element of Enhancing Sustainable Mobility—A Case Study based on a City in Poland. *Sustainability* **2020**, *12*, 3285. [CrossRef]
2. Song, M.; Wang, K.; Zhang, Y.; Li, M.; Qi, H.; Zhang, Y. Impact Evaluation of Bike-Sharing on Bicycling Accessibility. *Sustainability* **2020**, *12*, 6124. [CrossRef]
3. Alvarez-Valdes, R.; Belenguer, J.M.; Benavent, E.; Bermudez, J.D.; Muñoz, F.; Vercher, E.; Verdejo, F. Optimizing the level of service quality of a bike-sharing system. *Omega* **2016**, *62*, 163–175. [CrossRef]
4. Bruglieri, M.; Pezzella, F.; Pisacane, O. An Adaptive Large Neighborhood Search for relocating vehicles in electric carsharing services. *Discret. Appl. Math.* **2019**, *253*, 185–200. [CrossRef]
5. Bulhões, T.; Subramanian, A.; Erdoğan, G.; Laporte, G. The static bike relocation problem with multiple vehicles and visits. *Eur. J. Oper. Res.* **2018**, *264*, 508–523. [CrossRef]
6. Yang, X.-H.; Cheng, Z.; Chen, G.; Wang, L.; Ruan, Z.-Y.; Zheng, Y.-J. The impact of a public bicycle-sharing system on urban public transport networks. *Transp. Res. Part A Policy Pract.* **2018**, *107*, 246–256. [CrossRef]
7. Nogal, M.; Jiménez, P. Attractiveness of Bike-Sharing Stations from a Multi-Modal Perspective: The Role of Objective and Subjective Features. *Sustainability* **2020**, *12*, 9062. [CrossRef]
8. Fricker, C.; Gast, N. Incentives and redistribution in homogeneous bike-sharing systems with stations of finite capacity. *Euro J. Transp. Logist.* **2016**, *5*, 261–291. [CrossRef]
9. Hiermann, G.; Puchinger, J.; Ropke, S.; Hartl, R.F. The Electric Fleet Size and Mix Vehicle Routing Problem with Time Windows and Recharging Stations. *Eur. J. Oper. Res.* **2016**, *252*, 995–1018. [CrossRef]
10. Coelho, V.N.; Grasas, A.; Ramalhinho, H.; Coelho, I.M.; Souza, M.J.F.; Cruz, R.C. An ILS-based algorithm to solve a large-scale real heterogeneous fleet VRP with multi-trips and docking constraints. *Eur. J. Oper. Res.* **2016**, *250*, 367–376. [CrossRef]
11. Erdoğan, G.; Laporte, G.; Wolfler Calvo, R. The static bicycle relocation problem with demand intervals. *Eur. J. Oper. Res.* **2014**, *238*, 451–457. [CrossRef]

12. Dell'Amico, M.; Hadjicostantinou, E.; Iori, M.; Novellani, S. The bike sharing rebalancing problem: Mathematical formulations and benchmark instances. *Omega* **2014**, *45*, 7–19. [CrossRef]
13. Cruz, F.; Subramanian, A.; Bruck, B.P.; Iori, M. A heuristic algorithm for a single vehicle static bike sharing rebalancing problem. *Comput. Oper. Res.* **2017**, *79*, 19–33. [CrossRef]
14. Duan, Y.; Wu, J.; Zheng, H. A Greedy Approach for Vehicle Routing When Rebalancing Bike Sharing Systems. In Proceedings of the GLOBECOM 2018-2018 IEEE Global Communications Conference, Abu Dhabi, UAE, 9–13 December 2018.
15. Casazza, M.; Ceselli, A.; Wolfler Calvo, R. A route decomposition approach for the single commodity Split Pickup and Split Delivery Vehicle Routing Problem. *Eur. J. Oper. Res.* **2021**, *289*, 897–911. [CrossRef]
16. Papazek, P.; Kloimüllner, C.; Hu, B. Balancing Bicycle Sharing Systems: An Analysis of Path Relinking and Recombination within a GRASP Hybrid. In *Parallel Problem Solving from Nature–PSN XIII*; Bartz-eielstein, T., Ed.; Springer: Berlin/Heidelberg, Germany, 2014; pp. 792–801.
17. Blesa, M.J.; Blum, C.; Festa, P. A Hybrid ACO+CP for Balancing Bicycle Sharing Systems. *Hybrid Metaheuristics* **2013**, 198–212. [CrossRef]
18. Raviv, T.; Kolka, O. Optimal inventory management of a bike-sharing station. *Iie Trans.* **2013**, *45*, 1077–1093. [CrossRef]
19. Wang, Y.; Szeto, W.Y. Static green repositioning in bike sharing systems with broken bikes. *Transp. Res. Part D Transp. Environ.* **2018**, *65*, 438–457. [CrossRef]
20. Usama, M.; Shen, Y.; Zahoor, O. Towards an Energy Efficient Solution for Bike-Sharing Rebalancing Problems: A Battery Electric Vehicle Scenario. *Energies* **2019**, *12*, 2503. [CrossRef]
21. Pal, A.; Zhang, Y. Free-floating bike sharing: Solving real-life large-scale static rebalancing problems. *Transp. Res. Part C Emerg. Technol.* **2017**, *80*, 92–116. [CrossRef]
22. Zhang, D.; Yu, C.; Desai, J.; Lau, H.Y.K.; Srivathsan, S. A time-space network flow approach to dynamic repositioning in bicycle sharing systems. *Transp. Res. Part B Methodol.* **2017**, *103*, 188–207. [CrossRef]
23. Dell'Amico, M.; Iori, M.; Novellani, S.; Subramanian, A. The Bike sharing Rebalancing Problem with Stochastic Demands. *Transp. Res. Part B Methodol.* **2018**, *118*, 362–380. [CrossRef]
24. Maggioni, F.; Cagnolari, M.; Bertazzi, L.; Wallace, S.W. Stochastic optimization models for a bike-sharing problem with transshipment. *Eur. J. Oper. Res.* **2019**, *276*, 272–283. [CrossRef]
25. Yuan, M.; Zhang, Q.; Wang, B.; Liang, Y.; Zhang, H. A mixed integer linear programming model for optimal planning of bicycle sharing systems: A case study in Beijing. *Sustain. Cities Soc.* **2019**, *47*, 101515. [CrossRef]
26. Legros, B. Dynamic repositioning strategy in a bike-sharing system; how to prioritize and how to rebalance a bike station. *Eur. J. Oper. Res.* **2019**, *272*, 740–753. [CrossRef]
27. You, P.-S. A two-phase heuristic approach to the bike repositioning problem. *Appl. Math. Model.* **2019**, *73*, 651–667. [CrossRef]
28. Correia, G.H.A.; Antunes, A.P. Optimization approach to depot location and trip selection in one-way carsharing systems. *Transp. Res. Part E Logist. Transp. Rev.* **2012**, *48*, 233–247. [CrossRef]
29. Ons, S.; Wahiba, R.C.; Ammar, O. Vehicle Routing Problem with Mixed Fleet of Conventional and Heterogenous Electric Vehicles and Time Dependent Charging Costs. Technical Report. 2014. Available online: https://hal.archives-ouvertes.fr/hal-01083966 (accessed on 6 November 2020).
30. Goeke, D.; Schneider, M. Routing a mixed fleet of electric and conventional vehicles. *Eur. J. Oper. Res.* **2015**, *245*, 81–99. [CrossRef]
31. Macrina, G.; Di Puglia Pugliese, L.; Guerriero, F.; Laporte, G. The green mixed fleet vehicle routing problem with partial battery recharging and time windows. *Comput. Oper. Res.* **2019**, *101*, 183–199. [CrossRef]
32. Hansen, P.; Mladenović, N.; Todosijević, R.; Hanafi, S. Variable neighborhood search: Basics and variants. *Euro J. Comput. Optim.* **2016**, *5*, 423–454. [CrossRef]
33. Wei, L.; Zhang, Z.; Zhang, D.; Leung, S.C.H. A simulated annealing algorithm for the capacitated vehicle routing problem with two-dimensional loading constraints. *Eur. J. Oper. Res.* **2018**, *265*, 843–859. [CrossRef]
34. China's Seven Major Carbon Market K Line Chart. Available online: http://www.tanpaifang.com/tanhangqing/ (accessed on 25 October 2020).

sustainability

MDPI

Article

Spatial Cluster-Based Model for Static Rebalancing Bike Sharing Problem

Bahman Lahoorpoor [1], Hamed Faroqi [2], Abolghasem Sadeghi-Niaraki [3,4,*] and Soo-Mi Choi [4]

1 Department of Civil Engineering, K.N. Toosi University of Technology, Tehran 19697, Iran; b.lahoorpoor@mail.kntu.ac.ir

2 School of Civil Engineering, The University of Queensland, Brisbane, QLD 4072, Australia; h.faroqi@uq.edu.au

3 Geoinformation Tech. Center of Excellence, Faculty of Geomatics, K.N. Toosi University of Technology, Tehran 19697, Iran

4 Department of Computer Science and Engineering, Sejong University, Seoul 143-747, Korea; smchoi@sejong.ac.kr

* Correspondence: a.sadeghi.ni@gmail.com

Received: 23 April 2019; Accepted: 3 June 2019; Published: 8 June 2019

Abstract: Bike sharing systems, as one of the complementary modes for public transit networks, are designed to help travelers in traversing the first/last mile of their trips. Different factors such as accessibility, availability, and fares influence these systems. The availability of bikes at certain times and locations is studied under rebalancing problem. The paper proposes a bottom-up cluster-based model to solve the static rebalancing problem in bike sharing systems. First, the spatial and temporal patterns of bike sharing trips in the network are investigated. Second, a similarity measure based on the trips between stations is defined to discover groups of correlated stations, using a hierarchical agglomerative clustering method. Third, two levels for rebalancing are assumed as intra-clusters and inter-clusters with the aim of keeping the balance of the network at the beginning of days. The intra-cluster level keeps the balance of bike distribution inside each cluster, and the inter-cluster level connects different clusters in order to keep the balance between the clusters. Finally, rebalancing tours are optimized according to the positive or negative balance at both levels of the intra-clusters and inter-clusters using a single objective genetic algorithm. The rebalancing problem is modeled as an optimization problem, which aims to minimize the tour length. The proposed model is implemented in one week of bike sharing trip data set in Chicago, USA. Outcomes of the model are validated for two subsequent weekdays. Analyses show that the proposed model can reduce the length of the rebalancing tour by 30%.

Keywords: bike sharing; rebalancing; clustering; optimization; sustainability

1. Introduction

Quality of service in bike sharing systems is an important operational problem for authorities. In order to keep the quality of service at a certain level, these systems need rebalancing to compensate unbalances in the network. The static rebalancing problem is keeping the balance of all stations in the network at a predetermined level. The process of rebalancing usually is carried out by vehicles. The vehicles start from the deposit (depot) and deliver bikes to vacant docks at stations or pick up bikes from overloaded stations; the cost of running this process through the network can be very high, which is an important issue in the management of bike sharing systems. Therefore, finding a rebalancing strategy to keep the balance of network with the least possible cost can be an ideal case for operators.

Bike sharing systems are one of the effective and environmentally friendly solutions to the first/last mile problem; they notably reduce motorized trips, air pollution, fuel consumption and needs for parking spaces [1–7]. Since the last decade, these systems have become one of the main complementary modes of public transit systems in many urban areas. Different factors such as accessibility to stations, availability of bikes and vacant docks, and fares influence the effectiveness of these systems, among which the availability of bikes during the day is one of the most challenging issues in the designing, operating, and management of these services.

To a large extent, bike sharing systems are used for one-way trips, and such a trend leads to inappropriate bike distribution in time and space [7,8]. As a result, unmet demand and consequently user dissatisfaction occur in the system. To minimize the demand dissatisfaction (i.e., maximize usage and revenue), the redistribution of bikes over stations is required. This redistribution, also called rebalancing, is implemented by employing vehicles (usually trucks) to collect surplus bikes from overloaded stations and deliver to bike deficient stations [9]. The efficiency of rebalancing operations is important to the authorities to ensure that the number of bikes and vacant docks at each station are periodically restored to predefined target values [8].

Rebalancing of bikes in the network is carried out in two different manners: the user-based approach, in which users can rebalance a portion of bikes in the system by their daily trips or taking bonuses for specified trips; and the operator-based approach, where bikes are rebalanced by the system operators whether in a static or dynamic manner [4,6,8,10,11]. The static rebalancing is carried out during the night when the demand in the system is insignificant. The dynamic rebalancing is performed during the day when the system is active by taking into account the real-time usage of the bike sharing system. The operator-based bike rebalancing problem (BRP) is a *Pickup and Delivery Traveling Salesman*. The aim of rebalancing is minimizing unmet demand, minimizing total travel time and handling costs, minimizing tour length of the vehicles, and minimizing the number of tours.

In contrast to static rebalancing, dynamic rebalancing usually is complex and remarkably expensive enough not to be utilized in most cases where the system is relatively large in size and extent. In bike sharing systems, preferably, static rebalancing is performed by the operator only to rebalance bikes for the morning demand (morning peak). At the other times of day, users handle a dynamic bike rebalancing in the system. To exemplify, *person A* moves a bike from *station i* to *station j* at the morning peak-hour. In the evening, *person B* collects another bike at *station j* and drops it off at *station k*. As a late-night trip, *person C* picks a bike at *station k* and returns it back to *station i* where it needs a bike back for the morning demand. This trilogy of trips can be expanded to a more complex case on the whole network.

Ultimately, users' behavior and trip patterns have significant impacts on the distribution of bikes over the system and also determine how and how many bikes need to be rebalanced by the operator at nights. Therefore, the origin-destination (OD) matrix of undertaken trips is one of the main factors that needs to be considered in the static rebalancing. As a matter of fact, it is important to specify which stations need to be rebalanced. This will eliminate extra rebalancing tours between stations. In other words, it demonstrates the share of rebalancing done by users and how much needs to be carried out by the bike sharing service staff.

This paper proposes a bottom-up spatial cluster-based model to solve the static rebalancing problem in the bike sharing systems. First, spatial and temporal patterns of bike sharing trips in the network are investigated. Second, a similarity measure based on the trips between stations is defined to discover groups of correlated stations, using a hierarchical agglomerative clustering method. Third, two levels of rebalancing are assumed as intra-clusters and inter-clusters with the aim of keeping the balance of the network as the beginning of the day. The intra-cluster level keeps the balance of bike distribution inside each cluster, and the inter-cluster level connects different clusters in order to keep the balance between the clusters. Finally, rebalancing tours are optimized according to the positive or negative balance at both levels of intra-clusters and inter-clusters using a single objective genetic algorithm. The rebalancing problem is modeled as an optimization problem, which aims to minimize the tour length. The proposed model is implemented in one week of bike sharing trip data set in Chicago, USA.

The remainder of this paper is organized as follows. In Section 2, a literature review on bike rebalancing problem is provided. In Section 3, the mathematical problem definition of the clustering method, elements of the genetic algorithm for the static bike rebalancing problem, and the data are stated. In Section 4, the results of the computational experiments are provided. Finally, the concluding remarks and discussion are presented in Section 5.

2. Literature Review

Bike rebalancing problem is one of the main problems in operating the bike sharing systems [10,12]. The aim of rebalancing is to specify the number of bikes loading/unloading at each station and find the optimal truck routes subject to various constraints [7,9,13]. In order to have an overview on the status quo of the mentioned problem during the recent years, Table 1 gives an outline of some recent and most contributed studies based on their operation type, mathematical approach and algorithms, and objectives. The available studies could be divided into two separate groups as static or dynamic rebalancing problems. Also, choosing between appropriate mathematical approaches has been another challenge for the researchers. In this section, we investigate recent studies on both static and dynamic rebalancing problems in order to clearly present the current research gaps.

Most of the studies on bike sharing systems focus on the static bike rebalancing problem by proposing several optimization algorithms, while a few others are related to dynamic rebalancing by dividing the problem into some static sub-problems. The difference in the number of publications is mainly due to the availability of data, complexity, difficulty in managing the varying demand and setting time slots. In dynamic rebalancing, the vehicle loading/unloading and tours need to be updated regularly to determine the demand variations from time to time [9]. Most studies on both static and dynamic rebalancing adopt different objectives including minimizing vehicle travel time and cost, the tour length, vehicle emission and also minimizing user dissatisfaction and unmet demand.

Table 1. Summary of the rebalancing literature.

Reference	Type	Mathematical Approach	Objective
Dell'Amico et al. [1]	Static	mixed integer linear programming	Minimizing total traveling cost
Cruz et al. [11]	Static	iterated local search (ILS)	Minimizing total traveling cost
Chemla et al. [14]	Static	integer program	Minimizing total traveling cost
Forma et al. [10]	Static	mixed integer linear programming	Minimize unserved users and the total traveling distance
Liu et al. [15]	Static	mixed integer linear programming	Minimize the total traveling distance
Schuijbroek et al. [16]	Static	Constraint Programming	Optimal vehicle routes
Li et al. [4]	Static	mixed integer linear programming	Minimize the total cost
Alvarez-Valdes et al. [12]	Static	a heuristic algorithm	Minimizing the overall cost of unsatisfied demands
Pal and Zhang [5]	Static	mixed integer linear programming	Minimize the make-span of the fleet of rebalancing vehicles
Erdogan et al. [17]	Static	Exact method Greedy	Minimize rebalancing costs
Erdogan et al. [18]	Static	Exact method integer programming	Minimize travel and handling costs
Shui and Szeto [9]	Dynamic	artificial bee colony algorithm	Minimizes the total unmet demand and the fuel and CO_2 emission cost
Zhang et al. [6]	Dynamic	mixed-integer problem	Minimizes the total unmet demand and route
Caggiani et al. [8]	Dynamic	Travelling Salesman Problem	Minimize cost and maximization of user satisfaction

2.1. Static Rebalancing Problem

The static rebalancing problem mostly deals with solving the optimization problem with different mathematical approaches, which are briefly explained here. Studies in the rebalancing problem are also could be divided into studies with and without clustering steps.

Chemla et al. [14] proposed a math-heuristic to find a route that minimizes the total traveling distance. Their algorithm is based on a solution of a relaxation of the problem that is solved by a branch-and-cut procedure. Dell'Amico et al. [1] presented four mixed integer linear programming with the objective of minimizing the traveling cost. Their computational time efficiently solves networks with up to 50 vertices and extends exponentially with the larger networks. With different objective, Alvarez-Valdes et al. [12] optimized the level of service of bike sharing systems by forecasting the unsatisfied demand at each station and proposing an algorithm upon the estimated measures. Their tested procedure on a relatively small network demonstrates to be an extremely fast heuristic algorithm. Some research consider the problem as 1-Commodity Pickup and Delivery Traveling Salesman Problem (1-PDTSP) using a single vehicle and allowing multiple visits. Erdogan et al. [18] and Erdogan et al. [17] present an exact algorithm to minimize rebalancing costs. Pal and Zhang [5] present mixed integer linear program for solving the static complete rebalancing problem in a free-floating bike sharing (dock-less) system. Similarly, Li et al. [4] present a mixed-integer linear programming to minimize the total rebalancing cost for different types of bikes. They proposed a combined hybrid genetic algorithm in which the hybrid genetic search determines the optimal routes and a greedy heuristic to specify the loading/unloading number of bikes at stations. Their proposed method is efficient for small-sized networks. Cruz et al. [11] proposed a hybrid iterated local search (ILS) algorithm. The authors argue that their heuristic algorithm is capable of finding the optimal solution in 97% of cases compared with previous work. Therefore, the above-mentioned studies focused on proposing a better mathematical approach that could solve the real world size problems in an efficient way.

The other studies use clustering method to simplify the rebalancing problem. Schuijbroek et al. [16] proposed an inventory rebalancing and vehicle routing (cluster-first route-second heuristic) for polynomial size clustering problem and consider the service level feasibility and approximate routing cost simultaneously. Also, Dondo and Cerda [13] investigated the multi-depot routing problem with different time slots and heterogeneous vehicles using a three step heuristic and hybrid. Their approach aims to divide the problem into small mixed-integer linear programming (MILP) sub-problems by a heuristic clustering algorithm and then finding the optimize routing (minimum travel costs) for the multi-depot heterogeneous vehicle routing problem. With the similar approach, Forma et al. [10] geographically clustered stations and then, the vehicle routing is determined. Similarly, Liu et al. [15] proposed adaptive capacity constrained K-centers clustering algorithm to solve the mixed integer nonlinear programming with the objective of minimizing the tour length. Consequently, enriching the rebalancing problems by using clustering algorithms could successfully deal with the size of the problem and break it down into smaller problems.

2.2. Dynamic Rebalancing Problem

The main objective of the dynamic rebalancing is to determine the optimal inventory levels that need to be maintained at each time interval. The challenges, arising in the dynamic rebalancing, necessitate the dedicated and efficient methodologies [6]. However, considering the complexity of fully integrating three key components (the user dissatisfaction estimating, the bicycle rebalancing and the vehicle routing), several partially integrated or sequential solution methodologies of the dynamic rebalancing are proposed [6], which are investigated in the following.

Regue and Recker [19] proposed a sequential framework. First, the demand forecasting model and inventory levels model are solved, then redistribution demand is generated and finally, vehicle routing is calculated. Alongside the optimization of vehicle routing and unmet demand, Shui and Szeto [9] also consider the environmental aspect as their objective for a dynamic rebalancing problem to try to minimize the CO_2 emission and the associated costs. To handle the varying demand on bikes, they used a rolling horizon approach in which the dynamic rebalancing is broken down into several static rebalancing problems. An enhanced artificial bee colony algorithm and a route truncation heuristic are jointly used to optimize the route design in each stage, and the loading and unloading heuristic is used to tackle the loading and unloading sub-problem along the route in each stage. They show that the shorter stage length, the better is the solution; in addition to computation time, total unmet demand and the total fuel and CO_2 emission cost are higher with a longer stage length [9]. This reinforces splitting the problem into sub-problems. The smaller problem, the better and faster solution.

Zhang et al. [6] proposed a non-linear time-space network flow-base formulation to fully integrate three key components of the dynamic rebalandcing (the bicycle rebalancing, the vehicle routing and user dissatisfaction). They estimated the user dissatisfaction by decomposing the forecasting period into two sub-periods based on the vehicle arrival time. The experimental results revealed that significant improvements compared to several benchmark approaches are achieved. On the other hand, Waserhole and Jost [20] present a dynamic pricing mechanism that forces a balanced demand and hence saves the need for rebalancing altogether. However, such an approach clearly comes at the cost of reducing the level of service provided to the users of the system but nevertheless seems the only viable alternative to balancing a car sharing system where rebalancing is too expensive. Contardo et al. [21] proposed methodologies based on a time-space network, in which the estimation of the user dissatisfaction is simplified and consequently the estimation accuracy is sacrificed.

User satisfaction becomes crucial for free-floating bike sharing system and the dynamic rebalancing makes the problem more complicated. In order to reduce the complexity, Caggiani et al. [22] and Caggiani et al. [8] propose a dynamic clustering method in constant time gaps. After the spatio-temporal correlation pattern of different zones in a city, a decision support system is developed to achieve a high degree of user satisfaction and keep the costs of rebalancing operations as low as possible Therefore, the studies in the dynamic rebalancing problem have focused on different heuristic optimization methods to solve the problems in both spatial and temporal dimensions.

In conclusion, bike sharing rebalancing problem has attracted the interest of many researchers in recent years. A large portion of studies focuses on solving the optimization problem itself without considering users' behavior and their role in the rebalancing problem. Also, most studies conduct experiments on relatively small networks and the literature mostly simplifies and surveys the network as a vector graph. Few studies use clustering in order to reduce the complexity of large-scale optimization problem [8,10,13,15,23], while what is not considered in the most rebalancing studies is the users' behavior in bike sharing systems. Users' behavior is the result of the inherent interactions between demand and supply. Availability of bikes (supply) alongside the distance and the geographical terrains affect the users' behavior and consequently the demand for bikes and docks. Clustering the network by considering trips between stations makes the analysis of the network simpler and more efficient and also as it will be discussed in followings, it eliminates unnecessary tours.

3. Methodology

The state-of-the-art bike sharing systems mostly deals with the rebalancing problem. In the static rebalancing problems, the aforementioned studies in the literature try to simplify the rebalancing into an optimization problem and provide the best possible solution for that, while the effect of users' trips has been neglected. To overcome the research gap, this paper proposes a bottom-up cluster-based model with the aim of:

- Considering the users' behavior in the network and determining the user-based rebalancing portion;
- Clustering the stations needed to be rebalanced;
- Minimizing the rebalancing tour length (cost);
- Implementing in a real scale case study.

After clustering the stations, trips are divided into two different categories namely intra-clusters and inter-clusters. Intra-cluster trips are those occurring within a cluster, and inter-cluster trips refer to those occurring between two clusters. These trips unbalance the distribution of bikes inside and outside of each cluster and need to be rebalanced again. This rebalancing takes place in part through users activities and in part by a rebalancing vehicle. Therefore, it is necessary to determine the share of user-based rebalancing to specify the volume of bikes need to be rebalanced by the operator. Similar to trips, the rebalancing operations need to be done via inter-clusters and intra-clusters. A static rebalancing operation with a single vehicle of a specific capacity is assumed. The vehicle is allowed to visit a station multiple times in one tour. Also, the vehicle sets off from a deposit with some bikes and returns to the deposit with the same number of bikes at the final stage. This number of bikes is determined at the inter-cluster rebalancing optimization. The implemented procedure for rebalancing is summarized in Figure 1.

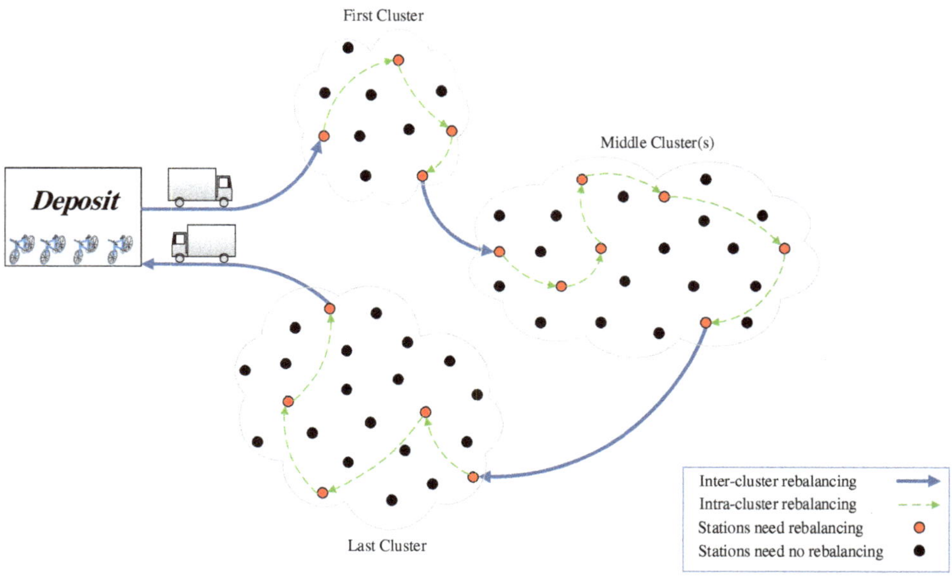

Figure 1. Rebalancing schematic scheme.

The goal of the rebalancing is keeping the balance of every single station at the predetermined level (initial balance). The initial status is computed by the data set of bike sharing transactions. The static rebalancing problem would be solved for each cluster in order to rebalance the stations in each cluster. Rebalancing each cluster would lead to a positive or negative balance for the whole cluster which is compensated by connecting that cluster with other clusters in the next steps. A genetic optimization algorithm is used to solve the rebalancing since it has positive performance in solving the routing problems in the literature [1,8,9]. As the remainder of this section, Section 3.1 displays the deployed notation and describes the problem and decision variables; Section 3.2 discusses the solution approach; and in Section 3.3 the case study and its characteristics are explained.

3.1. Notation

For clarity, the following describes the mathematical symbols adopted throughout the paper. They are notated into three main categories: indices, problem parameters, and decision variables.

- Indices

  i: id of stations

  j: id of vehicles

- Problem parameters

  I: set of id of all stations in the network

  IBI_i: Number of entering bikes into the station i

  OBI_i: Number of exiting bikes from the station i

  B_i: Balance of the station i at the end of the day; Bi = IBIi − OBIi

  Cap_j: Capacity of the vehicle

- Decision variables

  T: Tour for the vehicle; $T = S_1, \Delta B_1, S_2, \Delta B_2, \ldots, S_m, \Delta B_m$

  ΔB_m: Number of bikes delivered or picked up to/from m-th station

  S_m: id of the m-th station

The bike sharing network consists of number of stations, which contain docks that can be either occupied or vacant. During a day, a number of bikes enter or exit the stations by users, which lead to unbalance of stations at the end of the day. The difference between entering and exiting bikes for each station is considered as the balance of the station, which needs to be zero. In some stations, users keep the balance of the station as zero at the end of the day (which is notified as user-based rebalancing in this study); however, some other stations need to be rebalanced to zero by the vehicles. The authorities usually use vehicle(s) with specific capacity, to deliver/pick up the bikes in/out the stations in order to balance the network. The number of bikes on the vehicle at the beginning of the tour can be zero or a positive number of bikes inside the vehicle.

The cost of running the vehicles to rebalance the stations is a challenging issue for the authorities and needs to be as minimum as possible. The cost of rebalancing operation depends on many factors, among which the cost is heavily impacted by two factors as the number of vehicles and mileage. Hence, the decision variables are tours of the vehicle that rebalance the network. Every tour includes ID of the station and number of bikes that are needed to be loaded/unloaded (deliver/pick up) from specific stations. The objective function and constraints for the static rebalancing problem are presented in the following Equations (1)–(6); where D stands for the network distance between two stations in the network; m is the number of stations in the tour; and abs stands for the absolute values.

- Objective function

$$Min \sum_{i=1}^{m-1} D(S_i, S_{i+1}) \tag{1}$$

- Constraints

$$\Delta B_m \leq Cap \tag{2}$$

$$Cap_j \geq 0 \tag{3}$$

$$abs(\Delta B_m) \leq abs(B_m) \tag{4}$$

$$\Delta B_m \times B_m < 0 \quad (5)$$

$$\Delta B_m + B_m = 0 : i \in I \quad (6)$$

The objective function is minimizing the total network distance that a vehicle needs to traverse in order to rebalance all the stations. Equation (1) presents the total network distance that is the summation of the distances between every subsequent station in the tour. Also, there are five constraints. Equation (2) presents the number of moved bikes between stations and must be less or equal to the capacity of the vehicle. Equation (3) guarantees that the capacity of the vehicle cannot be a negative number. Equation (4) presents that the number of delivered/picked up bikes at every station cannot be greater than the balance of that station. Equation (5) ensures that the vehicle always helps to rebalance the station, not to strike the balance more; for example, if the balance of a station is negative then the vehicle can only deliver bikes to the station (not make it more negative). Equation (6) guarantees that the balance of every station in the network should be zero at the end of the tour.

3.2. Solution Approach

This section describes the proposed approach for solving the static rebalancing problem. The proposed approach begins with deriving the OD matrix between the stations in the network from the data set of bike sharing transactions. Then, stations are clustered according to the number of trips between them. Next, one OD matrix is reconstructed for each discovered cluster from the previous step. The static rebalancing problem is solved for each cluster in order to rebalance the stations in each cluster (intra-cluster level). Rebalancing each cluster would lead to a positive or negative balance for the whole cluster, which is compensated by connecting that cluster with other clusters in the next steps (inter-cluster level). An OD matrix is calculated by trips between the clusters, which actually focuses on the number of bikes trading by users between the clusters besides bikes from the first level (intra-cluster level) of rebalancing. Figure 2 presents the proposed approach to solving the rebalancing problem.

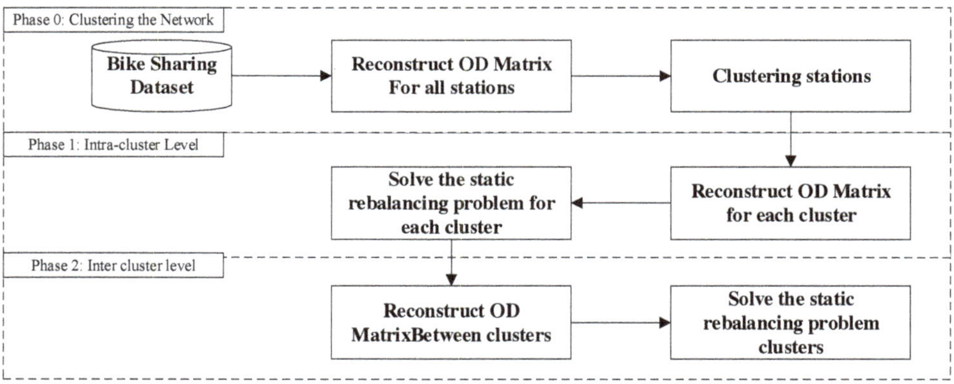

Figure 2. Solution approach.

The bike sharing data set include trips of users. Every trip has a unique ID. Every trip occurs between two stations as the origin and destination. Also, time for starting and ending the trip is included in the data set. It should be noted that the data sets usually need a clarification process in the case of missing data or unreasonable trip length or duration. Every station in the network has a unique ID that helps to reconstruct the OD matrix. Every cell in the OD matrix shows the number of trips between two stations considering the direction of the trip. Figure 3 presents an example for the data set and the derived OD matrix for five stations in the network; this example shows 17 trips between 5 stations in the network; and the derived OD matrix from these trips between these stations is presented at the right hand-side of the figure.

trip_id	start_time	end_time	tripduration	from_station_id	to_station_id
13550373	4/4/2017 23:26	4/4/2017 23:40	867	9	9
13549223	4/4/2017 19:11	4/4/2017 19:31	1182	4	4
13549015	4/4/2017 18:52	4/4/2017 19:14	1298	3	3
13548817	4/4/2017 18:34	4/4/2017 18:52	1071	6	3
13548299	4/4/2017 17:59	4/4/2017 18:24	1535	4	6
13548272	4/4/2017 17:58	4/4/2017 18:24	1599	4	6
13546231	4/4/2017 16:20	4/4/2017 17:18	3483	4	4
13546228	4/4/2017 16:20	4/4/2017 17:18	3499	4	4
13546225	4/4/2017 16:20	4/4/2017 17:18	3503	4	4
13545825	4/4/2017 15:46	4/4/2017 15:57	695	6	3
13545570	4/4/2017 15:17	4/4/2017 15:34	999	4	4
13545519	4/4/2017 15:10	4/4/2017 15:17	375	4	3
13545098	4/4/2017 14:08	4/4/2017 14:23	894	4	7
13544607	4/4/2017 12:40	4/4/2017 13:04	1435	3	3
13544602	4/4/2017 12:40	4/4/2017 13:04	1468	3	3
13543314	4/4/2017 9:13	4/4/2017 9:37	1437	6	4
13543312	4/4/2017 9:13	4/4/2017 9:37	1448	6	4

OD	3	4	6	7	9
3	3	0	0	0	0
4	1	5	2	1	0
6	2	2	0	0	0
7	0	0	0	0	0
9	0	0	0	0	1

Figure 3. An example for the bike sharing data set and origin-destination (OD) matrix.

The generated OD matrix for all stations in the network is used to cluster the stations. The similarity measure here is the number of trips between the stations. This paper utilizes the agglomerative hierarchical clustering algorithm using the *Ward* method that minimizes the total within-cluster variance. While the agglomerative hierarchical clustering algorithm can be implemented with various methods such as *Single*, *Average*, *Complete* and *Ward*, the *Ward* method is chosen according to the results of comparing these methods by [24]. It is chosen because it does not need to determine the number of clusters and is flexible with different similarity measures. It begins at the bottom where each object has its own cluster and merges them till all the objects form one cluster at the top. The result of the hierarchical agglomerative clustering is a dendrogram that shows how the objects are merged at each step [25]. According to the shape of the dendrograms and the Silhouette information [25], the dendrogram can be cut at a proper level. The Silhouette information refers to a method of interpretation and validation of consistency within clusters of data [26]. The initial cluster distances in Ward's minimum variance method are defined to be the squared Euclidean distance between points, which are shown in Equation (7), in which $d_q p$ stands for the squared Euclidean distance between two stations q and p [27]. Clustering the stations helps to discover groups of stations that have most interactions (trips) together. An obvious benefit of clustering is dividing the problem into several sub-problems that are computationally easier to solve.

$$d_{qp} = d(X_q, X_p) = \| X_q - X_p \|^2 \tag{7}$$

After discovering the clusters of the stations, the static rebalancing problem is solved for each single cluster in the network. The intra-cluster rebalancing problem just focuses on the imbalances that are caused by trips inside the clusters. Intra-cluster rebalancing aims at rebalancing the stations inside each cluster. However, trips between different clusters are unavoidable and must be considered as total positive or negative balance of the cluster, which is addressed at the next level (inter-cluster) rebalancing. The static rebalancing problem is solved for each cluster. The problem is a single objective optimization problem that can be solved, according to the size and complexity of the problem, with a variety of deterministic or heuristic algorithms. The utilized optimization algorithm in this study is the genetic optimization algorithm since it has positive performances in solving the routing problems in the literature [1,8,9].

The next step after the intra-cluster rebalancing is the inter-cluster rebalancing that focuses on trips between clusters and results from the previous step of the rebalancing. Similar to the intra-cluster level, an OD matrix is generated between all trips that have occurred between clusters; origin and destination of trips are not from the same cluster. Also, the static rebalancing problem at inter-cluster level is solved with a similar genetic algorithm as for the previous step.

The genetic algorithm is an evolutionary heuristic optimization one that produces a solution set resembling chromosomes of a population generation in biologic terms. Every chromosome consists of independent units called genes, which are in fact the components of the solution. This algorithm starts with an initialization step, in which the initial population of solutions is generated based on the constraints in the problems. Also, population size, cross over and mutation rates need to be set in the initialization level [28]. The primary steps of the main loop of the genetic algorithm are presented in Figure 4.

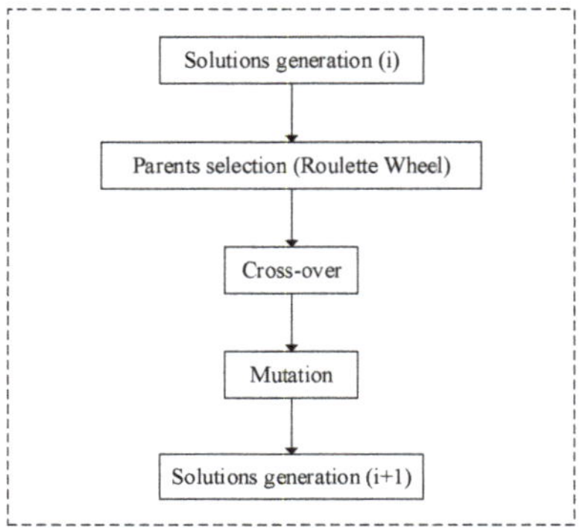

Figure 4. The genetic algorithm main steps.

Figure 5 shows a simple visualization for two chromosomes representing two possible tours with assumed length values (these values are for demonstration purposes only), cross-over and mutation operators between these two chromosomes. Each chromosome consists of separated cells, odd cells represent the ID of the stations, and even cells represent number of bikes for the rebalancing. The first two chromosomes are chosen for the cross-over and attached from station B. Also, the produced child from the cross-over is mutated into a new chromosome with a different tour length. Finally, ranking of these four chromosomes (regarding the tour length values) are presented that shows one of the parents (chromosome 2) is at the bottom of the ranking.

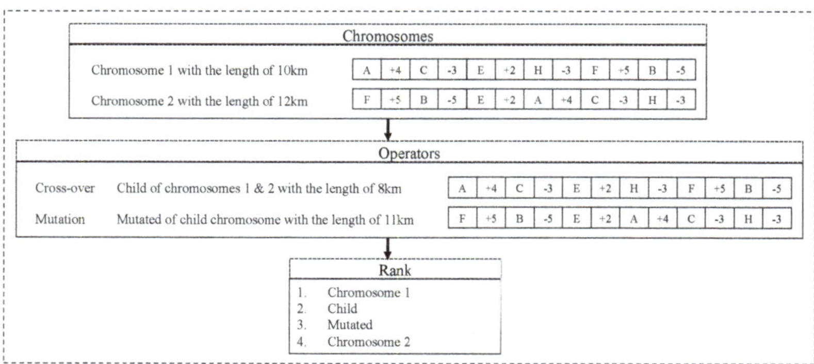

Figure 5. The genetic algorithm illustration.

Algorithms 1–4 present the pseudo code for implementing the genetic algorithm in the static rebalancing problem.

Algorithm 1: Genetic algorithm framework for the static rebalancing problem

```
Data: OD matrix
Result: tours
p ← population size;
gen ← maximum number of generations;
cr ← cross over rate;
mu ← mutation rate ;
P ← Initialize-Population ();
mutated ← Mutation (mut);
for i = 1 : gen do
    rank the solutions;
    parents ← select (cr × p) solutions using Roulette Wheel method;
    children ← Cross-over (parents);
    add children to the solutions;
    mut ← select (mu × p) solutions;
    mutated ← Mutation (mut);
    add mutated to the solutions;
    rank the solutions;
    eliminate ((cr + mu) × p)) of the worst solutions (highest cost);
```

Algorithm 2: Initialize-Population function

```
tour ← ∅;
while tour does not meet the constraints do
    randomly choose stations;
    randomly choose numbers of bikes;
    add stations and number of bikes to the tour;
return tour;
```

Algorithm 3: Cross-over function

```
while child tour does not meet the constraints do
    randomly cut parent tours;
    child ← attach cut parents;
return child;
```

Algorithm 4: Mutation function

```
while mutated tour does not meet the constraints do
    randomly choose number of stations in the selected tours;
    mutated ← replace the chosen stations with random stations;
return mutated;
```

3.3. *Data*

In order to evaluate the proposed rebalancing model, the data set from the Chicago bike sharing network called Divvy is selected in this study. The reason is twofold; first, Divvy is one of the biggest station-based bike sharing systems in the world which operates approximately 5800 bikes over 582 stations in the City of Chicago, USA; second, the data set is available online which enables researchers to compare their methods with the current approach with the same data. The trips' data set and stations' characteristics are available since 2013. Each trip occurs between two unique stations or similar stations (circle trips) with specific start and end times. Each station is identified with a unique ID. Also, the capacity of stations and their commencing dates are mentioned in the data set.

The second quarter of 2017 is selected for the primary network analysis. This period is the warm season in Chicago and is expected to have more bike trips compare to other seasons. Also, the proposed model is implemented for the second week of May 2017, when no specific weather condition was reported.

4. Results

In this study, first, a bottom-up clustering model is developed to cluster the stations and then divides the problem into two different levels. In the end, by using a single objective genetic algorithm, the objective function is to minimize the rebalancing vehicle' travel. In this section, after the primary analysis of the network, the temporal and spatial characteristics of the case study are studied. Then, the result of implementing the clustering method is discussed, followed by the rebalancing at two intra-cluster and inter-cluster levels. In the last part, the implemented method is validated by examining two different scenarios.

4.1. *Primary Network Analysis*

The distances between stations (density) in a bike sharing network, topographical and weather conditions affect the number of trips taken in the network. The primary analysis on the mentioned network data set demonstrates that first of all, the number of trips for Monday to Thursday is different from Friday to Sunday. Also, unexpected up and downs are seen in the diagram, which is examined with temperature and rainfall in Figure 6. A significant correlation exists between the overall average temperatures, and also a significant inverse relation between rainfall and number of trips is clear in the diagram.

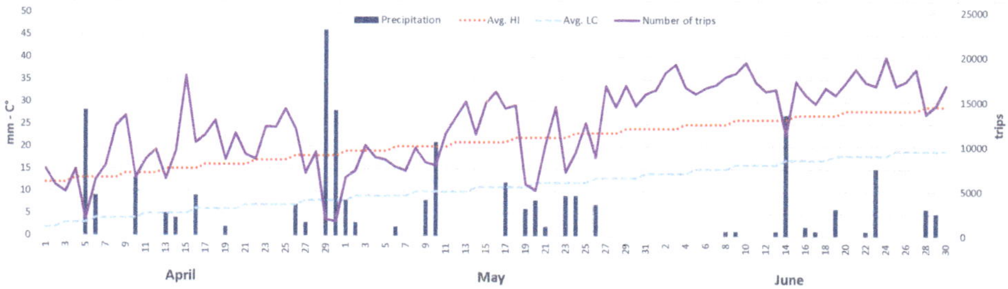

Figure 6. Weather condition vs. number of trips.

4.2. *Temporal and Spatial Analysis*

Temporal dimension of the trips can be summarized into two elements: start-end time and duration of trips. Results imply that the weekday patterns are similar and also weekends follow a unique pattern. The first two columns of Figure 7 presents the density plots of start-end time and trip duration for the second week of May 2017. Start and end time are mostly overlapped in the charts which is because of short durations. For Monday to Friday, there are clearly two peaks for morning and evening. Saturday and Sunday have one flat peak diagram. This difference can be originated from working days and the morning and evening demand of the facility. Trip durations for all weekdays and weekends present a one peak pattern around a 20-min travel time. This can be related to users' preferences and convenient biking duration which of course reinforce the result from Figure 8.

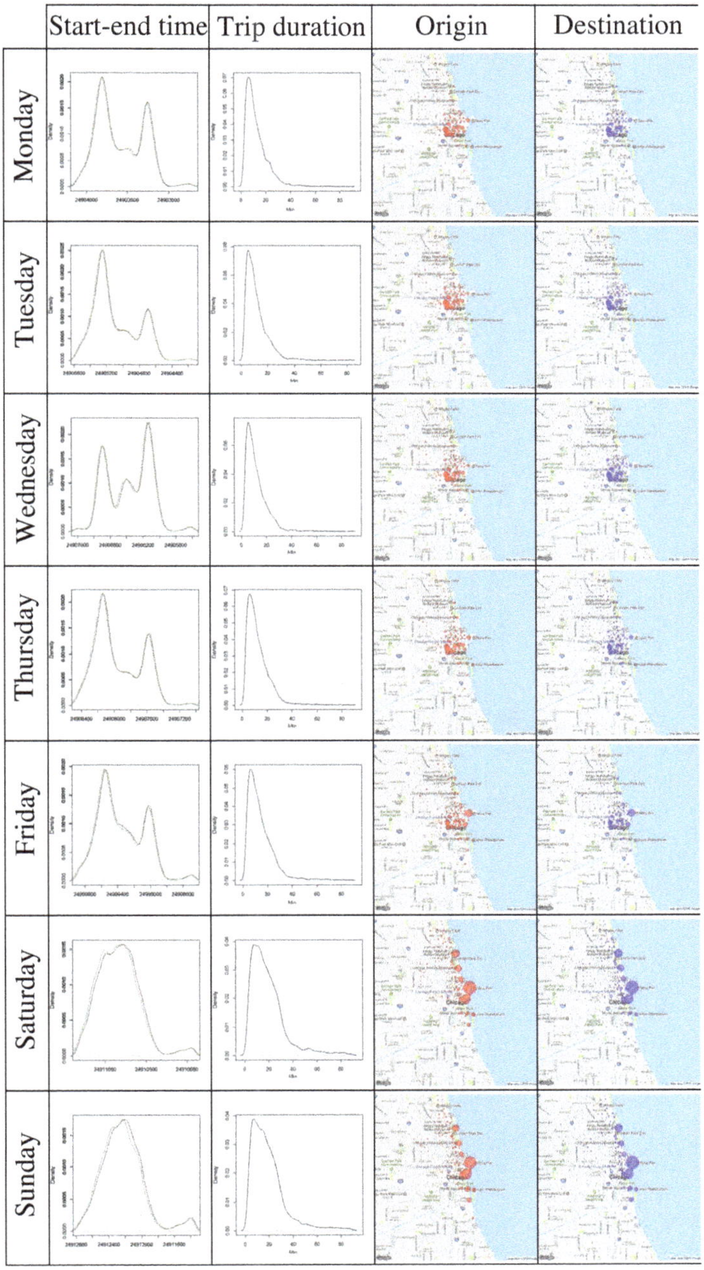

Figure 7. Temporal and spatial analysis of one week.

The frequency of trips over the network is studied. Results demonstrate that the most frequent origins and destinations of trips during the weekdays are different from weekends. On weekdays, the top five most frequent trips start from and end at central business district (CBD) while at weekends, trips mostly originate and destinate to coastal edges of Chicago. The *Origin* and *Destination* column of Figure 7 demonstrate the relative frequency trips on the map.

Another element of spatial analysis could be the length of trips occurring in the network. Results show that the density of trips' length is similar for the three months. The length of each trip is calculated based on network distance between the associated origin and destination. The density of travel length for June, which has the most number of trips, is illustrated in Figure 8. The result implies that about 80% of trips biked less than 3.5 km. Furthermore, with the increase in distance, the number of trips decreases exponentially. It is due to the users' preferences and convenient distance users are willing to bike. Also, no surprisingly, topological and topographical conditions and bike lane existence are important factors on number of trips and trips' length since Chicago is relatively a flat city and currently has more than 200 miles (320 km) of on-street protected, buffered and shared bike lanes.

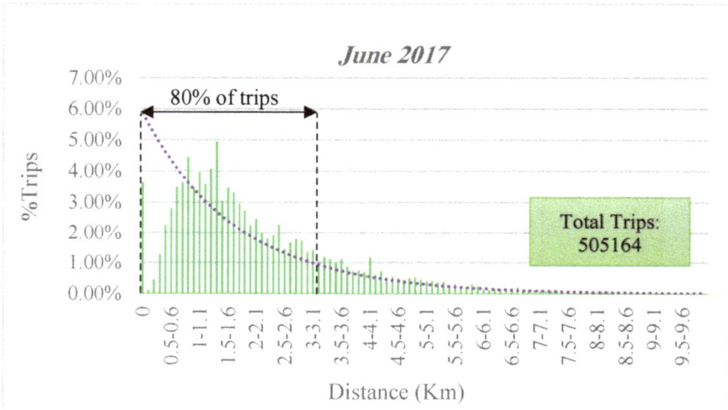

Figure 8. Density of traveled distance.

4.3. Clustering

By using the hierarchical agglomerative clustering method, the 582 stations in the Divvy bike sharing network are categorized into 13 clusters based on the OD matrix of the second week of May 2017. Although these clusters are different in size and distribution but mostly resemble each other and are scattered similarly through the network area. Clusters for each day are demonstrated in Figure 9. Cluster 2 (denoted as C2) is the largest cluster that has about 130 stations demonstrating a uniform distribution of trips among the center of Chicago. Also, there are 3 coastal clusters grouped in upper, middle and lower bounds of the coastal rim. Cluster 13 and Cluster 1 are showing two of Chicago's suburbs. It can be deduced that these clusters indicate 13 different districts (zones) in terms of number of trips, reflecting the underlying topographical layer of the Chicago city.

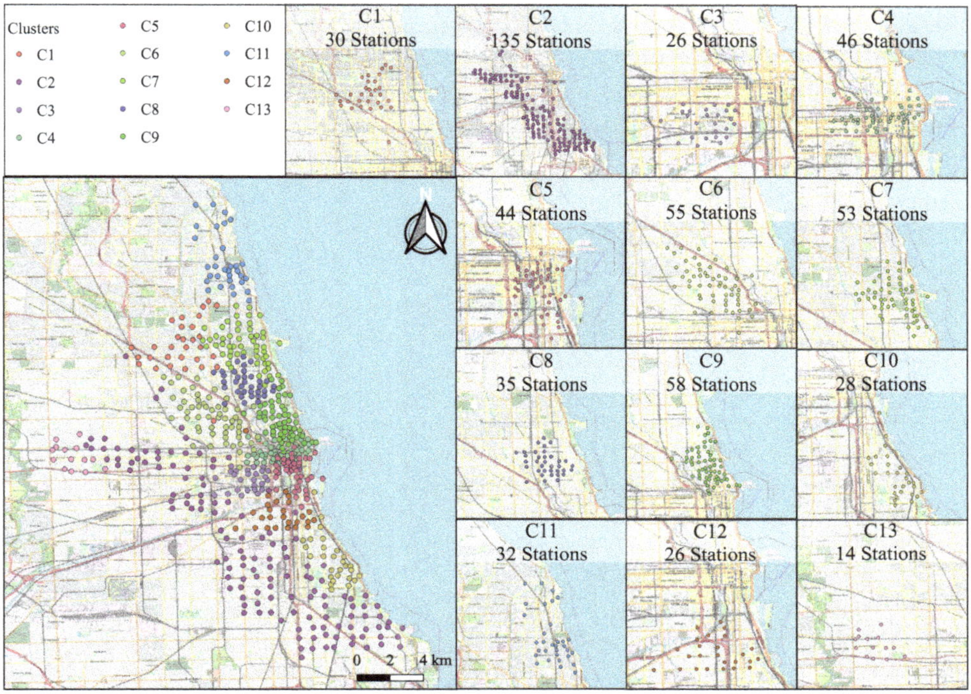

Figure 9. Clusters.

4.4. Rebalancing

Two levels, namely intra-cluster and inter-cluster are defined for rebalancing the bike sharing network. This is a bottom-up process in which rebalancing strategy begins from the lowest level of the network, which is trips and stations (i.e., OD matrix) to more aggregates level including clusters. Different values for population (5, 10, 30, 50, 75 and 100), cross over rate (10%, 30%, 60%, 70% and 80%), mutation rate (5%, 10%, 20%, 40% and 60%) are examined by running the algorithm 10 times for each specifications. Results of the sensitivity analysis indicate that population size of 50, cross over rate of 70% and mutation rate of 10% gives the best value of the objective function at the fewest number of iterations. Also, the maximum number of iterations is set to 10,000 (while the sensitivity analyses show that the algorithm is converged around the 100th iteration) for decreasing the chance of early convergence due to potential redundant solutions among the population. The higher number of the maximum iterations besides mutation and cross-over at each iteration produce new solutions that could compete with the existing solutions. The rebalancing process is carried out for one week (8–14 May) which is discussed as follows:

4.4.1. Intra-Cluster

After clustering the network and dividing it into 13 groups, the next step is determining which stations need to be rebalanced at each cluster and to specify the number of bikes needing to be loaded/unloaded at each station. Then, these bike deficiencies and the distance between them lead to the rebalancing scheme (i.e., tour route) at the intra-cluster level. Table 2 demonstrates the average number of stations needing to be rebalanced in the mentioned week which are visited in a rebalancing tour and also the number of stations satisfying the morning demand without a rebalancing tour. This is due to the user activity and shows the user-based rebalancing occurred in the bike sharing network. Cluster 2 has the highest number of stations which are scattered widely in the network. Among 135 stations on average, only 23 stations (17 percent) need to be rebalanced. It reinforces the fact that clustering by OD trips reduces the average tour length in proportion to the number of stations.

Table 2. Clusters rebalancing result.

Clusters	Visited Stations	Not-Visited Stations	Tour (km)
C1	19	11	33
C2	23	112	113
C3	23	3	24
C4	41	5	49
C5	37	4	47
C6	45	10	93
C7	45	8	82
C8	29	6	35
C9	52	6	82
C10	25	6	38
C11	25	7	46
C12	19	7	29
C13	9	5	8

Figure 10 shows the number of bikes loaded or unloaded at stations of each cluster. Similar patterns are observable for each day. Clusters 4, 5 and 9, which are mid-size clusters in terms of the number of stations, have the highest bike relocating among other clusters.

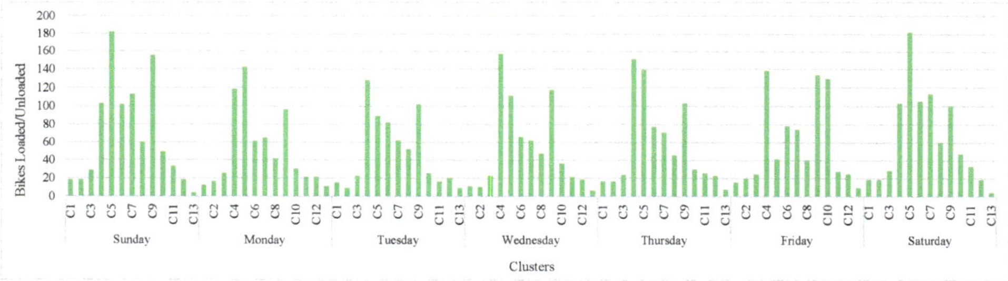

Figure 10. Bikes loaded/unloaded at each cluster.

Table 3 indicates the percent of stations visited, the number of bikes loaded/unloaded in a rebalancing tour and the tour length for each day. It can be realized that about 30 percent of stations do not have any bike deficiency. This is due to the user-based rebalancing that occurred in the network during the day. It can be concluded that the user-based portion of bike rebalancing in a bike sharing network is significant and has two important effects on the network rebalancing. First, it eliminates extra tours for rebalancing bikes over the bike sharing network. Second, it reduces the number of stations that need to be considered in a rebalancing process and consequently solving the problem becomes easier and much faster.

Table 3. Intra-Cluster rebalance tours.

Day	% Visited	Total Bikes Loaded/Unloaded	Tour (Km)
Monday	67	659	676
Tuesday	66	631	648
Wednesday	64	684	619
Thursday	70	724	720
Friday	68	751	727
Saturday	68	829	679
Sunday	68	883	689

4.4.2. Inter-Cluster

Inter-cluster trips lead to an unbalanced distribution of bikes over clusters. Cluster bike deficiency means the number of bikes that need to be entered to or taken out of each cluster. After determining the bike deficiency for each cluster, which is derived from the intra-cluster rebalancing level, the rebalancing tour will be calculated between clusters. Table 4 demonstrates the bike deficiency for each cluster for a week. It can be concluded that the grand total (cumulative sum) of bike deficiency is equal to zero since the overall system is balanced (unless a bike is broken or stolen). The significant result is the number of clusters with the zero deficiency which implies that they do not need inter-cluster rebalancing (as result of the clustering). This reinforces the fact that inward and outward trips to those clusters are equal and not leading to unbalances. Therefore, these clusters are reducing the total rebalancing tour length significantly as they do not attend in inter-cluster rebalancing process.

Table 4. Inter-cluster bike deficiency.

Day	C1	C2	C3	C4	C5	C6	C7	C8	C9	C10	C11	C12	C13	Tour (Km)
Sunday	0	0	0	2	13	0	0	0	−21	7	0	−1	0	28
Monday	1	4	−4	0	17	−1	0	0	−14	−3	0	0	0	46
Tuesday	0	0	−2	3	10	1	0	−1	−11	0	0	0	0	14
Wednesday	1	0	−2	1	5	−1	0	0	−6	2	0	0	0	19
Thursday	0	0	0	1	4	1	0	0	−6	−5	0	5	0	32
Friday	0	−1	−4	5	25	−1	0	0	−25	4	0	−3	0	40
Saturday	0	−1	0	1	37	3	0	−2	−41	5	0	−2	0	46

Figure 11 depicts the bike unbalances among clusters for a week. It demonstrates the portion of clusters which need (or do not need) to be balanced. Among clusters needing to be balanced, there is a logical trend for setting up a rebalance tour as the extra bikes at a cluster could compensate the other cluster needs. As can be seen from Figure 11, cluster 9 can compensate cluster 5 deficiency (clusters 9 and 5 are geographically close to each other according to Figure 9) since only these two clusters have significant bike unbalances. Besides, it can be concluded that the trends for bike loading/unloading has a similar pattern which could be predicted for other weeks.

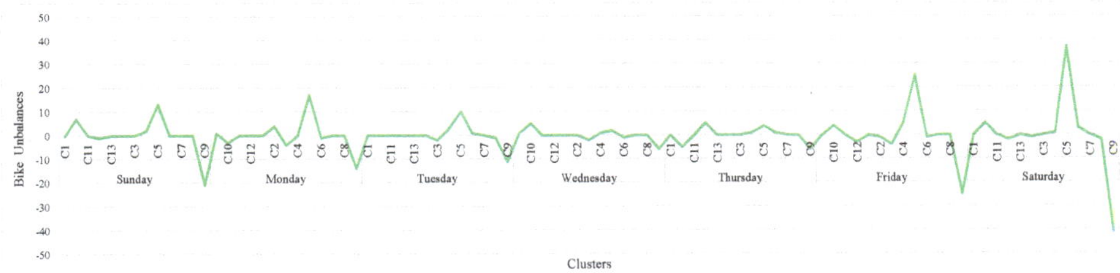

Figure 11. Clusters bike unbalances.

After determining bike unbalances at each cluster, the last step is to determine the routes between the clusters. Figure 12 demonstrates the rebalancing process for Monday. Clusters 4, 7, 8, 11, 12 and 13 have no deficiency at the inter-cluster level, therefore the rebalancing vehicle would not visit these clusters. It is worth mentioning that these clusters only need rebalancing at the intra-cluster level.

The rebalancing objective is to find a least-cost route that meets the demand of all stations and does not violate the minimum (zero) and maximum (vehicle capacity) load limits along the tour. Bikes loaded and unloaded at each stop determine the number of bikes at the beginning and end of each tour. Figure 13 plots the number of bikes on the vehicle for the inter-cluster rebalancing tour. It shows that the rebalancing vehicle sets off with 26 bikes from the deposit, visits 221 stations and finally goes back with the 26 bikes (the initial amount) to the deposit. As can be seen the number of bikes inside the truck are in the capacity range of the truck and does not violate the lower and upper bounds.

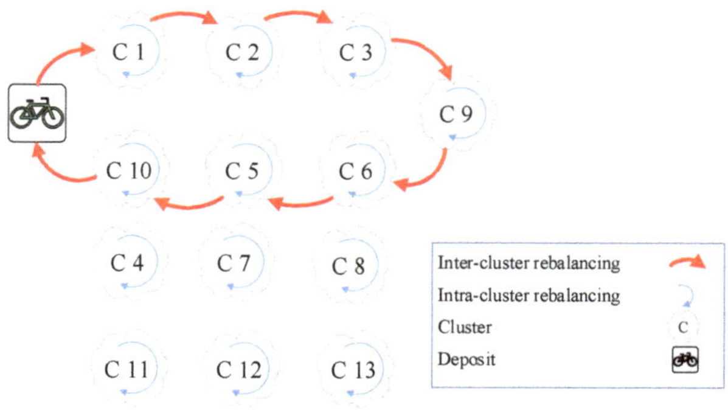

Figure 12. Inter-cluster rebalancing.

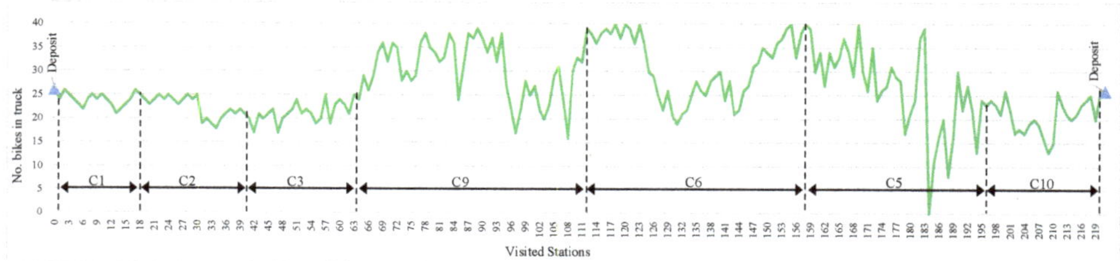

Figure 13. Number of bikes on the rebalancing vehicle.

4.5. Validation

One of the main purposes of rebalancing is to supply users' demand for bikes and vacant docks at each station. In order to evaluate the model efficiency, the model is run for two scenarios. *Scenario I* is independently implementing the model for Tuesday and conducting the rebalancing process based on the transactions for Tuesday in the data set. In *Scenario II*, rebalancing the network for the day before, that is, Monday is carried out first and then with the gained results, the model is implemented for rebalancing Tuesday. Table 5 shows the tour lengths in kilometer for all clusters both for scenario I and scenario II. Length of tours for all clusters is decreased by switching to the second scenario. Results indicate the proposed model can reduce the length of the total rebalancing tour length by 30%.

Table 5. Tour length before/after validation.

Cluster	Scenario I	Scenario II
C1	31	21
C2	75	50
C3	26	23
C4	51	35
C5	50	35
C6	96	61
C7	88	68
C8	40	27
C9	80	57
C10	32	22
C11	53	32
C12	19	17
C13	8	8
Total (km)	648	456

5. Discussion and Conclusions

From the operational viewpoint of bike sharing systems, one of the challenges is to keep the balance of bikes which are scattered in an unbalanced way over the stations due to the users' one-way daily trips. In order to keep providing service to users, operators try to rebalance bikes either once a day (i.e., static rebalancing) or many times a day (i.e., dynamic rebalancing). Finding a rebalancing strategy with the least possible cost (including operation, maintenance and environmental cost) is pivotal in the success of bike sharing systems.

In the literature, most of the studies deal with the static rebalancing problem [1,4,5,10,11,14,15,17]. The reason is three-fold. First, the dynamic rebalancing is expensive and as a result, most companies perform the static rebalancing in real cases. Second, the density of morning peak trips is usually more than the evening peak in most weekdays which reinforces the rationale to keep the network balance for the morning peak (static operations). Third, the dynamic rebalancing is the complex form of the static problem where the demand is varying and eventually can be converted into sub static problems for different time slots. For these reasons, it could be argued that static rebalancing still has its own currency both in literature and for operating companies.

Studies focusing on the static rebalancing problem, approach the optimization problem itself mathematically and consider it as a Pickup and Delivery Traveling Salesman problem [18] with different objective functions (from minimizing the general costs to maximizing the user satisfaction). Despite the rich optimization solutions in recent work, from a transportation point of view, there is less effort in considering the users' behavior in the network and determining the user-based rebalancing itself. This could be effectively carried out today with the availability of data.

To overcome the state of the art, this paper proposes a bottom-up clustering model to solve the static rebalancing problem. The aim of clustering is turning the problem into small size sub-problems. This has two advantages. First, it is easier to perform rebalancing on a small scale rather than the whole network. Second, the computational cost of optimizing hundreds of stations (the case of Chicago with 582 stations for example) would be significantly large [11,17] and clustering lightens the computational process. In the first step, the spatial analysis is done by the hierarchical agglomerative clustering method and a similarity measure based on OD trips. Then, stations are divided into separated clusters. In the second step, two rebalancing levels namely inter-cluster and intra-cluster are defined, where trips and rebalancing take place inside and outside the clusters.

Our method is quite different from typical clustering approaches. Unlike geographically clustered stations in previous studies, by clustering stations based on the OD trips, each cluster contains high intra-cluster trips. This gives a holistic perspective of how trips are concentrated between stations in the network and enables a new approach of rebalancing scheme. It is worth mentioning that with this method, no correlation has been found between the clusters and city functions. The reason is that clusters are defined by all day transactions as the static rebalancing occurs at the end of the day. The spatio-temporal analysis may reveal the inter-correlation between trips, activities and land use. This type of analysis is useful for dynamic rebalancing problems [8].

Finally, the rebalancing problem is modeled into tour finding problems using a single objective genetic algorithm at two different levels (inter and intra-cluster). The primary aim of the operators is to satisfy the demand and minimize the rebalancing costs (in terms of vehicle traveled distance) which both are addressed in this paper by balancing the number of bikes at each station to a predefined target value (enough vacant docks and bikes for the next morning demand) and setting the objective function to minimize the total tour length.

The proposed model is implemented in the Chicago bike sharing network. It is possible to enhance the efficiency of rebalancing by means of a clustering method. This conclusion is supported by Table 5 where the rebalancing of the network for one day has positive effects on a subsequent day. Results show a decrease in the length of tours for almost all clusters and the total tour length drops approximately by 30%. This illustrates the capability of the model to eliminate the extra tours and to reduce the length of the rebalancing tours significantly, which is ultimately a direct cost to the operator and indirect cost to the sustainable bike sharing systems.

Furthermore, in a large scale system, the rebalancing operation is difficult, expensive and needs an appropriate strategy. It has been observed that only a portion of stations (ranges from 64 to 70% in Table 3) needs rebalancing and roughly 33% of stations do not attend in the rebalancing process (which shows the users' role in rebalancing). In order to cope with the rebalancing problem in a large scale, using our proposed model would have two suggestions for the practitioners: (1) Inter-cluster rebalancing is not necessary due to the small number of deficiency outside the clusters. (2) Incentives could be arranged to encourage riders returning bikes to certain stations that would help with decreasing the intra-cluster rebalancing costs.

In the end, considering the network topology and available data of the case study, it would be possible to reduce the rebalancing tours, providing the operators with beneficial interventions on the system such as giving incentives or limiting the inter-cluster rebalancing. The future research may focus on optimization models with more details to solve the rebalancing problem in terms of one visit per station, the number of vehicles and rebalancing staff. Also, the effects of different clustering algorithms, considering both spatial and temporal dimensions [29], should be investigated on the outcome of the proposed model.

Author Contributions: Conceptualization, B.L. and H.F.; methodology, B.L.; software, H.F.; validation, A.S.-N.; formal analysis, H.F.; investigation, B.L. and H.F.; resources, A.S.-N.; data curation, B.L. and H.F.; writing–original draft preparation, B.L.; writing–review and editing, A.S.-N. and S.-M.C.; visualization, H.F.; supervision, A.S.-N.; project administration, S.-M.C.; funding acquisition, S.-M.C.

Funding: This research was supported by MSIT (Ministry of Science and ICT), Korea, under the ITRC (Information Technology Research Center) support program (IITP-2019-2016-0-00312) supervised by the IITP (Institute for Information & communications Technology Planning & Evaluation).

Conflicts of Interest: The authors declare no conflict of interest. The funders had no role in the design of the study; in the collection, analyses, or interpretation of data; in the writing of the manuscript, or in the decision to publish the results.

References

1. Dell'Amico, M.; Hadjicostantinou, E.; Iori, M.; Novellani, S. The bike sharing rebalancing problem: Mathematical formulations and benchmark instances. *Omega* **2014**, *45*, 7–19. [CrossRef]
2. Faroqi, H.; saadi Mesgari, M. Performance comparison between the multi-colony and multi-pheromone ACO algorithms for solving the multi-objective routing problem in a public transportation network. *J. Navig.* **2016**, *69*, 197–210. [CrossRef]
3. Faroqi, H.; Sadeghi-Niaraki, A. GIS-based ride-sharing and DRT in Tehran city. *Public Transp.* **2016**, *8*, 243–260. [CrossRef]
4. Li, Y.; Szeto, W.Y.; Long, J.; Shui, C.S. A multiple type bike repositioning problem. *Transp. Res. Part B Methodol.* **2016**, *90*, 263–278. [CrossRef]
5. Pal, A.; Zhang, Y. Free-floating bike sharing: Solving real-life large-scale static rebalancing problems. *Transp. Res. Part C Emerg. Technol.* **2017**, *80*, 92–116. [CrossRef]
6. Zhang, D.; Yu, C.; Desai, J.; Lau, H.Y.K.; Srivathsan, S. A time-space network flow approach to dynamic repositioning in bicycle sharing systems. *Transp. Res. Part B Methodol.* **2017**, *103*, 188–207. [CrossRef]
7. Wang, X.; Lindsey, G.; Schoner, J.E.; Harrison, A. Modeling bike share station activity: Effects of nearby businesses and jobs on trips to and from stations. *J. Urban Plan. Dev.* **2015**, *142*, 4015001. [CrossRef]
8. Caggiani, L.; Camporeale, R.; Ottomanelli, M.; Szeto, W.Y. A modeling framework for the dynamic management of free-floating bike-sharing systems. *Transp. Res. Part C Emerg. Technol.* **2018**, *87*, 159–182. [CrossRef]
9. Shui, C.S.; Szeto, W.Y. Dynamic green bike repositioning problem—A hybrid rolling horizon artificial bee colony algorithm approach. *Transp. Res. Part D Transp. Environ.* **2018**, *60*, 119–136. [CrossRef]
10. Forma, I.A.; Raviv, T.; Tzur, M. A 3-step math heuristic for the static repositioning problem in bike-sharing systems. *Transp. Res. Part B Methodol.* **2015**, *71*, 230–247. [CrossRef]

11. Cruz, F.; Subramanian, A.; Bruck, B.P.; Iori, M. A heuristic algorithm for a single vehicle static bike sharing rebalancing problem. *Comput. Oper. Res.* **2017**, *79*, 19–33. [CrossRef]
12. Alvarez-Valdes, R.; Belenguer, J.M.; Benavent, E.; Bermudez, J.D.; Muñoz, F.; Vercher, E.; Verdejo, F. Optimizing the level of service quality of a bike-sharing system. *Omega* **2016**, *62*, 163–175. [CrossRef]
13. Dondo, R.; Cerdá, J. A cluster-based optimization approach for the multi-depot heterogeneous fleet vehicle routing problem with time windows. *Eur. J. Oper. Res.* **2007**, *176*, 1478–1507. [CrossRef]
14. Chemla, D.; Meunier, F.; Calvo, R.W. Bike sharing systems: Solving the static rebalancing problem. *Discr. Optim.* **2013**, *10*, 120–146. [CrossRef]
15. Liu, J.; Sun, L.; Chen, W.; Xiong, H. Rebalancing bike sharing systems: A multi-source data smart optimization. In Proceedings of the 22nd ACM SIGKDD International Conference on Knowledge Discovery and Data Mining, San Francisco, CA, USA, 13–17 August 2016; pp. 1005–1014.
16. Schuijbroek, J.; Hampshire, R.; van Hoeve, W.J. Inventory rebalancing and vehicle routing in bike sharing systems. *Eur. J. Oper. Res.* **2017**, *257*, 992–1004. [CrossRef]
17. Erdoğan, G.; Battarra, M.; Calvo, R.W. An exact algorithm for the static rebalancing problem arising in bicycle sharing systems. *Eur. J. Oper. Res.* **2015**, *245*, 667–679. [CrossRef]
18. Erdoğan, G.; Laporte, G.; Calvo, R.W. The static bicycle relocation problem with demand intervals. *Eur. J. Oper. Res.* **2014**, *238*, 451–457. [CrossRef]
19. Regue, R.; Recker, W. Proactive vehicle routing with inferred demand to solve the bikesharing rebalancing problem. *Transp. Res. Part E Logist. Transp. Rev.* **2014**, *72*, 192–209. [CrossRef]
20. Waserhole, A.; Jost, V. Vehicle Sharing System Pricing Regulation: A Fluid Approximation. 2012. Available online: https://hal.archives-ouvertes.fr/hal-00727041v4/document (accessed on 8 June 2019).
21. Contardo, C.; Morency, C.; Rousseau, L.M. *Balancing a Dynamic Public Bike-Sharing System*; Centre Interuniversitaire de Recherche sur les Réseaux d'Entreprise, la Logistique et le Transport (CIRRELT): Montreal, QC, Canada; 2012; Volume 4.
22. Caggiani, L.; Camporeale, R.; Ottomanelli, M. A dynamic clustering method for relocation process in free-floating vehicle sharing systems. *Transp. Res. Procedia* **2017**, *27*, 278–285. [CrossRef]
23. Feng, Y.; Affonso, R.C.; Zolghadri, M. Analysis of bike sharing system by clustering: the Vélib'case. *IFAC-PapersOnLine* **2017**, *50*, 12422–12427. [CrossRef]
24. Ferreira, L.; Hitchcock, D.B. A comparison of hierarchical methods for clustering functional data. *Commun. Stat.-Simul. Comput.* **2009**, *38*, 1925–1949. [CrossRef]
25. Zaki, M.J.; Meira, W., Jr.; Meira, W. *Data Mining and Analysis: Fundamental Concepts and Algorithms*; Cambridge University Press: New York, NY, USA, 2014.
26. Rousseeuw, P.J. Silhouettes: A graphical aid to the interpretation and validation of cluster analysis. *J. Comput. Appl. Math.* **1987**, *20*, 53–65. [CrossRef]
27. de Amorim, R.C. Feature relevance in ward's hierarchical clustering using the L p norm. *J. Classif.* **2015**, *32*, 46–62. [CrossRef]
28. Deb, K.; Pratap, A.; Agarwal, S.; Meyarivan, T. A fast and elitist multiobjective genetic algorithm: NSGA-II. *IEEE Trans. Evol. Comput.* **2002**, *6*, 182–197. [CrossRef]
29. Faroqi, H.; Mesbah, M.; Kim, J. Comparing Sequential with Combined Spatiotemporal Clustering of Passenger Trips in the Public Transit Network Using Smart Card Data. *Math. Probl. Eng.* **2019**, *2019*, 5070794. [CrossRef]

sustainability

MDPI

Article

Why Shared Bikes of Free-Floating Systems Were Parked Out of Order? A Preliminary Study based on Factor Analysis

Qianling Jiang [1], Sheng-Jung Ou [2] and Wei Wei [2,*]

1 School of Design, Jiangnan University, Wuxi 214000, China; magicjql@gmail.com
2 School of Design, Chaoyang University of Technology, Taichung 41349, Taiwan; sjou@cyut.edu.tw
* Correspondence: weiwei925878620@gmail.com

Received: 20 May 2019; Accepted: 12 June 2019; Published: 14 June 2019

Abstract: Free-floating bicycle-sharing systems are an important component of sustainable transport. China's bicycle-sharing schemes have experienced ups and downs in the past three years, and there are a lot of related studies, but there are relatively few studies on the causes of disorderly parking of shared bikes. In this study, an open questionnaire is used to widely collect the causes of the disorderly parking of shared bicycles from users. Through factor analysis, six factors and 32 criteria for the causes of disorderly parking are constructed. Factor 1 'supervision and management of enterprises'; factor 2 'supervision and management of users'; factor 3 'parking space'; factor 4 'guidance of parking shared bikes'; factor 5 'user self-discipline'; factor 6 'operation and maintenance'. It requires the cooperation of multiple parties to solve the problem of disorderly parking of shared bicycles.

Keywords: Free-Floating Bike-sharing Systems; causes of disorderly parking; factor analysis

1. Introduction

As global warming intensifies, countries around the world have begun to save energy and reduce emissions to slow down global warming. At the Copenhagen Summit (2009 United Nations Climate Change Conference), the Chinese government made commitment to reduce carbon emissions and then opened the curtain for sustainable development. Alonso et al. [1] stated that in urban areas, where pollutants and consequently the impacts generated by unsustainable transport structure exist in a concentrated way, the sustainable mobility is a prerequisite of achieving sustainable cities. Transport systems are key elements of urban areas. Therefore, their sustainability has a pivotal role in achieving complex urban sustainability [2–4]. In China, where the population base is large, the amount of carbon dioxide produced by people in the daily transportation demand process should not be underestimated. Therefore, encouraging people to use public transport and develop a sustainable transportation system can effectively reduce carbon emissions.

Since the 1960s, Western countries have gradually realized the adverse factors such as traffic congestion, air pollution, and noise pollution caused by excessive motor vehicles, and they gradually began to realize the importance of bicycles in the transportation system. In the 1970s, the Dutch government improved bicycle transportation facilities in order to create a good bicycle traffic environment and return bicycles to people's lives. In the 1980s, the Federal Ministry of Transport and the Federal Ministry of Regional Planning, Building and Urban Development jointly promoted environmentally friendly traffic management strategies, including the integration of bicycles and public transportation systems [5]. In 2010, a National Cycling Plan has been drawn out with a vision of developing a cyclist-friendly, well-connected network, providing safe and healthy cycling for all [6]. Bicycle-sharing schemes are also called "Public-Use Bicycles" (PUBs), "Bicycle Transit", "Bikesharing" or "Smart Bikes", bicycle-sharing schemes comprise short-term urban bicycle rental schemes that enable bicycles to be picked up at any self-serve bicycle station and returned to any other bicycle station, which makes bicycle-sharing ideal for point-to-point trips [7,8]. China used to be named the "Kingdom of Bicycles" due to the nation's heavy reliance on cycling for mobility given the relatively low income of its citizens, compact urban development, and short trip distances in the 1970s [9]. However, China's bicycle ownership has declined year by year with its economic growth, popularity of motor vehicles, the longer distance of travel and the deterioration of the riding environment. For instance, average bicycle ownership in Chinese cities declined from 197 bikes/hundred households in 1993 to 113 bikes/hundred households in 2007 [10]. In light of growing traffic congestion and environmental concerns, the Chinese Ministry of Housing and Urban-Rural Development opposed bicycle use restrictions and supported tackling cycling barriers since 2007 [9]. The development of bicycle-sharing schemes in China has gone through three generations. The first-generation bicycle-sharing schemes were provided and managed by various local governments. It is required to apply for the designated IC card, and all fleets should be picked up and return to fixed stations. Most of the second-generation bicycle-sharing schemes were public-private partnership projects, the bicycle was used in a similar way to the first generation and the representative was Yonganxing. The third-generation bicycle-sharing schemes started since 2015. In addition to the advantages of flexible mobility, emission reductions, physical activity benefits, reduced congestion and fuel use, individual financial savings and support for multimodal transport connections [11], it also has the convenience features of digital registration, mobile payment, low cost and refundable deposit at any time. It provided a better solution to the problem of first/last mile of people's travel and became a popular travel mode in a short time (Figure 1 [12]).

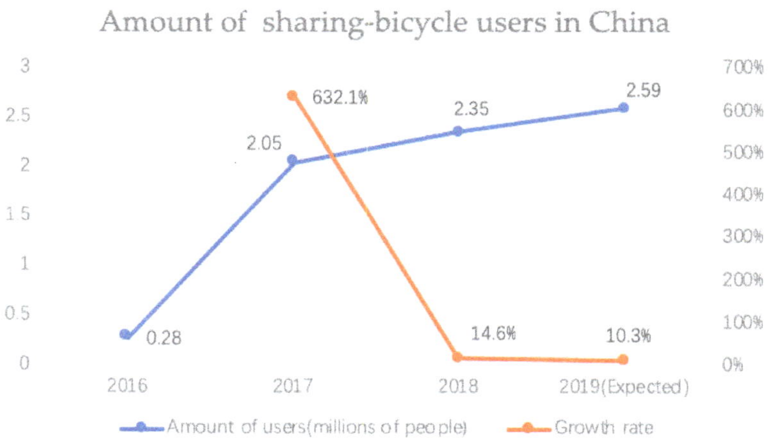

Figure 1. Amount of sharing-bicycle users in China [12].

In the past few decades, bicycle-sharing schemes around the world have been constantly evolving, but researches relative to bicycle-sharing schemes began roughly after 2000. The development of bicycle-sharing schemes is divided into four generations, their study sums the characteristics, problems and development process of the first three generations of shared bikes [11,13,14]. Then they point out the characteristics that the fourth generation shared bikes should have. DeMaio believes that the problems encountered by the first three generations of shared bikes include theft, inappropriate private use and anonymity. So the fourth generation bicycle-sharing schemes can put more effort into distribution of bikes, installation, powering of stations, tracking, offering pedalec (pedal assistance) bikes and new business models to improve efficiency, sustainability and usability [13]. Shasheen says that new bicycle-sharing schemes must pay attention to theft and vandalism, bicycle redistribution, information systems, insurance and liability consideration, and prelaunch consideration. The future bicycle-sharing schemes should focus more on: (a) flexible, clean docking stations; (b) bicycle redistribution innovations; (c) smartcard integration with other transportation modes, such as public transit and carsharing; and (d) technological advances including Global Positioning System (GPS) tracking, touch screen kiosks, and electric bikes [11]. Both of the seniors' researches gave suggestions on the macro direction and the detailed procedure as a development reference for the future bicycle-sharing schemes.

DeMaio [13] and Bührmann [15] divide bicycle-sharing schemes into different models depending on the business entity. The six business entities are: government, transport agency (quasi-governmental), university, non-profit organizations (e.g., foundation or advocacy group), advertising company, for-profit organizations. These business models can be divided into two orientations: public orientation and private orientation, the for-profit model is the most private one (Figure 2). In this model, service provided by private companies is should have limited or no government involvement. But if the private company used a fixed, versus flexible, system, it would need to have the local government's support to use the public space. This model is widely adopted by Chinese bicycle-sharing companies.

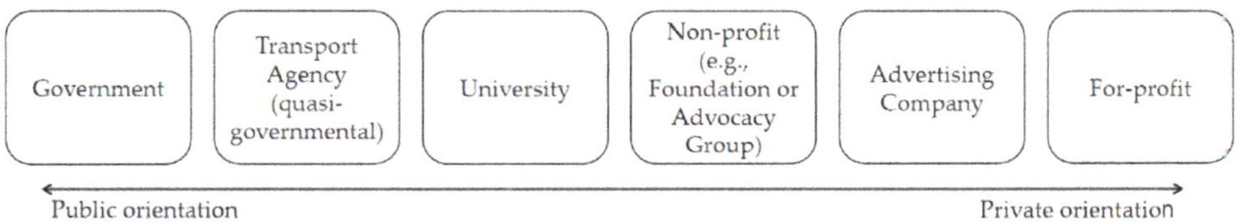

Figure 2. Different business models of bicycle-sharing schemes [13].

With the experience of previous bicycle-sharing schemes, China's bicycle-sharing schemes should be planned according to the standards of the fourth generation shared bicycles. However, the market results showed that China's shared bicycle-sharing schemes still encountered various problems in previous schemes. Due to social media's frequent exposure of disorderly parking, vandalism, theft, and wasting resources, the bicycle-sharing industry was pushed to the forefront and aroused the attention of all sectors of society. Among these problems, some of the problems have been analyzed by previous researchers and the corresponding countermeasures and suggestions are given. However, the problem of disorderly parking is rarely mentioned by scholars. Due to the failure of 'White Bikes' in Holland in 1965, most of the bicycle-sharing systems developed in early Europe and North America had parking stations, so they seldom encountered the problem of disorderly parking. In China, where free-floating bike-sharing systems are a big part of recent bicycle-sharing schemes, the problem of disorderly parking is particularly acute. In this research, the disorderly parking can be understood as users park the shared bikes out of the designated parking space. Therefore, the purposes of this study are as follows:

1. To widely collect the reasons for disorderly parking of shared bikes through the questionnaire survey of users, the results of the questionnaires are collated and analyzed as a basis for the subsequent research;
2. To construct the structure and factors of disorderly parking of shared bikes through factor analysis, and propose suggestions for improvement according to the results.

2. Methods

The first phase of the study focused on collecting the reasons for the disorderly parking of shared bikes as much as possible. Since the disorderly parking situation of shared bikes has appeared in China in recent years, and there are no foreign cases for reference, so the related research we can find is very limited. Therefore, this study used an open-ended questionnaire to ask respondents to fill in five reasons they thought would cause the disorderly parking of shared bikes in the first phase. During the period from December 20, 2018 to January 8, 2019, the users with experience of using shared bikes were invited to fill in an online questionnaire. The second online questionnaire was conducted between 8 May 2019 and 11 May 2019. The questionnaire design includes two parts: the reasons for disorderly parking of shared bikes and basic information of respondents. There are a total of 42 items of reasons, which are obtained from result of the first-phase questionnaire. Basic information of respondents includes gender, age, education level, city, area of use (downtown, outside the downtown but within the suburbs, the suburbs), frequency of using shared bikes [16], motivation of using shared bikes (commuting, leisure and social activities "outing, group travel, etc,.", exercise, personal affairs "shopping, eating, etc.," and others) [17], time of using shared bikes (weekdays or weekends), whether have the experience of disorderly parking, the frequency of disorderly parking. The 5-point scale was used in the questionnaire, with 1 being strongly disapproving and 5 being strongly approving. The respondents of both questionnaires are users who have the experience of using shared bikes in China. The respondents in this study are invited by the authors and their classmates. Respondents filled in the answers through the hyperlink of the online questionnaire. The basic variables of the questionnaire also investigated the frequency of using shared bikes, and the results showed that all respondents had experience in using shared bikes.

3. Results and Discussion

3.1. Descriptive Analysis of the First-Phase Questionnaire

A total of 240 questionnaires were collected, and 235 of them were valid questionnaires, the effective rate is 97.9%. The questionnaires with too many missing values and obvious deviation from the topic were considered as invalid. A total of 1004 valid items were obtained after collecting all the responses of respondents, 118 initial items were obtained after merging the similar-expression items. In the discussion part, all the items will be shown in the format of 'item description (frequency)'. All the items are sorted directly from the original content of the respondents' questionnaires.

After sorting out respondents' answers, we found that although some respondents' answers were different in expressions, they essentially described the same problem, so it is necessary to further combine such items. For better analysis of the items, we classified the items according to the main content discussed by the items after the semantic judgment of the items. All items were divided into six categories, including 'user-related problems', 'rules and regulations', 'rewards and punishment for users', 'supervision and management issues', 'parking facilities' and 'social advocacy', and the rest were discussed separately. The detailed process is as follows.

The first category is about users, including: for users' own convenience (75 times), users tended to follow the trend and crowd (45 times), users' didn't have time to find parking area(79 times), user quality problem(128 times), users don't care (6 times), personal habits (11 times), users are not conscious (12 times), no awareness of parking or no public awareness (29 times), users put shared bikes down stairs (once), laziness (15 times), bikes aren't their owns, so users don't care (8 times), users were not responsible (12 times), users forgot to park bikes in order (once), high user arbitrariness (twice), users were not familiar with the rules of parking (12 times), moral deficiency (13 times), users(who parked bikes out of order) were not punished (twice), user deliberate (16 times), users were not satisfied with the service of the bicycle-sharing companies, so they deliberately retaliated (once), parking bikes was inconvenient (twice), users didn't obey the regulations (once), without damaging the bikes, it was users' right and freedom to decide where and how to park the bike (once), users thought it was unnecessary to park bikes in order (twice), park nearby (once), and users were worried about shared bikes will be taken away by others, so they hid shared bikes for their own use (five times).

a. For users' own convenience is a statement covered a lot, so a lot of other items can be incorporated into it. In this study, 'parking bikes was inconvenient', 'park nearby', 'put shared bikes down stairs', and 'users were worried about shared bikes will be taken away by others, so they hid shared bikes for their own use' are all incorporated into for users' own convenience.

b. Users' didn't have time to find parking area. In addition to the situation that the users are really in a hurry, it may also because the user is unwilling to find a parking area or users can't find the parking area in a short time.
c. Users tended to follow the trend and crowd. This is one of the important factors that respondents believe to cause disorderly parking of shared bikes. Lots of users saw or felt that other users didn't park their shared bikes according to the regulations, so they imitated them, which eventually led to a large range of disorderly parking of shared bikes.
d. Shared bicycle users also need to bear certain responsibilities while enjoying the convenience brought by shared bicycles. The responsibilities can be divided into legal responsibility and moral responsibility. Bicycle-sharing schemes are newly developed things, and the development of corresponding laws and regulations is impossible to keep up with the development of the industry in a short time, which means the self-discipline of enterprises and users is required. 'Bikes don't belong to users', 'so users don't care', 'users were not responsible', and 'users thought it was unnecessary to park bikes in order' indicate that users subjectively believe that the shared bikes are no longer their business after use, and they don't need to bear corresponding responsibility. 'High user arbitrariness' and 'users forgot to park bikes in order' also reflect that users didn't pay enough attention to parking shared bikes in order after use. This study merges the above items into 'users don't attach enough importance to parking shared bikes in order'.
e. User quality problem. Bicycle-sharing schemes have been described by the media as the mirror for the quality of the civil, a large number of disorderly parking does reflect the national quality need to be improved.
f. Personal habits indicate that users may not change the old parking habits and they won't follow the current parking specifications of shared bikes. The habit of parking in order didn't fully spread indicates that the normative parking behavior of shared bicycles is not yet widespread, and users need some time to adapt to the new parking specifications. These two items are integrated into 'users need time to get rid of the old parking habits'.
g. No awareness of parking, no public awareness. Users themselves do not have the awareness to park shared bikes in order, nor have they been trained to do so, leading to a weak awareness of parking shared bikes in order.
h. Laziness. In many cases, users are not willing to park shared bikes in order just because they are lazy.
i. Users were not familiar with the rules of parking. On the one hand, it reflects that users, as the subject, didn't not take the initiative to understand the relevant rules of parking shared bikes; on the other hand, the bicycle-sharing companies didn't effectively promote the rules of parking shared bikes to users, or the rules of parking shared bikes were not perfect. This study focuses on the former here, and the latter will be discussed in a later section. So, this item is changed to 'users didn't take the initiative to understand the relevant rules of parking shared bikes'.
j. Users who parked bikes out of order were not punished. This item indicates that the punishment for users' illegal parking behavior was not completely implemented, and also reflects that users have some fluke mind. In this section, we focus on the latter, so this item is modified to 'users were not punished, so they had fluke mind'.
k. Without damaging the bikes, it was users' right and freedom to decide where and how to park the bike. Users of shared bikes must take corresponding responsibilities when obtaining the right to use shared bikes, so users should use and park shared bikes according to the regulations. Therefore, this item won't be discussed separately in this study.
l. Users couldn't find parking area. The reasons why users can't find the parking area can be roughly divided into the following categories. Firstly, there is no parking area near users' destination; secondly, the guidance system of parking area is not clear, which will be discussed

in the latter section; thirdly, users own reasons such as: lack of time, laziness or some other reasons. Therefore, this item is divided to discuss in other items.

After discussion, the first category reasons for disorderly parking are shown in Table 1.

Table 1. First category reasons for disorderly parking.

No.	Items
a	For users' own convenience
b	Users tended to follow the trend and crowd
c	Users' didn't have time to find parking area
d	User quality problem
e	Users need time to get rid of the old parking habits
f	No awareness of parking, no public awareness
g	Laziness
h	Users don't attach enough importance to parking shared bikes in order
i	Users didn't take the initiative to understand the relevant rules of parking shared bikes
j	Users were not punished, so they had fluke mind

The second category is about rules and regulations. Due to the rapid development of the bicycle-sharing industry, many local governments were unable to develop management regulations, and there were also no related regulations of parking shared bikes provided by bicycle-sharing companies. Items of second category are as follows: no relevant constraints (once), integrity system was not established completely (three times), lack of government management practice (three times), there was no corresponding management mechanism (13 times), insufficient regulations binding (once), there were no clear parking specifications (12 times), incomplete laws (8 times).

a. 'Lack of government management practice' was included in 'there was no corresponding management mechanism', so they were discussed together. To solve the problem of disorderly parking of shared bikes, the government needs to guide both bicycle-sharing companies and users of shared bikes, and bicycle-sharing companies also need to restrain users [18]. Therefore, these two items are divided into 'the government lacks enterprises management regulations' and 'bicycle-sharing companies lacks user management regulations'.
b. Same as above, 'insufficient regulations binding' can be divided into 'government's management regulations for enterprises has insufficient binding force' and 'bicycle-sharing companies' management regulations for users has insufficient binding force'.

After discussion, the second category reasons for disorderly parking are shown in Table 2.

Table 2. Second category reasons for disorderly parking.

No.	Items
a	There were no relevant constraints on parking shared bikes
b	Integrity system was not established completely
c	The government lacks enterprises management regulations
d	Bicycle-sharing companies lacks user management regulations
e	Government's management regulations for enterprises has insufficient binding force
f	Bicycle-sharing companies' management regulations for users has insufficient binding force
g	There were no clear parking specifications
h	Incomplete laws

The third category is about rewards and punishment for users, including the following: violation cost was low (twice), no punishment (39 times), the accountability of users was not well investigated and affixed (once).

a. 'No punishment' is a representation of 'violation cost was low'. These two items are combined into 'The penalty for users' violation of parking is not enough'.

After discussion, the third category reasons for disorderly parking are shown in Table 3.

Table 3. Third category reasons for disorderly parking.

No.	Items
a	The penalties for users' violation of parking is not enough
b	The accountability of users was not well investigated and affixed

The fourth category is supervision and management issues, including: No supervision and management (50 times), inadequate participation of managers (11 times), the government didn't pay enough attention (3 times), carrier management problem (10 times), the city administration is not in place (7 times), little follow-up guarantee was invested in bicycle-sharing schemes (3 times), no specific management staff (4 times), city management (once), governments' and bicycle-sharing companies' administration was not in place (12 times), disordered management (once), insufficient municipal management (twice), there were none staff managing parking area (4 times), lack of forward-looking for new things (once), immature management system (manual management and system management) (once), poor planning by providers (once), there were too many shared bikes (18 times), bicycle-sharing companies were not good at vehicle management (repair of damaged vehicle) (9 times), concentrated parking in certain periods (especially at noon and night near the school) (once), intense industry competition (4 times).

a. The expression of 'No supervision and management' is too succinct. The government has the responsibility to supervise and manage the bike-sharing companies, bike-sharing companies have the responsibility to supervise and manage users of shared bikes, and the government has a duty of supervision over the users of shared bikes. Insufficient municipal management, the government didn't pay enough attention, the city administration is not in place, city management, insufficient municipal management, carrier management problem, governments' and bicycle-sharing companies' administration was not in place, all these items are about government and business management and supervision. Three new items derived from the above items are as follows: The government has no supervision and management of bike-sharing companies, bike-sharing companies have no supervision and management of users of shared bikes, the government has no supervision of users of shared bikes.
b. 'Inadequate participation of managers' can be understood as there are not enough managers and managers in 'inadequate participation of managers' can be divided into managers in government and managers in enterprises. In addition, 'there were none staff managing parking area' also shows the problem of insufficient number of managers in government or bicycle-sharing companies. Therefore, this study will discuss this problem from the perspective of insufficient number of managers, which can be divided into 'insufficient number of government managers' and 'insufficient number of bicycle-sharing company managers'.
c. The meaning of 'little follow-up guarantee was invested in bicycle-sharing schemes' is the maintenance and management of shared bikes is not in place, which is consistent with the content of 'bicycle-sharing companies were not good at vehicle management (repair of damaged vehicle)' and 'concentrated parking in certain periods (especially at noon and night near the school)'. Therefore, the above items are sorted as follows: 'The bicycle-sharing companies can't maintain the damaged bikes in time' and 'the vehicle distribution policy of bicycle-sharing companies is unreasonable'.
d. 'Disordered management' and 'immature management system (manual management and system management)' not only point out the management of government and bicycle-sharing companies isn't in place, but also vaguely reveal that the division of responsibility for management of shared

bikes within the government is not clear. 'No specific management staff' is catered to the content of previous items. Therefore, these two parts are sorted into 'the division of responsibility for management of shared bikes within the government is not clear'.

e. The providers in 'poor planning by providers' can be the government or bicycle-sharing companies. If the provider referred to the government, it can be understood that the government's control measures for the scale of the bicycle-sharing market are not in time, which is consistent with 'lack of forward-looking for new things'. If the provider referred to bicycle-sharing companies, it can be understood combined with 'there were too many shared bikes' and 'intense industry competition' that bicycle-sharing companies put excessive amount of shared bikes in order to gain more market share. In this study, the items above are summarized as 'the government's control measures for the scale of the bicycle-sharing market are not in time' and 'bicycle-sharing companies put excessive amount of shared bikes'.

After discussion, the fourth category reasons for disorderly parking are shown in Table 4.

Table 4. Fourth category reasons for disorderly parking.

No.	Items
a	The government has no supervision and management of bike-sharing companies
b	Bike-sharing companies have no supervision and management of users of shared bikes
c	The government has no supervision of users of shared bikes
d	Insufficient number of government managers
e	Insufficient number of bicycle-sharing companies managers
f	The bicycle-sharing companies can't maintain the damaged bikes in time
g	The vehicle distribution policy of bicycle-sharing companies is unreasonable
h	The division of responsibility for management of shared bikes within the government is not clear
i	The government's control measures for the scale of the bicycle-sharing market are not in time
j	Bicycle-sharing companies put excessive amount of shared bikes

The fifth category of reasons center on parking facilities, including: no dedicated parking area (82 times), parking area was full (12 times), the bicycle-sharing companies didn't provide parking space (twice), there was little parking area (33 times), the parking area is far away (17 times), there were no more reasonable parking areas allocated (5 times), unreasonable setting of parking area (9 times), the parking space was occupied (once), street vendors occupied the parking area (once), poor design of bicycle parking (3 times), lack of fixed parking area (twice), the parking area was not clearly defined (5 times), the parking area was narrow (5 times), there were no suitable places for parking bikes (once), the parking area was too concentrated (3 times), the place was inappropriate for parking bikes (once), there was no specific position for parking bikes (once), urban planning problem (twice), the place was remote (twice), the sidewalk was narrow (once), the signs of parking area were not eye-catching (5 times), irregular parking area arrangement (once), the parking area is not big enough (twice), there was no electronic fence (once), the parking space was little (twice), parking facility (once), limitation of road conditions (once). In China, most cities take bicycle parking space into consideration at the beginning of planning. A small number of bicycles would be parked on the sidewalk when there is no bicycle parking space. Although it would not cause too much trouble when the amount of shared bikes is small, the original bicycle parking space in the city cannot meet the parking demand of users of shared bicycles after the rapid growth of bicycle-sharing industry, so a large number of Shared bicycles are parked out of order. Shared bicycle is actually a kind of private goods with public properties, whose ownership belongs to the for-profit bicycle-sharing company. However, shared bicycles play an active role in the public transportation system, especially in the slow traffic transportation system, which makes it have certain public attributes. Therefore, the government and bicycle-sharing companies should work together to solve the problems of facilities and parking area of shared bikes instead of either of them taking all the responsibilities.

a. 'No dedicated parking area', 'the bicycle-sharing companies didn't provide parking space', 'lack of fixed parking area', 'there were no suitable places for parking bikes', and 'there was no specific position for parking bikes' all show that 'there was no dedicated parking space for shared bikes'.
b. 'Parking area was full', 'there was little parking area', 'there were no more reasonable parking areas allocated', 'the parking area was narrow', 'the parking area is not big enough', and 'the parking space was little' all indicate that 'the capacity of shared bikes parking space is insufficient'.
c. 'The parking area is far away', 'poor design of bicycle parking', 'the parking area was too concentrated', 'the place was inappropriate for parking bikes', 'urban planning problem', 'the place was remote', and 'irregular parking area arrangement' all show that 'the setting of shared bikes parking area is unreasonable'.
d. 'Street vendors occupied the parking area' and 'the parking space was occupied' are telling the same truth.
e. 'The sidewalk was narrow' and 'limitation of road conditions': In those areas without designated parking space, it is customary for users to park shared bikes on sidewalks. These two items contains two aspects: on the one hand, the parking area has not been set up, on the other hand, the capacity of parking space is in sufficient.
f. The parking space of shared bikes should not only have clear mark of margin on the ground, but also have eye-catching guiding signs for users to find it. 'The signs of parking area were not eye-catching' is modified to 'the guiding signs of parking space is not eye-catching' for better understanding.
g. The meaning of 'parking facility' is vague, which is limited to the parking facilities of shared bikes in this study. Parking facilities of shared bicycle mainly include shared bicycle parking space, guiding signs of parking space, etc., which are similar to the above items. Therefore, this item will not be repeated in the follow-up study.

After discussion, the fifth category reasons for disorderly parking are shown in Table 5.

Table 5. Fifth category reasons for disorderly parking.

No.	Items
a	There was no dedicated parking space for shared bikes
b	The capacity of shared bikes parking space is insufficient
c	The setting of shared bikes parking area is unreasonable
d	The parking space was occupied
e	The parking area was not clearly defined
f	The guiding signs of parking space is not eye-catching
g	There was no electronic fence

The sixth category is about social advocacy, including: social atmosphere (3 times), influence of users' surrounding environment (twice), users are affected by people who have previously parked bikes out of order (twice), influence of the disorderly parking of bicycles and motorcycles (4 times), the habit of parking in order hasn't fully spread (twice), ineffective social supervision (once), the whole society doesn't care (once), the user wasn't discouraged by others (once), people turn a blind eye to the phenomenon of disorderly parking (twice), insufficient government and media promotion (26 times), insufficient education of parking in order (twice), bicycle-sharing schemes firstly promoted as 'shared bikes can be picked up and parked anywhere and anytime' (5 times), the bicycle-sharing companies didn't explicitly remind users to park in order (3 times), lack of reporting platform for easy reporting (once).

a. Wikipedia defines social atmosphere as the sum of the customs, cultural traditions, behavioral patterns, moral values, and fashion elements of a given society. In the past period of time, the phenomenon of disorderly parking of shared bikes can be seen everywhere, which is indeed a kind of behavior jointly presented by the whole society. 'Influence of users' surrounding environment', 'users are affected by people who have previously parked bikes out of order' and 'influence of the disorderly parking of bicycles and motorcycles' are incorporated into 'social atmosphere' due to similar meaning.

b. 'The whole city didn't care' is consistent with 'people turn a blind eye to the phenomenon of disorderly parking', which reflects the 'lack of social supervision'. Also, 'lack of reporting platform for easy reporting' is one of the causes of lack of social supervision.

c. After using shared bikes, users tend to park shared bikes in the place according to their old habits. They would park shared bikes in the place they thought right, they would also park the shared bikes nearby for their convenience. However, the users of shared bicycles do not have the ownership of the bicycles. Instead, they have obtained the right to use the shared bicycles within a certain period of time from the bicycle-sharing companies according to the rules of use, so they shall comply with the corresponding rules. The government, social media and bicycle-sharing companies should work together to carry out propaganda on parking shared bikes in order to help regulate users parking behavior. Therefore, 'insufficient government and media promotion' is modified to 'government's, social media's and bicycle-sharing companies' propaganda on parking shared bikes in order is not in place'.

d. "Bicycle-sharing schemes firstly promoted as 'shared bikes can be picked up and parked anywhere and anytime'" reflects the advantages of shared bikes for short trips, but there are limits to 'park anywhere and anytime', so there is misleading in initial propaganda of bicycle-sharing companies.

After discussion, the sixth category reasons for disorderly parking are shown in Table 6.

Table 6. Sixth category reasons for disorderly parking.

No.	Items
a	Influence of social atmosphere
b	Lack of social supervision
c	Lack of reporting platform for easy reporting
d	Government's, social media's and bicycle-sharing companies' propaganda on parking regulations is not in place
e	The bicycle-sharing companies didn't explicitly remind users to park in order
f	There is misleading in initial propaganda of bicycle-sharing companies

'Bikes were moved away by others' (4 times), 'there was something wrong with bicycle positioning' (once), 'the bicycle-sharing market was not optimistic' (once), 'weather problems such as strong winds' (4 times), 'users didn't put shared bikes back in place' (once), 'the bike was not parked in order at the beginning' (twice), accidental situation (once), 'the bike fleet was broken' (12 times), 'the bike was not comfortable to ride' (once), 'the bike is hard to ride and need to be changed' (twice), 'shared bikes are easy to get' (once), 'no cohesion' (once), "users' destinations were diverse" (once), 'the allocation of road rights of shared bikes was not clear' (once), 'limitation of road conditions' (once), 'the security guard was not responsible enough, so users could park the shared bikes in the community' (once), 'temporary parking' (once), 'poor traffic condition' (once), 'the bike was not parked steadily, then it toppled' (once), 'the market was not mature and perfect in all aspects' (once), 'the bikes should set reminders for users' (once) are difficult to be classified into one of the above categories, so they are discussed separately.

a. 'Weather problems such as strong winds' do cause trouble for users to use and park shared bikes. In addition to strong winds, heavy rains and snow weather will cause the same problem. Therefore, this item is adjusted to 'weather condition, which makes it impossible for users to use the shared bike as usual'.

b. 'The bike fleet was broken', 'there was something wrong with bicycle positioning', 'users didn't put shared bikes back in place', and 'the bike is hard to ride and need to be changed' can be summarized as 'the shared bike is damaged, so users can't park it in order'.
c. 'Temporary parking' is a common scenario when users using shared bikes.
d. The remaining items of this category will not be discussed in this study due to their trivial content and low relevance to the content of this paper.

After discussion, the seventh category reasons for disorderly parking are shown in Table 7.

Table 7. Seventh category reasons for disorderly parking.

No.	Items
a	Weather condition, which makes it impossible for users to use the shared bike as usual
b	The shared bike is damaged, so users can't park it in order
c	Temporary parking

In addition to the internal reasons for users cause the disorderly parking of shared bikes, the above part also discusses the external factors that lead to disorderly parking of shared bikes. In the special case of disorderly parking of shared bikes in China, Qin Zheng and Wang Qin [18] put forward the model of collaborative governance of government, market and society. They propose that the government should guide the market and society, and the market should restrain society, among which the market society respectively refer to bicycle-sharing companies and users. Therefore, this study firstly integrates and classifies the questions based on the content of the questions, and then reintegrates the answers of the interviewees through the relationship of the government, enterprises and users to remove the repeated content in the repeated questions, so as to construct the next questionnaire.

a. Considering that the bicycle-sharing industry is a recent development industry, 'insufficient number of government managers' can be recognized as the result of 'the division of responsibility for management of shared bikes within the government is not clear'. Therefore, these two items are merged into 'the division of responsibility for management of shared bikes within the government is not clear'.
b. 'The bicycle-sharing companies can't maintain the damaged bikes in time' are modified to 'the bicycle-sharing companies can't maintain the damaged vehicle properly' for better understanding.
c. Laws can be understood as stricter regulations. Thus, the meaning of 'incomplete laws' is included in 'the government lacks enterprises management regulations'. 'The government has no supervision and management of bike-sharing companies' has two scenarios. Firstly, the government has no rules to follow. Secondly, the government has rules to follow, but the enforcement is weak. Therefore, it is modified to 'the government's enforcement of business regulations is weak'.
d. 'The government has no supervision of users of shared bikes', 'incomplete laws', 'the government has no supervision of users of shared bikes' can be included in 'there were no relevant constraints on parking shared bikes', which is too vague for respondents to understand. So, 'there were no relevant constraints on parking shared bikes' is replaced by the three more specific items.
e. 'Bike-sharing companies have no supervision and management of users of shared bikes' was removed from this study because the content of this item is a summary of other items, such as 'bicycle-sharing companies lacks user management regulations', 'the bicycle-sharing companies didn't explicitly remind users to park in order' and so on.

All reasons for disorderly parking of shared bikes are shown in Table 8.

Table 8. All reasons for disorderly parking of shared bikes.

No.	Items
VAR1	For users' own convenience
VAR2	Users tended to follow the trend and crowd
VAR3	Users' didn't have time to find parking area
VAR4	User quality problem
VAR5	Users need time to get rid of the old parking habits
VAR6	No awareness of parking, no public awareness
VAR7	Laziness
VAR8	Users don't attach enough importance to parking shared bikes in order
VAR9	Users didn't take the initiative to understand the relevant rules of parking shared bikes
VAR10	Users were not punished, so they had fluke mind
VAR11	The division of responsibility for management of shared bikes within the government is not clear
VAR12	The government's control measures for the scale of the bicycle-sharing market are not in time
VAR13	The government lacks enterprises management regulations
VAR14	Government's management regulations for enterprises has insufficient binding force
VAR15	The government's enforcement of business regulations is weak
VAR16	Incomplete laws
VAR17	The government has no supervision of users of shared bikes
VAR18	The government's propaganda on parking shared bikes in order is not in place
VAR19	Insufficient number of bicycle-sharing companies managers
VAR20	The bicycle-sharing companies can't maintain the damaged vehicle properly
VAR21	The vehicle distribution policy of bicycle-sharing companies is unreasonable
VAR22	Bicycle-sharing companies put excessive number of shared bikes
VAR23	Bicycle-sharing companies lacks user management regulations
VAR24	There were no clear parking specifications
VAR25	The penalties for users' violation of parking is not enough
VAR26	The accountability of users was not well investigated and affixed
VAR27	The bicycle-sharing companies didn't explicitly remind users to park in order
VAR28	There is misleading in initial propaganda of bicycle-sharing companies
VAR29	There was no dedicated parking space for shared bikes
VAR30	The capacity of shared bikes parking space is insufficient
VAR31	The setting of shared bikes parking area is unreasonable
VAR32	The parking space was occupied
VAR33	The parking area was not clearly defined
VAR34	The guiding signs of parking space is not eye-catching
VAR35	There was no electronic fence
VAR36	Influence of social atmosphere
VAR37	Lack of social supervision
VAR38	Lack of reporting platform for easy reporting
VAR39	Social media's propaganda on parking regulations is not in place
VAR40	Weather condition, which makes it impossible for users to use the shared bike as usual
VAR41	The shared bike is damaged, so users can't park it in order
VAR42	Temporary parking

3.2. Results and Discussion of the Second-Phase Questionnaires

3.2.1. Descriptive Analysis of the Second-Phase Questionnaire

A total of 254 questionnaires were collected in the second phase, 245 of which are valid after screening, the effective rate was 96.5%. The questionnaires with too many missing values were considered as invalid. The sample size has met the requirement of factor analysis [19,20]. The Cronbach's α value is 0.951, 42 items have high internal consistency. Commuting (32.1%), personal affairs (27.7%) and leisure and social activities (24.2%) are the most common use scenarios, and the percentage of total cases showed that the respondents use shared bikes in various situations. Among the respondents, 156 of them are in the downtown area, accounting for 63.7%. 76 of them are in the suburbs outside the downtown area, accounting for 31%; 13 of them are in the suburbs, accounting for 5.3%. There are 209 users whose frequency of use is 5 times or less per week, accounting for 85.3%. 26 users' frequency of use between 6 to 10 per week, accounting for 10.6%; 10 users' frequency of use is 11 or more per week, accounting for 4.1%. Among the respondents, 96 have experience of disorderly parking, accounting for 39.2%, 149 of them don't, accounting for 60.8%.

3.2.2. Item Analysis of the Second-Phase Questionnaire

Since the items in this study are derived from the results of the previous questionnaire survey, it is necessary to judge whether the questions are relevant by item analysis before conducting factor analysis. The principal component analysis method is used to extract a factor. Using the common identity of 'composition matrix' less than 0.3 as standard, VAR1, VAR2, VAR3, VAR4, VAR5, VAR6, VAR22, VAR36, VAR38, and VAR41, a total of 10 questions are deleted. The remaining 32 items are used for factor analysis.

3.2.3. Factor Analysis of the Second-Phase Questionnaire

In this study, principal component analysis (PCA) and maximum axis method are used to extract and name the causes of disordered parking of Shared bicycles. Six factors are extracted from 32 items by factor analysis, factor 1 has 7 items, factor 2 has 8 items, factor 3 has 5 items, factor 4 has 5 items, factor 5 has 4 items, and factor 6 has 3 items. After the maximum rotation axis method, the eigenvalue of factor 1 is 4.409, factor 2 is 4.19, factor 3 is 3.572, factor 4 is 3.05, factor 5 is 2.877, and factor 6 is 2.459. Six factors explain the variable variation of 13.778%, 13.095%, 11.162%, 9.532%, 8.99% and 7.684% respectively, and the total explained variable is 64.241%. As can be seen from the table, the integrality of the factors after the rotation axis increases and the proportion of the factors that can be explained changes: factor 1 (39.375% 13.778%), factor 2 (6.727% 13.095%), factor 3 (6.282% 11.162%), factor 4 (4.605% 9.532%), factor 5 (3.801% 8.99%), factor 6 (3.451% 7.684%). The explainable specific gravity of factor 1 decreases, while the explainable specific gravity of factor 2, factor 3, factor 4, factor 5 and factor 6 increase. The commonality and relative positions of the six factors remain unchanged, and the synthesis of the characteristic values and the overall cumulative total variation remain unchanged, remaining at 64.241%. The Barrett sphericity test, KMO value, and the result of factor analysis is shown in Tables 9 and 10.

Table 9. Barrett sphericity test and KMO value.

Barrett Sphericity Test and KMO Value		
KMO sampling fitness measure	0.921	
Bartlett sphericity test	The approximate chi-square	4795.757
	Degrees of freedom	496
	Significance	0

After extraction, the criteria of factor 1 include: 'the government lacks enterprises management regulations', 'government's management regulations for enterprises has insufficient binding force', 'the government's enforcement of business regulations is weak', 'the government's control measures for the scale of the bicycle-sharing market are not in time', 'the division of responsibility for management of shared bikes within the government is not clear', 'the government's propaganda on parking shared bikes in order is not in place', 'there is misleading in initial propaganda of bicycle-sharing companies'. VAR13, VAR14, VAR15, VAR12, VAR11, VAR18, VAR28 are about the government's supervision and management of bicycle-sharing companies. In terms of solving the problem of disorderly parking of shared bikes, the government's management and supervision of companies are indeed an important link. Bicycle-sharing companies' misleading propaganda in the early stage can also be considered as the result of government's poor management. Therefore, factor 1 is named as 'supervision and management of enterprises'.

Table 10. Result of factor analysis.

No.	Factor 1	Factor 2	Factor 3	Factor 4	Factor 5	Factor 6
VAR13	**0.804**					
VAR14	**0.757**					
VAR15	**0.75**					
VAR12	**0.737**					
VAR11	**0.562**					
VAR18	**0.508**	0.494				
VAR28	**0.415**			0.412		
VAR37		**0.765**				
VAR39		**0.657**				
VAR26		**0.598**			0.458	
VAR25		**0.572**			0.475	
VAR23		**0.571**				
VAR16		**0.549**				
VAR17	0.49	**0.542**				
VAR24		**0.479**				
VAR31			**0.779**			
VAR30			**0.777**			
VAR29			**0.648**			
VAR32			**0.546**	0.511		
VAR27			**0.41**			
VAR33				**0.69**		
VAR42				**0.681**		
VAR34				**0.64**		
VAR40				**0.549**		0.508
VAR35				**0.489**		
VAR7					**0.76**	
VAR8					**0.695**	
VAR10					**0.684**	
VAR9					**0.559**	
VAR19						**0.785**
VAR20						**0.716**
VAR21			0.451			**0.599**
Eigenvalue	4.409	4.19	3.572	3.05	2.877	2.459
Extraction sums of squared loadings %	39.375	6.727	6.282	4.605	3.801	3.451
Rotation Sums of Squared Loadings %	13.778	13.095	11.162	9.532	8.99	7.684
Total explanatory variance %	64.241					
The overall-scale Cronbach's α	0.949					
Subscale Cronbach's α	0.883	0.887	0.830	0.808	0.797	0.815

The criteria of factor 2 include: 'lack of social supervision', 'social media's propaganda on parking regulations is not in place', 'the accountability of users was not well investigated and affixed', 'the penalties for users' violation of parking is not enough', 'bicycle-sharing companies lacks user management regulations', 'Incomplete laws', 'the government has no supervision of users of shared bikes', and 'there were no clear parking specifications'. VAR37 and VAR39 are about social supervision of users. After large quantities of exposure of disorderly parking of shared bikes, users began to realize the importance of parking shared bikes in order, showing the power of social supervision. VAR26, VAR25, VAR23 and VAR24 are about bicycle-sharing companies' supervision and management of users. VAR16 and VAR17 are about the government's supervision and management of users. Taking all three aspects of supervision and management into consideration, factor 2 is named as 'supervision and management of users'.

The criteria of factor 3 include: 'the setting of shared bikes parking area is unreasonable', 'the capacity of shared bikes parking space is insufficient', 'there was no dedicated parking space for shared bikes', 'the parking space was occupied', and 'the bicycle-sharing companies didn't explicitly remind users to park in order'. VAR31, VAR30, VAR29, VAR32 are about parking space. VAR27 are about bicycle-sharing companies' management of users. VAR27 is not strongly related to the previous criteria. Therefore, the naming of factor 3 is mainly based on the first four criteria, and factor 3 is named as 'parking space'.

The criteria of factor 4 include: 'the parking area was not clearly defined', 'temporary parking', 'the guiding signs of parking space is not eye-catching', 'weather condition, which makes it impossible for users to use the shared bike as usual', and 'there was no electronic fence'. VAR33 and VAR34 are about physical guiding system of parking shared bikes. The 'electronic fence' of VAR35 interacts with users through the mobile APP interface to guide users to park shared bikes. Electronic fence can be recognized as a kind of digital guiding system of parking shared bikes. VAR40 and VAR42 are all very special cases, and their content are very different from the previous criteria. Taking VAR33, VAR34 and VAR35 as the reference, and the study named factor 4 as 'guidance of parking shared bikes'.

The criteria of factor 5 include: 'laziness', 'users don't attach enough importance to parking shared bikes in order', 'users were not punished, so they had fluke mind', 'users didn't take the initiative to understand the relevant rules of parking shared bikes'. All criteria are about users' control of their own behavior and intention. Factor is named as 'user self-discipline'.

The criteria of factor 6 include: 'insufficient number of bicycle-sharing companies managers', 'the bicycle-sharing companies can't maintain the damaged vehicle properly', 'the vehicle distribution policy of bicycle-sharing companies is unreasonable'. All criteria are about bicycle-sharing companies' operation and maintenance of shared bikes. Factor 6 is named as 'operation and maintenance'.

3.2.4. Influence of Users' Basic Information on the Factors of Disorderly Parking of Shared Bikes

In this study, independent sample t test and one-way ANOVA are used to examine the influence of users' basic variables on the factors of disorderly parking of shared bikes. Users in different areas have no significant differences in factor 1, factor 2, factor 3, factor 4, and factor 5. Shared bike users on weekdays and weekends have no significant differences in factor 1, factor 2, factor 3, factor 4, factor 5, and factor 6. Users with or without experience of disorderly parking have no significant differences in factor 1, factor 2, factor 3, factor 4, factor 5, and factor 6.

But users in different areas have significant differences in factor 6. After post-event comparison, it was found that the users in the downtown area and users outside downtown but within the suburbs have much higher recognition of factor 6 than the users in the suburbs, indicating that bicycle-sharing companies should strengthen the operation and maintenance within the suburbs.

4. Conclusions and Suggestions

In the context of reducing emissions to achieve sustainable development, the development of a complete low-carbon transport system is crucial. As an important part of low-carbon transportation system, the development trend of bicycle-sharing industry has been stable. At this stage, solving the problem of disorderly parking of shared bicycles becomes the key to ensure the long-term development of the industry. This study collected the causes of disorderly parking of shared bicycles as comprehensively as possible through two-phase questionnaire survey, and constructed the factors of the causes of disorderly parking of shared bicycles through factor analysis, so as to facilitate the decision-making of the government and enterprises for reference. The conclusions of this study are as follows:

1. This study has extensively collected the causes of disorderly parking of shared bicycles. Combined with literature discussion, it is found that the three main objects involved in disorderly parking of shared bicycles are users, bicycle-sharing companies and the government. In addition, six factors and 32 criteria for the causes of disorderly parking of Shared bicycles are obtained, which are

as follows: factor 1 'supervision and management of enterprises'; factor 2 'supervision and management of users'; factor 3 'parking space'; factor 4 'guidance of parking shared bikes'; factor 5 'user self-discipline'; factor 6 'operation and maintenance'.

2. Users lack self-discipline; the supervision and management of users by bicycle-sharing companies, government and society is not in place; the government's management and supervision of enterprises are not in place, and bicycle-sharing companies' operation and maintenance of shared bikes are not in place; the failure of enterprises and governments to cooperate to provide sufficient parking space and to equip them with good parking guidance system, all show that the solution of disorderly parking requires multi-party cooperation. Therefore, in the formulation of solutions, we should take multiple factors into account in order to properly solve the problem of disorderly parking of shared bicycles.
3. Based on the tripartite cooperation mechanism and six factors, we propose the following suggestions: 1. The government needs to strengthen the supervision and management of bicycle-sharing companies. The government needs to formulate a complete, feasible and binding regulation for bicycle-sharing companies. Besides, the management responsibilities within the government should be clarified and management norms should be implemented effectively. 2. The government, enterprises and society should strengthen the supervision and management of users. The government and society should strengthen the publicity of parking shared bikes in order. The government and enterprises shall formulate clear parking standards or regulations of parking shared bikes, and violators shall be dealt with in strict accordance with the provisions. 3. The sharing-bicycle companies shall provide reasonable and sufficient parking space for users and shall clearly remind users to park shared bikes in accordance with parking regulations. The sharing-bicycle companies may cooperate with the government to provide parking space or adopt technical means to meet users' parking needs. 4. When setting up parking spaces, sharing-bicycle companies and governments should set up some guidance signs to facilitate users to find parking spaces. At the same time, the whole parking space and the single parking unit should be clearly demarcated to help users park shared bikes in order. 5. Users also need to correct their attitude, observe the usage and parking rules of Shared bikes to create a good environment for sharing bikes. 6. The sharing-bicycle companies should increase the investment in operation and maintenance and provide sufficient operation and maintenance personnel to ensure the good condition of vehicles. They should also constantly optimize the strategy of vehicle distribution according to users' demand to create a good experience for users.

Author Contributions: All authors contributed to the paper. Q.J., S.-J.O., and W.W. collected and organized data; W.W. wrote the manuscript with the supervision from S.-J.O. and Q.J.; and W.W. acted as a corresponding author.

Funding: This research received no external funding.

Conflicts of Interest: The authors declare no conflict of interest.

References

1. Alonso, A.; Monzón, A.; Cascajo, R. Comparative analysis of passenger transport sustainability in European cities. *Ecol. Indic.* **2015**, *48*, 578–592. [CrossRef]
2. Murphy, E.; Killen, J.E. Commuting economy: An alternative approach for assessing regional commuting efficiency. *Urban Stud.* **2011**, *48*, 1255–1272. [CrossRef]
3. Buehler, R.; Pucher, J. Cycling to work in 90 large American cities: New evidence on the role of bike paths and lanes. *Transportation* **2012**, *39*, 409–432. [CrossRef]
4. Buzási, A.; Csete, M. Sustainability indicators in assessing urban transport systems. *Period. Polytech. Transp. Eng.* **2015**, *43*, 138–145. [CrossRef]
5. Liu, L.; Lu, W.; Cai, J. German urban green traffic management strategy based on green traffic system. *Urban Dev. Res.* **2012**, *3*, 71–77. (In Chinese)

6. Lim, C.H.C.; Liak, T.K.; Jin, H.H. Singapore's Cycling Plan. Available online: https://trid.trb.org/view/1538625 (accessed on 20 April 2019).
7. Midgley, P. Bicycle-sharing schemes: Enhancing sustainable mobility in urban areas. *United Nations Dep. Econ. Soc. Aff.* **2011**, *8*, 1–12.
8. New York City Department of City Planning. Available online: https://www1.nyc.gov/assets/planning/download/pdf/plans/transportation/bike_share_complete.pdf (accessed on 20 May 2019).
9. Guzman, S.; Shaheen, S.; Zhang, H.; Martin, E. Hangzhou public bicycle: Understanding early adoption and behavioral response to bikeshare in Hangzhou, China. *Transp. Res. Rec.* **2011**, *2247*, 33–41.
10. National Bureau of Statistics. *China Statistic Yearbook (1989–2009)*; China Statistical Press: Beijing, China, 2010.
11. Shaheen, S.A.; Guzman, S.; Zhang, H. Bikesharing in Europe, the Americas, and Asia: Past, present, and future. *Transp. Res. Rec.* **2010**, *2143*, 159–167. [CrossRef]
12. iiMedia. Research on the Current Situation of China's Shared Bicycle Development in 2018. Available online: https://www.iimedia.cn/c400/63243.html (accessed on 20 May 2019).
13. DeMaio, P. Bike-sharing: History, impacts, models of provision, and future. *J. Public Transp.* **2009**, *12*, 3. [CrossRef]
14. Fishman, E.; Washington, S.; Haworth, N. Bike share: A synthesis of the literature. *Transp. Rev.* **2013**, *33*, 148–165. [CrossRef]
15. Bührmann, S. Bicycles as Public-Individual Transport–European Developments. Available online: https://docplayer.net/22521099-Bicycles-as-public-individual-transport-european-developments.html (accessed on 20 May 2019).
16. Buck, D.; Buehler, R. Bike lanes and other determinants of capital bikeshare trips. In Proceedings of the 91st Transportation Research Board Annual Meeting, Washington, DC, USA, 22–26 January 2012.
17. Buck, D.; Buehler, R. Bike Lanes and Other Determinants of Capital Bikeshare Trips. Available online: https://nacto.org/wp-content/uploads/2012/02/Bike-Lanes-and-Other-Determinants-of-Capital-Bikeshare-Trips-Buck-et-al-12-3539.pdf (accessed on 20 May 2019).
18. Qin, Z.; Wang, Q. Synergy Mechanism in the Vision of Sharing Economy: Taking Shared Bikes for Example. *Reform* **2017**, *5*, 124–134. (In Chinese)
19. Arrindell, W.A.; Van der Ende, J. An empirical test of the utility of the observations-to-variables ratio in factor and components analysis. *Appl. Psychol. Meas.* **1985**, *9*, 165–178. [CrossRef]
20. MacCallum, R.C.; Widaman, K.F.; Zhang, S.; Hong, S. Sample size in factor analysis. *Psychol. Methods* **1999**, *4*, 84. [CrossRef]

sustainability

MDPI

Article

Shifting to Shared Wheels: Factors Affecting Dockless Bike-Sharing Choice for Short and Long Trips

Ioannis Politis *, Ioannis Fyrogenis, Efthymis Papadopoulos, Anastasia Nikolaidou and Eleni Verani

School of Civil Engineering, Aristotle University of Thessaloniki, 54124 Thessaloniki, Greece; fyrogeni@civil.auth.gr (I.F.); efthympg@civil.auth.gr (E.P.); nikolaid@civil.auth.gr (A.N.); verani@civil.auth.gr (E.V.)

* Correspondence: pol@civil.auth.gr

Received: 5 August 2020; Accepted: 1 October 2020; Published: 5 October 2020

Abstract: In this paper, we explore users' intentions to use bike-sharing systems (BSS) compared to traditional competitive transport modes—private car, bus and walking. Fueled by the increasingly rampant growth of shared economy and Information and Communication Technology (ICT), shared mobility is gaining increasing traction. The numbers of shared mobility schemes are rapidly growing worldwide and are accompanied by changes in the traditional vehicle ownership model. In order to pinpoint the factors that strongly affect the willingness to use BSS, a stated preference survey among car and bus users as well as pedestrians was designed and conducted. Binary logit models of the choice between the currently preferred transportation modes and BSSs were developed, for short and long-duration trips, respectively. The results highlight a distinctive set of factors and patterns affecting the willingness to adopt bike-sharing: choice is most sensitive to travel time and cost of the competitive travel options. In general, users are more willing to make the switch to a BSS, especially for short trip durations, when their typical mode of transport becomes more expensive. Bike-sharing also seems to be a more attractive option for certain user socio-demographic groups per mode and trip duration (age, education level, employment status, household income). Trip characteristics such as trip purpose and frequency were also found to affect the willingness to choose BSS. In general, BSS seem to mainly attract bus users and pedestrians, while car users may use BSS more sparingly, mainly for commuting purposes.

Keywords: sharing economy; bike-sharing; stated preference; discrete choice models

1. Introduction

In the last decade, a tremendously intense transition from an "ownership" model to a "shareship" status has occurred in all aspects of global economy. This adaptation is primarily observed in one of the economy's main pillars, that of transport and mobility. Cohen and Shaheen [1] defined shared mobility as "an innovative transportation strategy that enables users to have short-term access to a mode of transportation (vehicle, bicycle, or other low-speed travel mode) on an as-needed basis". In a broader context, shared mobility is an umbrella term that encompasses several service models, including bike-sharing, car-sharing, ride-sharing (carpooling, vanpooling), ride-hailing, scooter-sharing, shared parking, public transit services, courier network services (shared trucks, electric vehicles, electric cargo bikes), etc. [1,2]. Shared mobility services are constantly expanding and improving, often by integrating new innovative technologies like autonomous vehicles [3].

Within this framework, over the last two decades, the bike-sharing concept has gradually turned into a mainstream form of urban mobility in numerous cities around the world, providing a viable alternative mode of transport for short or medium urban distances [4,5]. To describe bike-sharing, several definitions have already been provided in the literature [4,6,7]. In this context, Mátrai and Tóth [8] reviewed the already existing definitions of the bike-sharing concept, all of which include some of its core characteristics: bike rental schemes, shared mobility, short-term access, point-to-point urban trips. The anticipated benefits and positive effects of bike-sharing include, traffic congestion alleviation, reduced fuel consumption, emissions reduction and air quality improvement, road safety improvement, physical activity increase and public health improvement, accessibility and multi-modality enhancement, reduced individual mobility costs and quality of urban life, to name a few [9–13]. Considering its expected contribution to a set of key objectives, bike-sharing is broadly believed to be a significant component of sustainable urban mobility [1,5,8,14]. Some studies recognized that bike-sharing, along with an effective public transport system and a demand management policy formulation, may be an important driver for achieving sustainability goals [3,14].

Although bike-sharing systems have existed for about sixty years, the last two decades have witnessed a significant growth and spread of such schemes in many cities across the globe [9]. During their historical evolution, bike-sharing schemes have passed through different stages, having undergone several changes regarding their core characteristics. The evolution of bike-sharing systems could be categorized into four stages, also called "generations" [3,8,15]. The first generation of bike-sharing, also known as "white bikes", emerged in 1965 in the Netherlands. A few regular bikes were painted white, randomly distributed across the city and provided for public use, free of charge. In the absence of any security measure to prevent bicycles from being stolen or vandalized, the general failure of these systems was inevitable [9,16]. The problems experienced by the first generation bike-sharing systems, stressed the need for a more structured and secure approach, adopted by the second generation systems. The latter, broadly known as "coin-deposit systems", were initially developed in Denmark in the early 1990s. Although the incorporation of specially designed bikes and coin-deposit docking stations made second generation systems more reliable, bike theft remained a major problem, resulting from low deposit fees and user anonymity [9,16].

The transition to the third-generation bike-sharing, is inseparably linked with the rapid development of Information and Communication Technology—ICT. Third generation systems became increasingly popular, incorporating advanced technologies, such as RFID (Radio-Frequency Identification) and GPS (Global Positioning System), that enabled bicycle and user information tracking. The utilization of such technologies not only helped systems deter bike theft, but also made them capable of monitoring and controlling bike usage. The substantial contribution of ICT to the evolution of the third bike-sharing generation, is reflected in the term "IT-based systems", which is broadly used to describe such programs. The first typical example of this type of bike-sharing system, was developed in France [5,9,16]. The knowledge gained so far, has already set the scene for an emerging fourth generation bike-sharing model. Such a concept was initially introduced by Shaheen et al. [9], referring to demand-responsive, multi-modal systems with innovative characteristics: electric bicycles, enhanced user interface, integration with public transport, bicycle redistribution innovations, GPS tracking, smartphone applications for real-time information, etc. [4,9,17].

During recent years, a consistently increasing trend towards sustainability and reduced energy consumption has been pursued by existing and developing urban transport policy regulations [18]. Multiple, competing or cooperating solutions have been examining regarding this problem in lieu of car usage, that has been dominating the urban transport landscape for decades with repercussions for environmental, social and economic sustainability. Bike sharing is considered one of the most promising solutions to this problem.

While bike-sharing thrives around the world, its low usage gives a definitive cause for concern and further analysis. For the development of smarter and viable bike-sharing systems, so as to be consolidated as convincing mode choice options, it is important to recognize those factors affecting their usage. Since future demand and long-term sustainability of such systems are in doubt [9], a better understanding of those factors could provide valuable insights for the improvement of their efficiency and promotion of their usage [19].

Literature Review

Several studies have already attempted to identify the factors influencing bike-sharing usage, mainly using revealed preference data, including system-use data and data from specifically designed user surveys. Within this context, various factors broadly considered to affect bike-sharing usage were examined, such as individual socio-demographic characteristics (gender, age, occupation, education level, monthly income, household bicycle ownership, etc.), individual travel patterns (trip mode, travel time, trip purpose, etc.), transportation infrastructure, land-use and built environment characteristics, bike-sharing facilities, as well as environmental conditions.

Largely based on system-use data, a certain number of studies sought to recognize the effect of individual socio-demographic characteristics and travel patterns on bike-sharing usage. Major examples of such papers are presented next.

Shaheen et al. [20] conducted a survey in Hangzhou, China, with the overall aim of capturing the determinants of bike-sharing usage and adoption. Having developed two different questionnaires addressed to bike-sharing members and non-members, the authors examined the potential influence of several factors on bike-sharing usage, which could be grouped into four main categories: travel behavior (travel patterns), sociodemographics, psychographics (attitudes towards cycling conditions and environmental issues), and bike-sharing perception and satisfaction degree. The results suggested that bike-sharing members were likely to be less than 45 years old and have a moderate household income, indicating the potential influence of these two variables on bike-sharing usage. Moreover, bike-sharing membership was found not to be negatively affected by high car ownership rates, while bicycle ownership was found to be positively related to greater interest in bike-sharing.

Fuller et al. [21] used data collected by a random-digit dialing telephone survey, seeking to detect correlations between several factors and BIXI (a bike-sharing scheme in Montreal, Canada) usage. A multi-variate logistic regression analysis was conducted, which led to the identification of significant positive correlations between bike-sharing usage and (a) the closer proximity of home addresses to docking stations, (b) the 18–24 age group, (c) higher levels of education (university educated), (d) the return from work trip purpose and (e) the use of bicycle as the primary mode of transport to work.

Ogilvie and Goodman [22] used system-registration data, seeking to detect inequalities in Barclays Cycle Hire usage, in London UK. The authors examined the relationship between bike-sharing usage levels and various explanatory variables, including gender, income deprivation, etc. To that end, a GIS (Geographic Information System)-based, linear regression as well as a logistic regression analysis were performed, leading to the following outcomes: compared to the general population, system members were more likely to be male and live in relatively wealthy areas or in areas of high cycling prevalence. Considering the lower docking station density found in the deprived areas and the trip frequency of users living there, the lower rates of bike-sharing adoption among these areas were attributed to the docking station location. The demographics of system members were also found to be different to those of the general population, in another study concerning Capital Bikeshare system in Washington, DC, USA [23]. The report came up with significant findings, recognizing that when compared to all commuters in the region, bike-sharing members were more likely to be noticeably younger, male, highly educated and slightly less affluent than regional population. Furthermore, bike-sharing members were more likely to live and work within urban areas.

Based on data coming from a pre-existing household travel survey and CaBi (Capital Bikeshare) system-use data, Buck et al. [24] found that, in regard to demographics, socio-economics and travel patterns, significant differences do exist not only between bike-sharing users and the general population, but also between bike-sharing users and traditional cyclists. The analysis concluded that, when compared to traditional cyclists, bike-sharing users were more likely to be younger and female, belong to lower income groups, own fewer cars and bicycles and cycle for personal and work trips. Moreover, the analysis recognized that bike-sharing users mainly shifted from public transport and walking trips.

Guo et al. [19] applied a bivariate ordered probit modelling approach to explore factors affecting bike-sharing usage and satisfaction among the bike-sharing user population in Ningbo, China. A questionnaire survey was used to collect bike-sharing usage and satisfaction data as well as other variables that include socio-economic characteristics and travel patterns. The survey was carried out among bike-sharing members. The statistical analysis of the results indicated that the usage of bike-sharing was affected by gender, household bicycle/ebike ownership, trip mode, travel time, bike-sharing stations location and users' perception of bike-sharing. Additionally, the degree of satisfaction with bike-sharing was affected by household income, bike-sharing station location and users' perception of bike-sharing.

Analyzing system-usage data collected by a survey of users, Yang et al. [25] compared the bike-sharing systems of Beijing, Shanghai and Hangzhou in China, aiming to explore potential differences between trip purpose frequencies. The study did not find a specific trip purpose to be related to increased bike-sharing usage; on the contrary, significant differences in trip purpose were identified across the three cities examined. In Beijing, nearly 45% of users reported bike-sharing usage for journeys to work, compared to 18% for both Shanghai and Hangzhou. On the other hand, in Shanghai, almost 50% of the respondents reported bike-sharing usage for the return from work trip, compared to 29% for Beijing and 23% for Hangzhou. Furthermore, convenience and integration with public transit (metro), were found to be significant factors positively related to bike-sharing usage.

Jensen et al. [26] used bike-sharing ridership data provided by the operator of Lyon's system, in an attempt to understand cyclists' average speed and travel characteristics. The authors identified that trip distances between the system's stations appeared to be shorter compared to the distances that a car user would need to cover in order to travel between these two points. Furthermore, bike-sharing ridership data revealed that the average cyclists' speed ranged between 10 km/h and 14.5 km/h, which made bicycle travel faster than car travel in inner city areas. The study concluded that travel time was a significant factor influencing bike-sharing usage and consequently, the desired modal shift to bike-sharing is heavily dependent on creating conditions favourable to bike-sharing route choice.

Martin and Shaheen [27] used data from a survey conducted in collaboration with Nice Ride Minnesota and Capital Bikeshare schemes, to explore the shift towards public transit as a consequence of bike-sharing. Data collected were analyzed in conjunction with geospatial data and the respondents were mapped depending on their modal shift towards or away from bus and rail. The study also analyzed respondents' socio-demographic characteristics related to modal shift (age, gender, household income, education level, etc.), performing cross-tabulation and ordinal regression analysis. A number of different factors were found to be associated with shifting towards public transit, including increased age, being male, living in lower density areas and longer commute distances.

In their review paper, Fishman et al. [28] critically examined previous studies related to bike-sharing, in order to identify knowledge gaps and provide an outline of the global research on bike-sharing. Through an extended literature review, the authors recognized that convenience and value for money were the most significant components in members' motivation to use bike-sharing schemes. Additionally, private bicycle ownership was identified to be positively related to bike-sharing membership. Lastly, this paper reported that bike-sharing was far from substituting car usage and highlighted a literature gap regarding the perceptions and attitudes of bike-share non-users (especially car users) towards bike-sharing.

Mostly using system-use data, several studies focused on exploring the effect of transport infrastructure, bike-sharing facilities and operations, land use characteristics and weather conditions on bike-sharing usage.

Conducting focus groups with members and non-members and carrying out a thematic analysis for the collected data interpretation, Fishman et al. [29] sought to identify the major barriers and facilitators towards using CityCycle, a bike-sharing system in Brisbane, Australia. The study recognized several factors leading to low bike-sharing usage, including the lack of accessibility/spontaneity, overnight cease of operations of the system, sign-up procedure complexity and safety issues that stemmed from the perceived lack of car driver awareness towards cyclists and poor bicycle infrastructure.

Analyzing system-use data provided by the operator, Buck and Buehler [30] explored determinants of bike-sharing usage regarding the Capital Bikeshare system in Washington, DC, USA. A GIS-based, bivariate correlation as well as a multiple regression analysis were performed, resulting in the identification of a significant correlation between the existence of bicycle lanes and bike-sharing usage. In addition, research findings suggested that population density and mixed land uses were major factors towards encouraging bike-sharing usage.

Cervero and Duncan [31] used factor analysis to take into consideration the urban design and land-use diversity dimensions of the built environment and estimated discrete choice models for bicycles. Data were obtained from the 2000 Bay Area Travel Survey (BATS) and contained information regarding socio-economic characteristics of all household members, as well as their everyday activities, including travel and out-of-home activities. Data on built-environment, density and land-use composition were collected for the year 2000 to match up with BATS travel records. The results from the discrete choice models indicated that weekend and shopping trips were weakly related to bicycling. It was also found that rainfall did not dissuade people from bicycling, while nightfall was more of a barrier. Furthermore, the likelihood of bicycling was found to be increased with the number of bicycles in a person's household. Finally, mixed land use and balances of residences, jobs and retail services seemed to work in favor of bicycling.

Zhao et al. [32] indicated that bike-sharing ridership and turnover rate tended to increase with urban population, government expenditure and the number of bike-sharing members and docking stations. Rixey [33] also suggested that bike-sharing network-based factors, including access to a comprehensive network of stations, bikeways and proximity to bike stations, were highly important to bike-sharing ridership levels, with the other demographic and built environment variables controlled for. Population and retail employment density, as well as middle income levels, were also critical factors in assessing bike-sharing demand.

The proximity of homes to docking stations and the increase in the number of docking stations in residential neighborhoods, appeared also to have the greatest effect on the likelihood of using a bike-sharing system, based on the study of Bachand-Marleau et al. [34]. Similarly, Wang et al. [35] reported that the number of trips made with bike-sharing systems was associated with the proximity to the central business district, accessibility to trails and distance to other bike-share stations. In line with the aforementioned findings, Faghih-Imani et al. [36] concluded that transportation infrastructure, bike-sharing facilities and weather conditions were all significant factors affecting bike-sharing ridership.

The literature review indicated that previous research on factors influencing bike-sharing usage was largely based on revealed preference data, namely, system-use data. On the contrary, little research evidence exists on the identification of the major factors affecting bike-sharing usage, by using stated preference data.

Campbell et al. [37] employed a stated preference survey to model those factors influencing the choice to switch from an existing transportation mode to bike-share or e-bike-share in Beijing. To that end, a mode choice survey was conducted and the collected data were used to develop a multinomial logit model. The study examined trip characteristics and attributes, as well as environmental and weather conditions in order to answer questions about bike-sharing adoption in Beijing. The results of the multinomial logit model indicated that demand was mainly influenced by measures of effort and comfort (trip distance, temperature, precipitation, poor air quality), whereas user demographics were not found to strongly affect the mode choice of the respondents. Research also concluded that the bike-sharing market in Beijing would mostly attract users from other sustainable modes of transport (walking and public transport), rather than private car.

Using data collected by a combined revealed preference and stated preference survey, Shengchuan and Yuchuan [38] developed structural equation models to explore the major factors affecting mode choice and bike-sharing user satisfaction. The overall aim was to identify potential differences of behavior between bike-sharing users and non-users. The study identified the environment of bike-sharing stations and their proximity to home or metro stations as major factors affecting peoples' choice to use bike-sharing schemes. On the contrary, bike-sharing usage was found not to be affected by trip purpose, occupation, income and car ownership. Moreover, the discrete selection model developed showed that when compared to environment and distance, cost was found to play a much less important role in peoples' choice.

Table 1 summarizes the findings of the literature review by grouping factors into six categories; sociodemographic, spatial or infrastructure characteristics, BSS (Bike-Sharing System) characteristics, user behavioral attributes, trip and mobility characteristics and weather/environment characteristics. It also includes a short description of the methods of analysis that were used in each study and the study area.

From the above-mentioned Table, it can be seen that a large part of the literature has exhaustively gone over factors that affect bicycle choice and the characteristics of existing BSS users. On the other hand, much less has been done to examine the incentives that would be necessary for users of other modes of transport to shift to BSS; this examination poses a different question that might be crucial towards shaping a more sustainable urban mobility future. This research gap has also been identified by the pertinent literature. Are users who belong to certain sociodemographic groups or have certain predispositions more prone towards using a BSS? What levels of cost and time gains would persuade users of different modes of transport to switch to a BSS? Several of the studies have found, or hypothesized, that users that shift towards BSS are mainly pedestrians or public transport users and not car users. Are those values different for car or public transport users? Do they change based on the duration of the trip? This paper aims to answer the above in the shape of three research questions:

1. How likely are users with an existing mode choice behavior to shift to a BSS? Does this differentiate among the users with different mode choice?
2. Does and to what extent trip duration affect the probability of choosing a BSS? Should urban transport planning policy be reformulated/adapted to the new challenges?
3. Which individual factors affect the willingness to choose the BSS in favor of currently preferred (and competitive) modes of transport and in what way?

Table 1. Review of studies on factors affecting willingness to use Bike Sharing Systems.

Authors	Year	Study Area	Sociodemographic	Spatial/ Infrastructure	System Characteristics	Behavioral	Mobility and Trip Characteristics	Weather/ Environmental	Method of Analysis
Cervero & Duncan	2003	Bay Area, USA		X	X		X	X	• Discrete choice model that used data from the Bay Area Travel Survey and spatial data
Jensen et al.	2010	Lyon, France					X		• Analysis of BSS users' average speed and trip characteristics using BSS ridership data
Shaheen et al.	2011	Hangzhou, China	X			X			• Questionnaires that compared BSS members to non-members
Fuller et al.	2011	Montreal, Canada	X	X	X		X		• Multi-Variate Logistic Regression using random-digit dialing telephone surveys
Yang et al.	2011	Beijing, Shanghai & Hangzhou, China			X		X		• Comparison between different cities using system-usage data collected via user surveys
Ogilvie & Goodman	2012	London, UK	X	X					• Linear and logistic regression using system-registration data
LDA consulting	2012	Washington DC, USA	X	X					• Comparison between BSS members and general population
Fishman et al.	2012	Brisbane, Australia		X	X	X		X	• Thematic groups of focus groups data with members and non-members
Buck & Buehler	2012	Washington DC, USA		X					• GIS-based, bivariate correlation and a multiple regression analysis using system-use data provided by the operator
Bachand-Marleau et al.	2012	Montreal, Canada		X	X				• Binary logistic model and linear regression model using data from an online survey
Buck et al.	2013	Washington DC, USA	X				X		• Differences between BSS members, general population and traditional cyclists using pre-existing household travel surveys and CaBi system-use data
Fishman et al.	2013		X			X	X		• Literature Review

Table 1. *Cont.*

Authors	Year	Study Area	Sociodemographic	Spatial/ Infrastructure	System Characteristics	Behavioral	Mobility and Trip Characteristics	Weather/ Environmental	Method of Analysis
Rixey	2013	Washington DC, Minneapolis–St. Paul and Denver, USA	X	X	X			X	• Regression analysis that includes demographic and infrastructure characteristics and compares data from three BSS
Shengchuan & Yuchuan	2013	Shanghai, China	X	X	X		X		• Structural equation models using combined revealed and stated preference data
Zhao et al.	2014	China		X	X				• Regression and comparison of data from 69 BSS
Faghih-Imani et al.	2014	Montreal, Canada		X	X			X	• Linear mixed models using minute-by-minute availability data from BSS stations
Wang et al.	2015	Minneapolis–St. Paul, USA		X	X				• Log-linear and negative binomial regression using data from the BSS operator and the 2010 U.S. Census, regional planning agencies and local government
Campbell et al.	2016	Beijing, China	X				X	X	• Multinomial choice model using stated preference data
Guo et al.	2017	Ningbo, China	X	X		X	X		• Bivariate ordered probit model using survey among BSS members data

2. Materials and Methods

2.1. Case Study Area

Located in Northern Greece, Thessaloniki is the second largest city of Greece and one of the major cities in Balkans and the Mediterranean. The Metropolitan area of Thessaloniki covers a geographic area of 1455 km^2 and its population exceeds the 1,000,000 inhabitants [39,40]. The only available mass transit option is the bus, while at the same time bicycle usage is very low (less than 5%) and the bicycle infrastructure is limited (almost 12 km of cycleways). Over the last decades, the modal share of private vehicles has increased from 58% to 68% (+10%), while the modal share of public transport has decreased from 40% to 28% (−12%) [41].

The BSS in Thessaloniki began its operation in 2013. To date, it remains private and includes 200 bikes and eight stations, mainly located along the city's waterfront. The system provides access to its users with an electronic subscriber card and the charge for renting a bicycle includes a cost for accessing the system and a cost for using a bicycle, depending on the usage period. The minimum charge for renting a bicycle is 1€ (for non-registered users) and the maximum permissible duration of each rental is 24 h. The BSS has so far recorded more than 20,000 subscribers. However, in recent years, there has been a reduction in the number of subscribers (−5%) as well as in the average journey time of bicycle trips (−25%) [42,43].

In 2018, a pilot run of a dockless BSS was launched in Thessaloniki. The bicycle fleet is entirely composed of electric bikes which do not require a docking station and can be locked/unlocked using a smartphone app [44]. The promotion of bike-sharing networks has emerged as a priority action for the city of Thessaloniki, as emphasized in the recently published action guide of the "100 Resilient Cities" network entitled "Resilient Thessaloniki: A strategy to 2030" [39].

2.2. Methodology

2.2.1. Data Collection

In order to collect the necessary data to quantify the identified research questions, a stated preference survey was designed and created in the limesurvey platform [45]. The survey's structure is shown in Figure 1 and targeted the users of the two dominant modes of transport in Thessaloniki, the private car and the public bus as well as the pedestrians, which account for the vast majority of trips in the city. The design of the study focuses separately on the 3 main travel choices of the city, since they reflect completely different mobility needs, purposes, flexibility levels, safety and security demands, etc. Multimodal trips were not taken into consideration, since scarcely any multimodal activity takes place in the city (the multimodal transfer rate is equal to 1.0). Users of privately owned bikes were also not included in the survey as they account for an extremely small percentage of the trips in the city [46]. In the first section, respondents were asked about their personal sociodemographic characteristics and whether they own a private bicycle or not. In the second, they were asked about the trip characteristics of their most recent frequently repeated trip, including their estimated or perceived total duration, In Vehicle Time (IVT), Out of Vehicle Time (OVT) and cost. As those terms might have been confusing for many of the respondents, the terms were explained to them in detail in the relevant questions' descriptions. Afterwards, they are provided with a thorough but concise description of the dockless BSS. Finally, they were presented with a stated preference game, in the form of conjoint tasks and were asked to choose between the BSS and their revealed, currently primary mode of transport (car, public bus or walking) for their most recent frequently repeated trip.

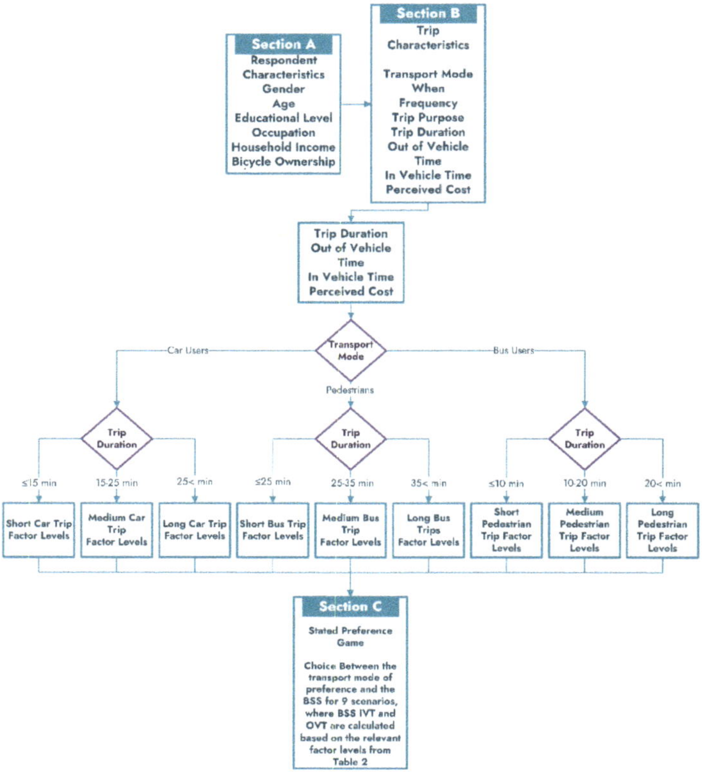

Figure 1. Survey Structure.

As can be seen in Figure 1, the survey's stated preference game was adaptively designed and incorporated different scenarios based on the respondent's answers in the second section of the survey. Depending on their preferred mode of transport and their total trip duration, the stated preference games that were displayed to the respondents were calculated dynamically. Differences between expected OVT, IVT and cost values for the three dominant city transport modes and the BSS alternative were estimated for three trip durations (short, medium and long), by taking into consideration typical elements of a generalized trip cost such as average private car speed in the city, fuel consumption of the average car in the city fleet, average fuel cost, parking fees, maintenance fees and depreciation rates of private cars, average commercial speed of the city buses, average bus stop waiting time, bus fare, average walking speed and average cycling speed. The estimation of those elements, for the specific case of Thessaloniki, was made possible by previously performed traffic analyses, case studies and a macroscopic traffic demand model of the city [46]. Based on those estimations, the factor levels of the BSS trip characteristics were calculated. The BSS IVT and OVT were calculated as percentages of the revealed IVT and OVT for the respondents' transport mode of preference, while the flat charging rates were used for the BSS cost. The full combination of 3 factors with 3 levels each would be $3^3 = 3 \times 3 \times 3 = 27$ different choice combinations. In order to reduce the amount of presented choice scenarios, a fractional factorial design with orthogonality and dominance criteria was applied and resulted in 9 games (choice sets). The factor levels used in the stated preference game are shown in Table 2. The respondents were asked to choose between repeating the same trip with their current mode of preference (with the IVT, OVT and Cost they had already revealed) or the alternative option of the dockless BSS choice (with IVT, OVT and cost that change with each game). The differing factor levels of the BSS IVT and OVT, as well as the different pricing of the BSS, based on duration, were mainly used to adjust BSS characteristics and have a much smaller effect on the outcome than different available choices based on previous answers would have. An example of the stated preference game questions can be seen in Figure 2 (The choice the respondents were put up to in the stated preference game, referred to their most recent frequently repeated trip, for which the respondents have already chosen their currently primarily selected mode of transport. So, it is a choice between an option they have already made and a not-yet-implemented mode of transport that would have been soon added to the city's mobility ecosystem. Only the BSS characteristics changed between the different scenarios of the games). The same type of illustration was chosen for competitive modes, in order to avoid bias. The respondents could see values of the mode characteristics, in terms of time and cost.

Table 2. Factor Levels of the Stated Preference Game.

		Car			Bus			Walk		
		≤15 min	15–25 min	25< min	≤25 min	25–35 min	35< min	≤10 min	10–20 min	20< min
IVT (% of revealed In Vehicle Time)	Level 1	70	100	110	90	80	70	80	80	70
	Level 2	50	80	90	80	70	50	70	70	60
	Level 3	30	60	70	70	60	40	60	60	50
OVT (% of revealed Out of Vehicle Time)	Level 1	100	100	100	80	80	80	N/A	N/A	N/A
	Level 2	80	80	80	60	60	60	N/A	N/A	N/A
	Level 3	60	60	60	40	40	40	N/A	N/A	N/A
Cost (€)	Level 1	1.5	2	2.5	1.5	2	2.5	1.5	1.5	1.5
	Level 2	1	1,5	2	1	1,5	2	1	1	1
	Level 3	0.5	1	1.5	0.5	1	1.5	0.5	0.5	0.5

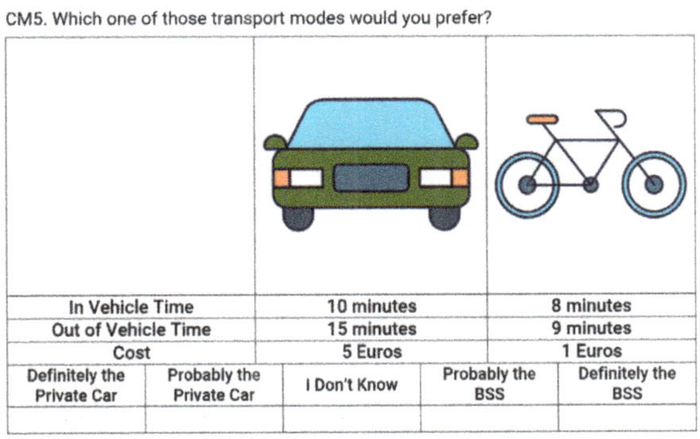

Figure 2. Example of the stated preference games questions as they were presented to the respondents.

Due to the demanding nature of the survey—as it included both terms that many respondents might have been unfamiliar with and a stated preference game—it was decided that data collection should take place on field. Passers-by were randomly approached by interviewers at the city's main intersections and poles of attraction. The interviewers were carefully trained and were equipped with tablets that loaded the online survey. After the interviews, answers were immediately submitted and stored. In order to avoid early morning peak hour, when the vast majority of the respondents would be too busy on their way to and from work, the interviews took place from 10:00 to 20:00. The interviews took place from April to May 2019 and 500 questionnaires were considered as valid for further analysis. Table 3 shows a comparison between sociodemographic characteristics of the collected sample (gender and age group) and the population of the Thessaloniki regional unit. The ages of the respondents are shifted towards the younger age groups as older age groups were more unwilling to answer the survey.

Table 3. Sociodemographic sample and population comparison.

Variable	Factor Levels	Sample Count	Sample Percentage	Population Percentage
Gender	Male	245	49%	47.8%
	Female	255	51%	52.2%
Age Group	18–24	139	28%	10.9%
	25–34	154	31%	17.8%
	35–44	88	18%	18.8%
	45–54	66	13%	17.1%
	55–64	36	7%	13.8%
	>64	17	3%	21.7%

2.2.2. Data Manipulation Based on Trip Duration

For the purposes of this paper the "short" and "medium" trip durations, as they can be seen in Table 2, were unified into one category. This was done in order to achieve a more concise and immediate interpretation of the results and because the "short" and "medium" trip duration categories were found to have more coherent mode choice behaviors, compared to the "long" trip category. The thresholds for the short and long trip durations for each mode of transport were decided based on data from a revealed preference survey from the city of Thessaloniki [46]. A duration threshold that split the number of trips with a 2:1 short to long duration ratio was chosen for each mode of transport. Figure 3 shows the distribution of trips by trip duration, while Table 4 shows the relevant ratios. A slightly larger ratio was eventually chosen for the pedestrian trips due to the higher concentration of trips in shorter durations.

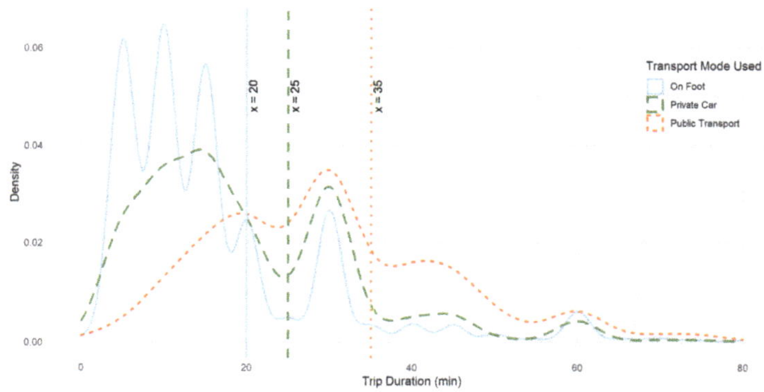

Figure 3. Trip Distribution based on Trip Duration..

Table 4. Ratio of short duration to long duration trips.

	Car Trips (25 min Threshold)	Bus Trips (35 min Threshold)	Pedestrian Trips (20 min Threshold)
Short to Long Duration Ratio	1.84	1.94	2.59

2.2.3. Sample Sizes and Analysis Tools

Out of the 500 responses, 4500 observations/choices were derived and 4167 were eligible to be included in the choice models, as they were the only ones that made a definitive choice of either the BSS or the previously preferred transport mode. Observations that either answered "I don't know" or "probably" for either one of the choices were not used (representing less than 8% of the total responses).

Table 5 shows the respondents and the observations that were included in each mode- and duration-based sub-sample. Each sub-sample's size, as can be seen in the table, was considered adequate for model fitting.

Table 5. Sub-sample sizes.

Sub-Sample	Respondents	Observations/Choices
Car User Short Duration	113	923
Car User Long Duration	101	853
Bus User Short Duration	71	586
Bus User Long Duration	70	574
Pedestrians Short Duration	91	774
Pedestrians Long Duration	54	457

Binomial choice modeling techniques and, more specifically, a binary logit model [47], were utilized to explore the data. The binary nature of the choice has already mentioned in this section; since responders were asked to choose between their current mode of transport (without any alterations at the trip characteristics) and a BSS alternative, there is no possibility to have shift choices among the available/current modes of transport in the city. The analysis was performed with the use of the R programming language [48]. The data handling, manipulation and the subsequent analysis were performed with the following R packages: Dplyr [49], Plyr [50], Stringr [51], Pscl [52], generalhoslem [53], ROCR [54], epiR [55] and ResourceSelection [56].

3. Results

This section presents the results of the binary logit choice models that have been developed within the framework of the study for each for the three discrete population segments: car users, bus users and pedestrians. Additionally, the above three population segments were further divided into short and long segments based on the trip's travel time, as stated by the respondents. So, overall, six discrete datasets were examined and a separate binary logit model was developed for each one. Three types of factors were examined; mode specific (cost, time), trip characteristics and socioeconomic. In Tables 6–8, the six binary logit choice models are presented, for car users, bus users and pedestrians, respectively. The statistically significant variables in the Tables (those with p-value less than 0.05) are highlighted with bold font. For nominal and ordinal factors, the reference category was set as follows: For Sex, the reference was set as the "Male" category; for Age, the reference category was the interval "18–24 years old"; for Trip Frequency, the reference category was the "Daily" trips; for Trip Purpose, the reference category was the "Work" purpose; and for Household Income, the reference was the interval "0–400 euros". Finally, two dummy variables were also considered; the first to examine the preferences of higher educated responders (Bsc, Msc and Phd awarded) against those who had primary and secondary education. The second one to examine differences between those who may be considered as having a stable daily trip schedule (university student, employee, business owner) against those who may not (freelancers, pensioner, unemployed, etc.).

Table 6. Binary Logit Model of Mode Choice between private car and BSS for short and long trip durations for car users.

		Short Trips (≤25 min)					Long Trips (>25 min)				
		Estimate	Std. Error	z Value	Pr(>|z|)	OR	Estimate	Std. Error	z Value	Pr(>|z|)	OR
(Intercept)		−0.903	0.604	−1.497	0.134	0.405	6.536	1.082	6.043	<0.001	689.300
IVT.BSS (min)		−0.258	0.031	−8.402	<0.001	0.772	−0.080	0.018	−4.447	<0.001	0.923
OVT.BSS (min)		−0.099	0.130	−0.759	0.448	0.906	0.065	0.027	2.457	0.014	1.068
Cost.BSS (€)		−1.463	0.196	−7.457	<0.001	0.232	−0.854	0.259	−3.303	0.001	0.426
IVT.Car (min)		0.173	0.028	6.289	<0.001	1.189	0.045	0.016	2.787	0.005	1.046
OVT.Car (min)		0.118	0.105	1.127	0.260	1.126	−0.041	0.021	−1.989	0.047	0.960
Cost.Car (€)		0.221	0.064	3.444	0.001	1.247	0.065	0.020	3.246	0.001	1.067
Frequency	2–3 Times a Day	0.413	0.308	1.341	0.180	1.512	−1.134	0.385	−2.946	0.003	0.322
	3–5 Times a Week	−0.470	0.223	−2.105	0.035	0.625	−0.132	0.280	−0.473	0.636	0.876
	3–5 Times a Month	0.702	0.290	2.424	0.015	2.018	−0.556	0.354	−1.569	0.117	0.574
Purpose	Other Reasons	−0.074	0.481	−0.153	0.878	0.929	−2.539	0.831	−3.055	0.002	0.079
	Education	−0.197	0.328	−0.599	0.549	0.821	−18.665	905.897	−0.021	0.984	0.000
	Entertainment	−0.861	0.288	−2.986	0.003	0.423	−2.304	0.482	−4.777	<0.001	0.100
Sex		−0.436	0.187	−2.330	0.020	0.646	−0.467	0.233	−2.009	0.045	0.627
Age Group	25–34	0.139	0.258	0.537	0.591	1.149	−3.183	0.507	−6.278	<0.001	0.041
	35–44	−0.056	0.284	−0.197	0.843	0.945	−2.942	0.521	−5.643	<0.001	0.053
	45–54	0.359	0.293	1.227	0.220	1.432	−2.444	0.517	−4.724	<0.001	0.087
	55–64	−0.458	0.455	−1.006	0.315	0.633	−5.510	1.142	−4.823	<0.001	0.004
	>64	−15.027	458.368	−0.033	0.974	0.000	−18.076	601.093	−0.030	0.976	0.000
Higher Education		1.043	0.468	2.226	0.026	2.837	−1.964	0.648	−3.030	0.002	0.140
Stable Schedule		0.520	0.214	2.435	0.015	1.682	−0.818	0.245	−3.332	0.001	0.441
Goodness of Fit Metrics		Null deviance: 1209.31 on 922 degrees of freedom; Residual deviance: 927.22 on 902 degrees of freedom; AIC: 969,22; Number of Fisher Scoring iterations: 14; McFadden R^2: 0.230; Hosmer and Lemeshow goodness of fit (GOF) test; X-squared = 7.063, df = 8, p-value = 0.530					Null deviance: 793.36 on 852 degrees of freedom; Residual deviance: 587.51 on 832 degrees of freedom; AIC: 629.51; Number of Fisher Scoring iterations: 16; McFadden R^2: 0.259; Hosmer and Lemeshow goodness of fit (GOF) test; X-squared = 9.134, df = 8, p-value = 0.331				

Table 7. Binary Logit Model of Mode Choice between bus and BSS for short and long trip durations for bus users.

		Short Trip (≤35 min)					Long Trip (>35 min)				
		Estimate	Std. Error	z Value	Pr(>|z|)	OR	Estimate	Std. Error	z Value	Pr(>|z|)	OR
(Intercept)		0.991	0.851	1.165	0.244	2.695	4.587	1.381	3.323	0.001	98.216
IVT.BSS (min)		−0.325	0.067	−4.840	<0.001	0.722	−0.099	0.023	−4.321	<0.001	0.906
OVT.BSS (min)		−0.208	0.054	−3.841	<0.001	0.812	−0.043	0.026	−1.672	0.094	0.958
Cost.BSS (€)		−3.045	0.304	−10.021	<0.001	0.048	−2.122	0.303	−7.006	<0.001	0.120
IVT.Bus (min)		0.213	0.047	4.479	<0.001	1.237	0.036	0.016	2.277	0.023	1.036
Cost.Bus (€)		1.409	0.426	3.304	0.001	4.091	0.240	0.070	3.418	0.001	1.271
Frequency	2–3 Times a Day	−0.505	0.387	−1.304	0.192	0.604	0.412	0.417	0.988	0.323	1.510
	3–5 Times a Week	−1.703	0.446	−3.817	<0.001	0.182	−0.754	0.365	−2.064	0.039	0.470
	3–5 Times a Month	−1.622	0.532	−3.049	0.002	0.198	−0.110	0.434	−0.253	0.800	0.896
Purpose	Other Reasons	0.286	0.506	0.566	0.572	1.331	0.500	0.809	0.618	0.537	1.649
	Education	0.291	0.372	0.782	0.434	1.338	0.387	0.377	1.026	0.305	1.473
	Entertainment	1.008	0.459	2.197	0.028	2.740	−0.765	0.453	−1.688	0.091	0.465
Sex		0.973	0.312	3.117	0.002	2.647	−0.148	0.284	−0.520	0.603	0.863
Age Group	25–34	−0.598	0.325	−1.841	0.066	0.550	0.369	0.325	1.136	0.256	1.446
	35–44	−0.394	0.465	−0.849	0.396	0.674	0.375	0.466	0.805	0.421	1.455
	45–54	−1.235	0.626	−1.973	0.048	0.291	−14.892	737.153	−0.020	0.984	0.000
	55–64	−0.246	0.615	−0.400	0.689	0.782	−2.764	0.759	−3.640	<0.001	0.063
	>64	−15.625	793.954	−0.020	0.984	0.000	-	-	-	-	-
Higher Education		2.586	0.500	5.167	<0.001	13.279	2.248	0.810	2.773	0.006	9.466
Stable Schedule		−0.423	0.304	−1.389	0.165	0.655	−2.261	0.370	−6.108	<0.001	0.104
Household Income	401–800 €	0.661	0.414	1.597	0.110	1.936	−0.898	0.439	−2.048	0.041	0.407
	801–1200 €	0.439	0.461	0.954	0.340	1.552	−0.651	0.462	−1.409	0.159	0.521
	1201–1600 €	0.899	0.466	1.930	0.054	2.457	−1.808	0.516	−3.506	<0.001	0.164
	1601–2000 €	1.425	0.582	2.447	0.014	4.160	1.398	0.778	1.797	0.072	4.046
	2001–2400 €	0.812	0.601	1.350	0.177	2.251	−2.089	0.695	−3.006	0.003	0.124
	More than 2400 €	0.838	0.845	0.992	0.321	2.312	−1.016	0.537	−1.890	0.059	0.362
Goodness of Fit Metrics		Null deviance: 794.52 on 585 degrees of freedom; Residual deviance: 509.28 on 560 degrees of freedom; AIC: 561.28; Number of Fisher Scoring iterations: 15; McFadden R^2: 0.359; Hosmer and Lemeshow goodness of fit (GOF) test; X-squared = 4.900, df = 8, *p*-value = 0.768					Null deviance: 657.37 on 573 degrees of freedom; Residual deviance: 494.46 on 549 degrees of freedom; AIC: 544.46; Number of Fisher Scoring iterations: 15; McFadden R^2: 0.248; Hosmer and Lemeshow goodness of fit (GOF) test; X-squared = 16.693, df = 8, *p*-value = 0.033				

Denotes that no responders were allocated at that segment.

Table 8. Binary Logit Model of Mode Choice between walking and BSS for short and long trip durations for pedestrians.

		Short Trip (≤20 min)					Long Trip (>20 min)				
		Estimate	Std. Error	z Value	Pr(>|z|)	OR	Estimate	Std. Error	z Value	Pr(>|z|)	OR
(Intercept)		−2.227	0.902	−2.469	0.014	0.108	6.678	1.589	4.201	<0.001	794.572
T.BSS (min)		−0.404	0.086	−4.699	<0.001	0.668	−0.178	0.053	−3.328	0.001	0.837
T.Walk (min)		0.517	0.069	7.507	<0.001	1.678	0.051	0.037	1.401	0.161	1.053
Cost.BSS (€)		−3.138	0.342	−9.181	<0.001	0.043	−4.206	0.491	−8.560	<0.001	0.015
Frequency	2–3 Times a Day	0.433	0.367	1.178	0.239	1.541	0.976	0.894	1.092	0.275	2.654
	3–5 Times a Week	0.278	0.376	0.738	0.460	1.320	2.048	0.620	3.303	0.001	7.752
	3–5 Times a Month	−0.384	0.440	−0.873	0.383	0.681	2.017	0.930	2.170	0.030	7.514
Purpose	Other Reasons	1.903	0.447	4.254	<0.001	6.708	−0.005	1.149	−0.004	0.997	0.995
	Education	1.464	0.446	3.284	0.001	4.322	1.754	0.819	2.142	0.032	5.778
	Entertainment	0.631	0.379	1.665	0.096	1.879	−1.612	0.774	−2.082	0.037	0.200
Sex		−0.074	0.262	−0.281	0.779	0.929	1.585	0.503	3.154	0.002	4.880
Age Group	25–34	0.874	0.364	2.402	0.016	2.397	0.196	0.672	0.292	0.771	1.216
	35–44	1.356	0.452	3.003	0.003	3.881	−0.552	0.911	−0.606	0.545	0.576
	45–54	−1.240	0.538	−2.304	0.021	0.289	2.252	1.394	1.616	0.106	9.508
	55–64	−1.743	0.686	−2.539	0.011	0.175	−3.790	1.232	−3.076	0.002	0.023
	>64	−1.377	0.728	−1.892	0.058	0.252	0.408	0.907	0.449	0.653	1.503
Higher Education		−0.770	0.371	−2.075	0.038	0.463	0.842	0.521	1.617	0.106	2.321
Stable Schedule		−0.860	0.375	−2.296	0.022	0.423	−3.217	0.587	−5.476	<0.001	0.040
Household Income	401–800 €	−0.177	0.425	−0.416	0.677	0.838	−2.847	0.837	−3.400	0.001	0.058
	801–1200 €	−0.105	0.421	−0.250	0.803	0.900	−2.842	0.797	−3.564	<0.001	0.058
	1201–1600 €	0.253	0.455	0.555	0.579	1.287	−3.267	0.993	−3.289	0.001	0.038
	1601–2000 €	−0.473	0.534	−0.887	0.375	0.623	−2.843	0.899	−3.164	0.002	0.058
	2001–2400 €	−0.143	0.614	−0.233	0.816	0.867	−0.584	1.105	−0.528	0.597	0.558
	More than 2400 €	1.943	0.648	2.999	0.003	6.978	−6.939	1.534	−4.522	<0.001	0.001
Goodness of Fit Metrics		Null deviance: 758.27 on 773 degrees of freedom; Residual deviance: 495.71 on 750 degrees of freedom; AIC: 543.71; Number of Fisher Scoring iterations: 6; McFadden R^2: 0.346; Hosmer and Lemeshow goodness of fit (GOF) test; X-squared = 5.215, df = 8, p-value = 0.734					Null deviance: 530.31 on 456 degrees of freedom; Residual deviance: 276.40 on 433 degrees of freedom; AIC: 324.4; Number of Fisher Scoring iterations: 7; McFadden R^2: 0.479; Hosmer and Lemeshow goodness of fit (GOF) test; X-squared = 6.209, df = 8, p-value = 0.624				

3.1. Car Users Datasets

All the mode-specific variables were found to be statistically significant for the long-duration model, while only OVT of the car and BSS trips were not found to be statistically significant for the short-duration model. Increased BSS cost harshly reduces the probability of preferring the BSS for both trip durations (0.2 and 0.4 of the odds of preferring a car per increased Euro of BSS cost, respectively). Increased car cost makes it more likely to prefer the BSS, which is more profound for the short-duration model (1.19 and 1.07 of the odds of preferring the car per increased Euro of car cost, respectively). Increased IVTs of the BSS decrease the probability of preferring the BSS (0.77 and 0.92 of the odds of preferring a car per increased minute of BSS IVT for the short and long duration model, respectively). Increased IVT of the car increases the probability of preferring the BSS (1.19 and 1.05 of the odds of preferring a car per increased minute of car IVT for the short and long duration model, respectively). On the other hand, increased OVT of the BSS slightly seems to increase the probability of preferring the BSS for the long duration model (1.068 of the odds of preferring the car per increased minute of BSS OVT) and increased OVT of the car seems to decrease the probability of preferring the BSS for the long duration model (0.96 of the odds of preferring a car per increased minute of car OVT).

Regarding trip characteristic variables, both trip frequency and trip purpose were found to be statistically significant for both trip durations. For short-duration trips, car users that repeat the trip 3–5 times a week were less likely to prefer the BSS compared to users that repeat the trip daily. On the other hand, users that repeat the trip less frequently (3–5 times a month) are more likely to prefer the BSS for that trip duration. For long-duration trips, car users that repeat the trip multiple time a day are less likely to prefer the BSS compared to users that repeat the trip daily. For both trip durations, car users are less likely to choose the BSS for trips with entertainment as the trip purpose, compared to commuters. For long-duration trips, trips with "other reasons" as a purpose are also less likely to be done with the BSS rather than the car.

Out of the variables describing the socioeconomic characteristics of the car users, the variables for sex, education level and occupation were statistically significant for both the short and long-duration model, while the users' age group was statistically significant for the long-duration model. Female users are less likely to prefer the BSS for both trip durations. Having a higher level of education and a stable form of occupation was found to increase the probability of choosing the BSS for short-duration trips but decrease it for long-duration ones. For long-duration trips, all age groups were much less likely to prefer the BSS compared to the reference age group "18–24" (odds ratios ranging from 0.004 to 0.087).

3.2. Bus Users Datasets

IVTs and cost of the BSS and the bus were found to be statistically significant for both trip durations, while only the OVT of the BSS was found to be statistically significant for short-duration trips. Increased BSS cost intensely reduces the probability of preferring the BSS for both trip durations (0.05 and 0.12 of the odds of preferring the bus per increased Euro of BSS cost, respectively). Increased bus cost increases the probability of preferring the BSS for both durations, but to a greater degree for short-duration trips (4.09 and 1.27 of the odds of preferring the bus per increased Euro of bus cost, respectively). Increased IVT of the BSS acts as a deterrent towards bus users preferring it, both for short and long duration trips (0.72 and 0.91 of the odds of preferring the bus per increased minute of BSS IVT, respectively). Increased OVT of the BSS seems to reduce the probability of preferring the BSS for short duration trips (0.81 of the odds of preferring the bus per increased minute of BSS OVT). Increased IVT of the bus increases the probability of the BSS being preferred for both trip durations but more intensely for short duration trips (1.24 and 1.04 of the odds of preferring the bus per increased minute of bus IVT, respectively).

Trip frequency was included in both trip-duration models, while trip purpose was found to be statistically significant only for short-duration trips. Bus users that repeat the trip 3–5 times a week are less likely to prefer the BSS for both short and long-duration trips, compared to users that repeat the trip daily. Bus users that repeat the trip 3–5 times a month were found to be less likely to prefer the

BSS just for the short-duration model. Bus users are more likely to prefer the BSS for short-duration trips and for trips with entertainment as the purpose.

Regarding the variables describing the socioeconomic characteristics of the bus users, the variables for the users' age group, education level and household income were found to be statistically significant for both the short and long-duration models, while the variables of the users' sex and occupation schedule stability were only statistically significant for the short and long-duration models, respectively. Female bus users have a higher probability of preferring the BSS for short-duration trips. For both trip lengths, certain older age groups of users are less likely to choose the BSS compared to the reference age group "18–24". More specifically, for short duration trips, the age group "45–54" has 0.29 of the odds of the reference age group "18–24" regarding preferring the BSS. For long duration models, the age group "55–64" has 0.06 of the odds of the age group "18–24" regarding preferring the bus. Having a higher education level makes it more likely to prefer the BSS. Having an occupation with a stable schedule seems to decrease the probability of preferring the BSS for long-duration trips. Users with a higher household income seem more likely to prefer the BSS for short-duration tips and less likely for long-duration trips. For the short duration trips, the household income groups "1201–1600 €" and "1601–2000 €" have 2.46 and 4.16 of the odds of the group "0–400 €", respectively, regarding preferring the BSS. For longer duration trips, the household income groups "401–800 €" and "1201–1600 €" have 0.41 and 0.16 of the odds of the group "0–400 €", respectively, regarding preferring the BSS.

3.3. Pedestrian Datasets

Regarding mode-specific variables, the cost and the duration of the BSS trip were found to be statistically important for both the short and long-duration models, while the duration of the trip by foot was only found to be statistically significant for the short-duration model. Increased BSS cost harshly reduces the probability of pedestrians preferring the BSS both for short and long duration trips (0.04 and 0.015 of the odds of preferring to walk per increased Euro of BSS cost, respectively). The increased duration of the BSS trip reduces the probability of being preferred, especially for the short-duration model (0.67 and 0.84 of the odds of preferring to walk per increased minute of BSS time for short and long duration trips, respectively). Increased duration of the trip by foot increased the probability of the BSS being chosen for the short-duration model (1.68 of the odds of preferring to walk per increased minute of walking time).

Trip frequency was only found to be statistically significant for long-duration trips, while trip purpose was found to be statistically significant for both short and long trip duration models. Pedestrians that repeat the trip less frequently (3–5 times a week or 3–5 times a month) are much more likely to prefer the BSS for long-distance trips compared to pedestrians that repeat the trip daily. Pedestrians with education as a trip purpose are more likely to prefer the BSS compared to pedestrians with work as a trip purpose for both trip durations, something that is also observed for pedestrians with "other reasons" as a trip purpose for short-duration trips. For long-duration trips, pedestrians with entertainment as a trip purpose are less like to prefer the BSS.

Regarding the variables describing the socioeconomic characteristics of the pedestrians, the variables for the pedestrians' age group, occupation schedule stability and household income were statistically significant for both trip length models, while the variables for the users' level of education and sex were only found to be statistically significant for the short- and long-duration models, respectively. For the short-duration model, pedestrians within the age group of 25–44 have increased probability of preferring the BSS compared to the reference age group of "18–24" (2.4 and 3.89 of the odds of preferring to walk for the age groups "25–34" and "35–44", respectively, compared to the age group "18–24"), while pedestrians in the age group "45–64" have decreased probability in comparison (0.29 and 0.18 of the odds of preferring to walk for the age groups "45–54" and "55–64", respectively, compared to the age group "18–24"). For long duration trips, only the age group "55–64" has decreased probability of preferring the BSS compared to the age group "18–24" (0.023 of the odds of preferring to walk). For short duration trips, household incomes of more than 2400 € appear to have increased probabilities of choosing the BSS compared to the reference household income group "0–400 €" (6.98 of the odds of preferring to walk). For longer duration trips, all household income categories except the "2001–2400 €" have increased probabilities of preferring the BSS compared to the reference group (0.001 to 0.058 of the odds of preferring to walk). Female pedestrians are more likely to prefer the BSS for long-duration trips, while the pedestrian's gender did not seem to affect short-duration trips. Having a stable occupation schedule decreases the chance of preferring the BSS for both trip durations but more intensely so for long-duration trips. In addition, pedestrians with a higher education were less likely to prefer the BSS for short-duration trips but education was found to have no statistically significant effect for long-duration trips.

The variables found to affect the probability of preferring the BSS over each competitive transportation mode, for short and long duration trips, respectively, is visualized in the Venn diagrams of Figures 4 and 5.

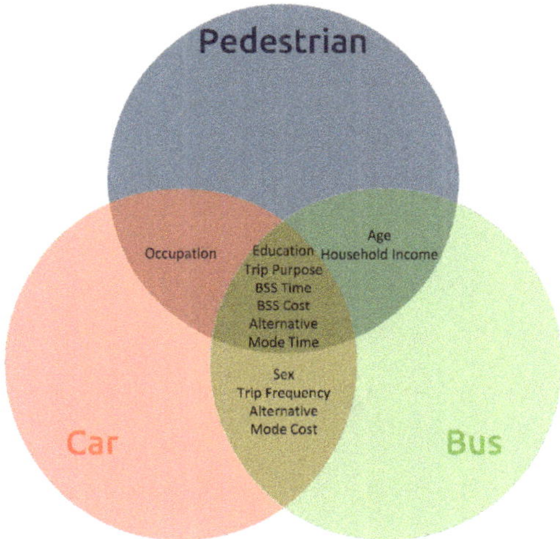

Figure 4. Short distance models' significant factors per current transport mode.

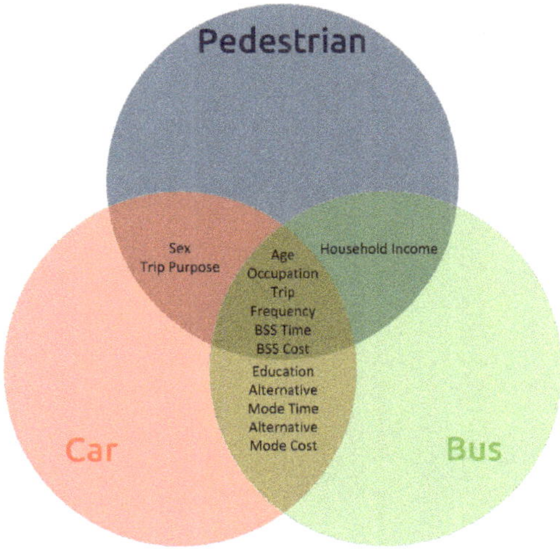

Figure 5. Long distance models' significant factors per current transport mode.

3.4. Models' Goodness of Fit Tests

The statistical tests undertaken show a good fit of the proposed models. The Hosmer and Lemeshow Goodness of Fit test failed for the long-distance bus users model but literature has shown that the test's results can be inaccurate for data sets with a number of covariate patterns less than the number of subjects, as is the case for the data of the long-duration model [57].

Figure 6 shows the Receiver Operating Characteristic (ROC) curves that have been plotted for all six models and consist of the true positive rate plotted against the false positive rate. The closer the plotted curves are to the left and top borders of the plot and the bigger the Area Under the Curve (AUC) is, the better the predictive capabilities of the model. The models' AUCs are displayed in Table 9 and show that all of the models are very predictively efficient.

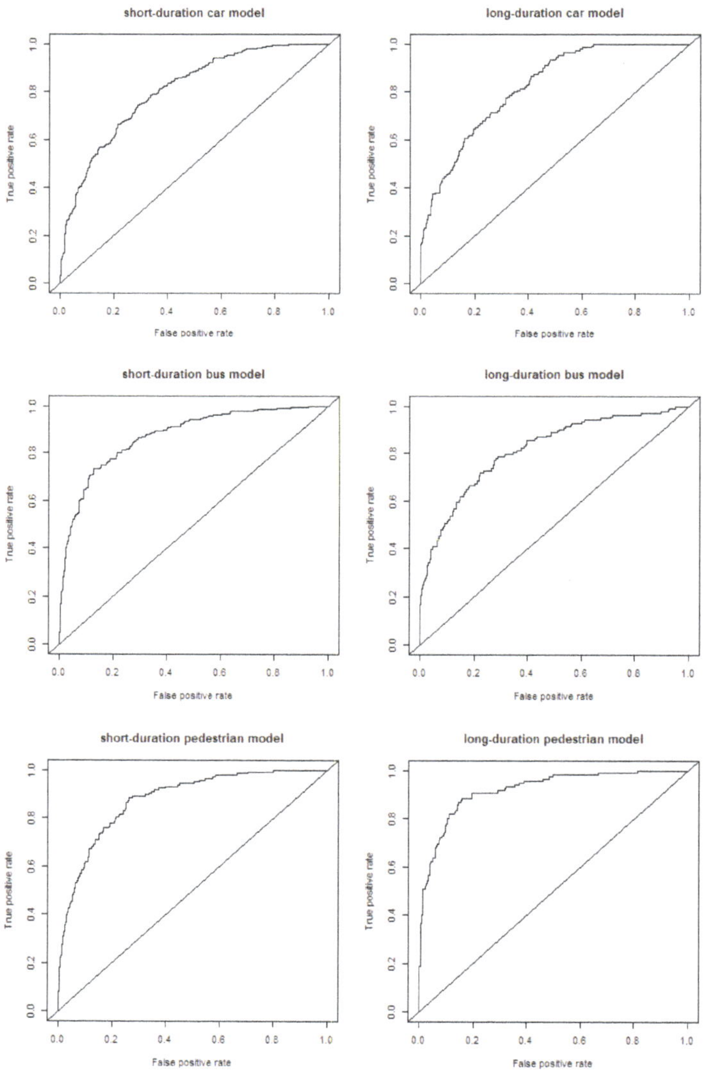

Figure 6. Receiver Operating Characteristic (ROC) curves of the models.

Table 9. Model Area Under the Curve (AUC) values.

Model	AUC
Short-Duration Car	80.7%
Long- Duration Car	82.4%
Short- Duration Bus	87.5%
Long- Duration Bus	82%
Short- Duration Pedestrian	87.9%
Long- Duration Pedestrian	92.3%

4. Discussion and Conclusions

4.1. Main Findings

This paper attempts to identify the crucial factors that contribute towards BSS choice, by setting trip duration as a vital stratification parameter for analysis. Through that, various outcomes have been derived that demonstrate BSSs' potential to substantially become a part of the larger urban ecosystem and replace or supplement traditionally dominant transport modes, both for shorter and longer trips.

The cost of the BSS was found to be statistically significant across all six datasets. While increased BSS cost radically decreases the probability of choosing the BSS across all three modes of transport, the decrease is much more intense for the pedestrian models and the short-duration bus users model, while it is less for the long-duration car users model. For car users, car cost was included in both short and long duration models but was more of a deterrent in the short-duration model. The same can be observed for the bus users' models but its effect on the short-duration model is larger in magnitude. An increase in bus cost heavily increases the probability of choosing the BSS for short trips. While increased car cost for short trips also increases the probability of choosing the BSS, the increase is much more moderate compared to the bus, showing that car users are more hesitant to make the switch to the BSS due to increased costs compared to bus users and it also indicates that users are more willing to make the switch to an alternative mode of transport for short-duration trips when their typical mode of transport becomes more expensive.

IVT or total time of the BSS trips are included in all models and its increase is a meaningful deterrent to choose the BSS, especially for short-duration trips. At the same time, increased IVT and OVT of car and bus, respectively, increases the probability of choosing the BSS to a smaller degree, more prominently for short duration trips.

Regarding socioeconomic characteristics, older users seem to be more reluctant towards choosing the BSS across modes, while higher household income does not seem to affect different groups of users in the same way. Women car users are less likely to make the switch to the BSS compared to women that use the bus or travel on foot. Having an occupation with a stable schedule increases the probability of preferring the BSS only for short-duration car trips and decreases the probability for all other modes and durations except for short-duration trips by foot, where it was not found to be statistically significant. Bus users and pedestrians, with higher household income, are both more likely to prefer the BSS for short-duration trips and less likely to prefer it for long duration trips, possibly showing that, despite income, the BSS is a more attractive alternative for shorter trips.

Car users are more willing to switch to bike-sharing for commuting, while pedestrians are more willing choose it for entertainment, especially for short-duration trips. Pedestrians are very likely to choose the BSS for trips with education as a purpose.

If bike-sharing is to play an increasingly enhanced role towards a more sustainable urban transportation landscape, it is essential to understand what makes choosing it an attractive alternative. Different and discrete groups of users need to be identified and their separate needs and views of the mode evaluated and taken into consideration. The current paper manages to offer a deeper look into the profile of potential BSS users and the mechanisms behind their decision-making. The potential BSS user is more likely to choose the BSS for short duration trips but is very conscious of the BSS's cost, especially when walking is a zero-cost alternative. Given enough incentives, including competitive cost and improved level of service—something that requires investments in dedicated bicycle lane infrastructure—they would be more willing to make the switch, and substitute their current mode of preference for a wide range of trip purposes, including both commuting and less frequently repetitive trips.

BSSs' contribution to increased sustainability or urban mobility is two-fold; it is an active mode of transport, and it is a shared one. The external benefits of bicycle use have been consistently found to be heavily increased compared to the private car and they extend to health benefits, noise reduction, increased safety and environmental pollutants [58]. This difference becomes even more prominent by taking into consideration the heavy usage of cars for short trips in the urban environment. More specifically, a recent traffic study of Thessaloniki showed that approximately 45% of all trips done by car were shorter than 2.5 km [46]. This is an even more pronounced concentration of car trips on the shorter end of the distance spectrum than has been observed elsewhere [59]. Following a quantified approach that highlights the necessary incentives that would convince short trip car users to switch to a more sustainable alternative, provides the necessary insights towards formulating appropriate tools and mechanisms that would convert a good amount of those car trips to bicycle trips. These outcomes can be very productively utilized by policy makers and transportation planners towards formulating new regulations and incentives and setting up an integrated transport and mobility plan. Pricing BSS competitively and providing the necessary infrastructure makes them a time-saving option especially in the most time-sensitive short trips in the dense urban center, will increase usage and make them an even more approachable mobility option, through economies of scale. Municipality or corporation-supported actions that support or even partially fund regular use for commuting for their workers will promote frequent trips during peak-hours, when the increased congestion will make them even more competitive compared to heavily mechanized traffic.

Without a clear and concise image of the way potential users view bike-sharing—which factors affect their willingness to regularly use it and which make it a more daunting alternative—effectively planning for a future that promotes it and strives for its optimal utilization becomes challenging and uncertain. This challenge is getting even greater where a number of new players (e.g., bikes, scooters, etc.) are directly competing with each other without strict and structured regulations.

4.2. Limitations of the Study and Future Research Directions

This study is prone to limitations, such as the longitudinal limited area of the study and the focus on trip and sociodemographic characteristics. This does not take into consideration the behavioral predispositions that make BSS various levels of appealing or even a non-option, all other things considered equal, or practical challenges, such as unpredictable weather conditions and the want or need (depending on trip purpose and health) to abstain from physical exercise. In addition, the collected sample is shifted towards younger age groups compared to the city's population, due to the relative unwillingness of older age groups to take part in the survey. The difference is more pronounced in the "55–64" and especially in the ">64" age groups, also because Greece is a country with a high average population age. The representativeness of the age group "35–44" is satisfactory as the percentage of the sample is close to that of the population. The younger age groups "18–24" and "25–34" are overrepresented in the sample, possibly because it was easier for younger respondents to take part in the online survey. The stated survey data collection that was used is less reliable than revealed preference alternatives, due to the hypothetical nature of the games in it. The data was collected by trained interviewers and while that method of collection offers some benefits, like better guidance of the respondents through the questions, it is also accompanied by threats, such as social desirability bias that could methodically affect the results, if the respondents felt they were put in a position where they were positively predisposed towards one choice. Furthermore, only the observations that made a non-neutral ("Definitely the Private Car" and "Definitely the BSS") choice were utilized in the current study. While the observations with a neutral choice ("Probably the Private Car", "I Don't Know",

"Probably the BSS") were a low percentage of the overall sample (less than 8%), it is a limitation of the current study that they were not included. Lastly, some endogeneity bias might have been introduced in the model due to the adaptive nature of the survey and merged the short and medium trip durations into one [60].

It would be beneficial for future research to focus, in greater detail, on the disincentives and deterrents or practical challenges that would make BSS services less appealing for car users, especially for short urban trips. That would allow for potential hurdles on the way to be removed and even alternative business and operating plans to be developed, which are more closely fitted to the users' needs. Towards that, a multinomial model could be utilized, which makes full use of the possible choices available to the respondents. Moreover, as the city's mobility ecosystem becomes richer, the interactions between currently dominant modes of transport and the BSS with new modes, like shared ridehailing or shared e-scooters, is very promising.

Author Contributions: Conceptualization, I.P. and I.F.; methodology, I.P. and I.F.; formal analysis, I.F. and A.N.; investigation, I.P., I.F. and E.P.; resources, E.P. and E.V.; data curation, A.N. and E.V.; writing—original draft preparation, I.P., I.F., E.P., A.N. and E.V.; writing—review and editing, I.P., I.F., E.P., A.N. and E.V.; visualization, I.F. and E.P.; supervision, I.P. All authors have read and agree to the published version of the manuscript.

Funding: Part of this study was funded within the framework of the C.H.A.N.G.E. project: "EnhanCing tHe bicycLe-shAre EcoNomy throuGh InnovativE Services & Applications". This research has been co-financed by the European Regional Development Fund of the European Union and Greek national funds through the operational program Competitiveness, Entrepreneurship and Innovation, under the call RESEARCH – CREATE – INNOVATE (project code: T1EDK- 04582).

Acknowledgments: The authors would like to thank the three anonymous reviewers that evaluated our paper through the peer review system. The time they invested, their constructive criticism and thoughtful and fair comments on the submitted version of our paper made a much-improved version of it possible.

Conflicts of Interest: The authors declare no conflict of interest.

References

1. Cohen, A.; Shaheen, S. *Planning for Shared Mobility*; American Planing Association: Chicago, IL, USA, 2018.
2. Shared-Use Mobility Center Shared-Use Mobility Reference Guide. Available online: https://sharedusemobilitycenter.org/wp-content/uploads/2015/09/SharedUseMobility_ReferenceGuide_09.25.2015.pdf (accessed on 19 July 2019).
3. Turoń, K.; Kubik, A. Economic aspects of driving various types of vehicles in intelligent urban transport systems, including car-sharing services and autonomous vehicles. *Appl. Sci.* **2020**, *10*, 5580. [CrossRef]
4. Midgley, P. *Bicycle-Sharing Schemes: Enhancing Sustainable Mobility in Urban Areas*; Commission on Sustainable Development: New York, NY, USA, 2011.
5. Parkes, S.D.; Marsden, G.; Shaheen, S.A.; Cohen, A.P. Understanding the diffusion of public bikesharing systems: Evidence from Europe and North America. *J. Transp. Geogr.* **2013**, *31*, 94–103. [CrossRef]
6. ECF. *Factsheet—The Rise of Bicycle Sharing Schemes*; EFC: Brussels, Belguim, 2012.
7. New York City Department for City Planning. *Bike-Share. Opportunities in New York City*; New York City Department for City Planning: New York, NY, USA, 2009.
8. Mátrai, T.; Tóth, J. Comparative Assessment of Public Bike Sharing Systems. *Transp. Res. Procedia* **2016**, *14*, 2344–2351. [CrossRef]
9. Shaheen, S.A.; Guzman, S.; Zhang, H. Bikesharing in Europe, the Americas, and Asia. *Transp. Res. Rec. J. Transp. Res. Board* **2010**, *2143*, 159–167. [CrossRef]
10. Fishman, E.; Washington, S.; Haworth, N.; Watson, A. Factors influencing bike share membership: An analysis of Melbourne and Brisbane. *Transp. Res. Part A Policy Pract.* **2015**, *71*, 17–30. [CrossRef]
11. Ricci, M. Bike sharing: A review of evidence on impacts and processes of implementation and operation. *Res. Transp. Bus. Manag.* **2015**, *15*, 28–38. [CrossRef]
12. Yang, T.; Li, Y.; Zhou, S. System dynamics modeling of dockless bike-sharing program operations: A case study of mobike in Beijing, China. *Sustainability* **2019**, *11*, 1601. [CrossRef]
13. Ban, S.; Hyun, K.H. Designing a User Participation-Based Bike Rebalancing Service. *Sustainability* **2019**, *11*, 2396. [CrossRef]

14. Nikitas, A. How to save bike-sharing: An evidence-based survival toolkit for policy-makers and mobility providers. *Sustainability* **2019**, *11*, 3206. [CrossRef]
15. Shi, J.G.; Si, H.; Wu, G.; Su, Y.; Lan, J. Critical factors to achieve dockless bike-sharing sustainability in China: A stakeholder-oriented network perspective. *Sustainability* **2018**, *10*, 2090. [CrossRef]
16. DeMaio, P. Bike-sharing: History, impacts, models of provision, and future. *J. Public Transp.* **2009**, *12*, 41–56. [CrossRef]
17. Winslow, J.; Mont, O. Bicycle sharing: Sustainable value creation and institutionalisation strategies in Barcelona. *Sustainability* **2019**, *11*, 728. [CrossRef]
18. European Commission Urban Mobility Package. Available online: https://ec.europa.eu/transport/themes/clean-transport-urban-transport/urban-mobility/urban-mobility-package_en (accessed on 20 July 2019).
19. Guo, Y.; Zhou, J.; Wu, Y.; Li, Z. Identifying the factors affecting bike-sharing usage and degree of satisfaction in Ningbo, China. *PLoS ONE* **2017**, *12*, 1–19. [CrossRef] [PubMed]
20. Shaheen, S.A.; Zhang, H.; Martin, E.; Guzman, S. China's Hangzhou Public Bicycle: Understanding Early Adoption and Behavioral Response to Bikesharing. *Transp. Res. Rec. J. Transp. Res. Board* **2011**, *2247*, 33–41. [CrossRef]
21. Fuller, D.; Gauvin, L.; Kestens, Y.; Daniel, M.; Fournier, M.; Morency, P.; Drouin, L. Use of a new public bicycle share program in Montreal, Canada. *Am. J. Prev. Med.* **2011**, *41*, 80–83. [CrossRef]
22. Ogilvie, F.; Goodman, A. Inequalities in usage of a public bicycle sharing scheme: Socio-demographic predictors of uptake and usage of the London (UK) cycle hire scheme. *Prev. Med.* **2012**, *55*, 40–45. [CrossRef]
23. LDA Consulting. *Capital Bikeshare 2011 Member Survey Report*; LDA Consulting: Washington, DC, USA, 2012.
24. Buck, D.; Buehler, R.; Happ, P.; Rawls, B.; Chung, P.; Borecki, N. Are Bikeshare Users Different from Regular Cyclists? *Transp. Res. Rec. J. Transp. Res. Board* **2013**, *2387*, 112–119. [CrossRef]
25. Yang, T.; Haixiao, P.; Qing, S. Bike-sharing systems in Beijing, Shanghai and Hangzhou and their impact on travel behaviour. In Proceedings of the Transportation Research Board 90th Annual Meeting, Washington, DC, USA, 23–27 January 2011.
26. Jensen, P.; Rouquier, J.B.; Ovtracht, N.; Robardet, C. Characterizing the speed and paths of shared bicycle use in Lyon. *Transp. Res. Part D Transp. Environ.* **2010**, *15*, 522–524. [CrossRef]
27. Martin, E.W.; Shaheen, S.A. Evaluating public transit modal shift dynamics in response to bikesharing: A tale of two U.S. cities. *J. Transp. Geogr.* **2014**, *41*, 315–324. [CrossRef]
28. Fishman, E.; Washington, S.; Haworth, N. Bike Share: A Synthesis of the Literature. *Transp. Rev.* **2013**, *33*, 148–165. [CrossRef]
29. Fishman, E.; Washington, S.; Haworth, N. Barriers and facilitators to public bicycle scheme use: A qualitative approach. *Transp. Res. Part F Traffic Psychol. Behav.* **2012**, *15*, 686–698. [CrossRef]
30. Buck, D.; Buehler, R. Bike lanes and other determinants of capital bikeshare trips. In Proceedings of the 91st Transportation Research Board Annual Meeting, Washington, DC, USA, 22–26 January 2012.
31. Cervero, R.; Duncan, M. Walking, Bicycling, and Urban Landscapes: Evidence from the San Francisco Bay Area. *Am. J. Public Health* **2003**, *93*, 1478–1483. [CrossRef] [PubMed]
32. Zhao, J.; Deng, W.; Song, Y. Ridership and effectiveness of bikesharing: The effects of urban features and system characteristics on daily use and turnover rate of public bikes in China. *Transp. Policy* **2014**, *35*, 253–264. [CrossRef]
33. Rixey, R.A. Station-Level Forecasting of Bikesharing Ridership. *Transp. Res. Rec. J. Transp. Res. Board* **2013**, *2387*, 46–55. [CrossRef]
34. Bachand-Marleau, J.; Lee, B.H.Y.; El-Geneidy, A.M. Better Understanding of Factors Influencing Likelihood of Using Shared Bicycle Systems and Frequency of Use. *Transp. Res. Rec. J. Transp. Res. Board* **2012**, *2314*, 66–71. [CrossRef]
35. Wang, X.; Lindsey, G.; Schoner, J.E.; Harrison, A. Modeling Bike Share Station Activity: Effects of Nearby Businesses and Jobs on Trips to and from Stations. *J. Urban Plan. Dev.* **2015**, *142*, 04015001. [CrossRef]
36. Faghih-Imani, A.; Eluru, N.; El-Geneidy, A.M.; Rabbat, M.; Haq, U. How land-use and urban form impact bicycle flows: Evidence from the bicycle-sharing system (BIXI) in Montreal. *J. Transp. Geogr.* **2014**, *41*, 306–314. [CrossRef]
37. Campbell, A.A.; Cherry, C.R.; Ryerson, M.S.; Yang, X. Factors influencing the choice of shared bicycles and shared electric bikes in Beijing. *Transp. Res. Part C Emerg. Technol.* **2016**, *67*, 399–414. [CrossRef]

38. Shengchuan, J.; Yuchuan, D. Combining Revealed and Stated Preference Methods to Evaluate the Use of Bicycle-Sharing Systems. In Proceedings of the Fourth International Conference on Transportation Engineering, Chengdu, China, 19–20 October 2013.
39. Thessaloniki, C. *Resilient Thessaloniki: A Strategy for 2030*; Thessaloniki: Thessaloniki city, Greece, 2017.
40. Pozoukidou, G. Land use transport interaction models: Application perspectives for the city of Thessaloniki. *Spatium* **2014**, *1*, 7–14. [CrossRef]
41. Thessaloniki Public Transportation. *Socio-Economic Study in the Thessaloniki Metro extension to Kalamaria*; Thessaloniki: Thessaloniki City, Greece, 2011.
42. Thessbike. Website of Thessaloniki's Bike Sharing System. Available online: http://www.thessbike.gr/en/ (accessed on 20 July 2019).
43. Katkadikas, R. *No TitleManagement and Analysis of Massive Data (Big Data) of the Public Bicycle Users in the City of Thessaloniki by Using the Programming Language R*; Aristotle University of Thessaloniki: Thessaloniki, Greece, 2016.
44. Bikeazy. Website of Bikeazy System. Available online: https://bikeazy.gr/ (accessed on 20 July 2019).
45. Limesurvey. Available online: https://www.limesurvey.org/ (accessed on 20 July 2019).
46. Politis, I.; Papadopoulos, E.; Fyrogenis, I.; Nikolaidou, A.; Verani, E.; Delivopoulos, G. *CHANGE Project Transport Model—Report 2.2*; Aristotle University of Thessaloniki, Transport Engineering Laboratory: Thessaloniki, Greece, 2019.
47. Garson, G.D. *Logistic Regression: Binary and Multinomial*; Statistical Associates: Asheboro, NC, USA, 2014.
48. R Core Team. *R: A Language and Environment for Statistical Computing*; R Core Team: Vienna, Austria, 2018.
49. Wickham, H.; Francois, R.; Henry, L.; Muller, K. dplyr: A Grammar of Data Manipulation, R Package Version 0.8.3. Available online: https://cran.r-project.org/package=dplyr (accessed on 20 July 2019).
50. Wickham, H. The Split-Apply-Combine Strategy for Data Analysis. *J. Stat. Softw.* **2011**, *40*, 1–29. [CrossRef]
51. Wickham, H. Stringr: Simple, Consistent Wrappers for Common String Operations, R Package Version 1.3.0. 2018. Available online: https://CRAN.R-project.org/package=stringr (accessed on 20 July 2019).
52. Jackman, S. Pscl: Classes and Methods for R Developed in the Political Science Computational Laboratory, R Package Version 1.4.9. Available online: http://pscl.stanford.edu/ (accessed on 20 July 2019).
53. Jay, M. Generalhoslem: Goodness of Fit Tests for Logistic Regression Models, R Package Version 1.3.3. Available online: https://cran.r-project.org/ (accessed on 20 July 2019).
54. Sing, T.; Sander, O.; Beerenwinkel, N.; Lengauer, T. ROCR: Visualizing classifier performance in R. *Bioinformatics* **2005**, *21*, 3940–3941. [CrossRef] [PubMed]
55. Stevenson, M.; Nunes, T.; Heuer, C.; Marshall, J.; Sanchez, J.; Thorn-, R.; Reiczigel, J.; Robison-cox, J.; Sebastiani, P.; Solymos, P.; et al. Epir: Tools for the Analysis of Epidemiological Data, R Package Version 0.9–79. Available online: https://cran.r-project.org/package=epiR (accessed on 20 July 2019).
56. Lele, S.R.; Keim, J.L.; Solymos, P. ResourceSelection: Resource Selection (Probability) Functions for Use-Availability Data. Available online: https://www.google.com/url?sa=t&rct=j&q=&esrc=s&source=web&cd=&cad=rja&uact=8&ved=2ahUKEwiz6--n9pTsAhVScCsKHXbSDK8QFjAAegQIARAC&url=https%3A%2F%2Fcran.r-project.org%2Fweb%2Fpackages%2FResourceSelection%2FResourceSelection.pdf&usg=AOvVaw16eU8ZwC_4aQ6tDtDzfF-g (accessed on 20 July 2019).
57. Bertolini, G.; D'Amico, R.; Nardi, D.; Tinazzi, A.; Apolone, G. One model, several results: The paradox of the Hosmer-Lemeshow goodness-of-fit test for the logistic regression model. *J. Epidemiol. Biostat.* **2000**, *5*, 251–253. [PubMed]
58. Meschik, M. Reshaping City Traffic Towards Sustainability Why Transport Policy should Favor the Bicycle Instead of Car Traffic. *Procedia Soc. Behav. Sci.* **2012**, *48*, 495–504. [CrossRef]
59. Cole, R.; Turrell, G.; Koohsari, M.J.; Owen, N.; Sugiyama, T. Prevalence and correlates of walkable short car trips: A cross-sectional multilevel analysis. *J. Transp. Health* **2017**, *4*, 73–80. [CrossRef]
60. Liu, Q.; Otter, T.; Allenby, G.M. Investigating endogeneity bias in marketing. *Mark. Sci.* **2007**, *26*, 642–650. [CrossRef]

sustainability

MDPI

Article

Optimization of a Bikeway Network with Selective Nodes

C. S. Shui [1,*] and W. L. Chan [2]

[1] Department of Transportation and Logistics Management, The National Chiao Tung University, Hsinchu City 300, Taiwan

[2] Department of Civil Engineering, The University of Hong Kong, Hong Kong; wengcy@connect.hku.hk

* Correspondence: csshui@nctu.edu.tw; Tel.: +886-3-5712121-57217

Received: 9 October 2019; Accepted: 19 November 2019; Published: 20 November 2019

Abstract: Setting up a bikeway network has been recognized as one of the most effective measures to motivate cycling. In fact, a highly connected, exclusive bikeway network that covers all demand sources can be an attractive and time-saving measure, but it requires very high setup costs. The planner often needs to have a trade-off between demand coverage and travel time under a given construction cost. This paper introduces a novel bikeway design problem which determines an optimal bikeway network that covers all potential cycling demand sources with minimal total travel time and under budget constraints. In the context of designing a bike sharing system, the resultant node set of the bikeway network can be interpreted as the locations of the shared bike stations which can cover all cycling demands. A two-stage solution method, by combining the genetic algorithm and a novel elimination heuristic, is proposed to solve the problem by firstly determining the subset of nodes (selected nodes) that can cover all the demand sources and then designing the bikeway network that connects all selected nodes within a given budget. Numerical studies illustrate the advantages of elimination heuristics in solving the proposed problem and the effect of the budget towards the solution fitness with or without a solution. Case studies of two proposed new towns in Hong Kong are provided to illustrate the applicability and effectiveness of the method in bikeway design. This optimization model can be applied to bike-sharing system design problems which aims to cover all demand sources by providing bike stations that are also well connected with exclusive bikeways subject to budget constraints.

Keywords: bikeway network design; selective nodes; elimination heuristic; demand coverage

1. Introduction

Cycling has been receiving attention in the past decade due to its associated benefits, including improving public health, reducing greenhouse gas emissions, alleviating traffic congestion, and increasing the catchment area of public transportation [1]. It is competitive with cars for short trips in urban regions, as it has a lower total travel cost and sometimes a shorter travel time as well [2]. These benefits have motivated a worldwide development of bicycle networks and public bike-sharing systems (PBSs) [3]. At present, there are more than 1,500 PBSs with more than 17 million bicycles in operation [4].

The provision of a connected bikeway network has been proven to be one of the main measures to motivate cycling because it greatly improves cyclists' safety (e.g., references [1,5]). Buehler and Pucher [6] revealed three critical elements that a bikeway network can use to increase cycling levels, including (1) separation with the roadway traffic, (2) high continuity and connectivity, and (3) high bikeway density. These elements can also be internalized in other complex measures such as a bicycle level of service (BLOS), bicycle compatibility index, or level of traffic stress. In other words, a bikeway network that can motivate cycling should be a dense network of exclusive and continuous bikeways, in which the cyclists can ride safely and encounter fewer detours. It is noted that these elements are also implicitly covered in the existing bikeway network design models. The bikeways (and the intersections) are required to meet a predefined level of service, and the trip lengths for the origin-destination pairs should be minimized or below a predefined upper bound ([2,7]). Though the setup of exclusive bikeways offers a safer cycling environment, its adverse impacts on the existing traffic (e.g., reducing the roadway capacity or on-street parking spaces) should be addressed in order to achieve a balance between the performance of the bikeway and the roadway networks. This introduces bikeway network design problems involving a co-existing roadway network (e.g., references [8,9]).

In addition to these elements, the bikeway network design needs to consider the trade-off between the travel time between every origin-destination (O-D) pair and the bikeway construction costs. By increasing the bikeway density, the travel times between the O-D pairs can be reduced, and more cycling activities can thus be motivated ([2,10]). However, higher bikeway density incurs a higher bikeway setup cost. Since the setup costs for a bikeway network are often limited, there is a question on how a bikeway design that minimizes total travel times for all cyclists can be achieved under the budget constraint. However, previous bikeway design studies have not addressed this question and only focus on either cost minimization or travel time minimization (e.g., references [2,7,9]). On the other hand, a bikeway network needs to handle its service coverage. A common approach is to assume an aggregated demand at each potential bike station, while the objective of the bikeway network design is to connect all potential stations with minimal costs in order to cover all demands. However, this approach neglects the fact that the demand points (e.g., transport interchanges, tourist spots, and schools) are not located exactly at but instead close to all potential nodes: as long as there is one station proximate to a demand point, the bikeway network can cover that demand point. In other words, it is sufficient to cover all demands by selecting a set of nodes among all potential nodes (this set of nodes can be named as a selected node set), and the bikeway layout that connects these selected nodes must have a *lower* construction cost than that connects all potential nodes. This concept of demand coverage is modeled in bike station location design (e.g., reference [11]) but never in bikeway network design, which is used to determine the selected node set and connect these nodes by using bikeways. This study, therefore, proposes a new bikeway design problem that minimizes the total system travel time between the selected nodes under both the budget and demand coverage constraints.

As the above novel problem involves two types of discrete decisions (i.e., locations of the selected nodes and built bikeways), this study prefers to develop a two-stage heuristic to solve the proposed problem. The first stage node selection is solved by the classic Genetic Algorithm, which has proved to be powerful in solving many combinatorial problems (e.g., reference [12]). Meanwhile, this stage can also be solved by other meta-heuristics (e.g., a tabu search and large neighborhood search). Nevertheless, no existing heuristic is applicable to solve the second-stage bikeway construction. As a congestion-free transport mode, the travel time of the bikeway is always assumed to be flow-independent and constant (e.g., reference [7]), and therefore the optimal route for all cyclists of each O-D pair is equivalent to the lowest cost route. Nevertheless, this problem is not equivalent to the shortest path problem because of the budget constraint. The overall bikeway construction cost is limited and sometimes needs to be minimized, which is similar to the minimum spanning tree problem or Steiner tree problem. Table 1 compares the common shortest path problems and minimum spanning tree problems with respect to the number of origins and destinations considered in these problems and lists out their design objectives, and the common heuristics used to solve these problems (i.e., references [13–18]). Regarding the problem type, "1", "all", and "some" represent the number of nodes chosen to be the origins/ destinations, in which "some" indicates that a subset of nodes among all nodes are selected to be the origins/ destinations (which cannot be "one" or "all"). It is noted that unselected nodes can be included in the "selective" case if they can reduce the total cost and maintain the connectivity of the selected nodes. The design objectives can be classified into two classes, cost (weight) minimization, and travel time minimization. The former aims to determine a network that connects the nodes with minimal weight, and the latter aims for a set of minimal cost routes between the origin(s) and the destination(s). In other words, the former guarantees a minimal overall construction cost while the latter provides minimal travel costs between the O-D pairs, but none of them can provide an intermediate solution, which is a network with minimal total travel time under a given budget, especially in the case with a selected set of nodes. In addition, the studied problem should be distinguished from the Generalized Minimum Spanning Tree (G-MST) problem [19] which aims to connect part of the nodes in the network with minimum construction cost. In G-MST, each node has been firstly allocated to a group, and the resultant network requires that all groups should be connected, which implies that at least one node in each group is required to be connected. However, this study does not have any predefined group as not all nodes are selected, so the requirement that at least one element in each group cannot be held. As the studied problem differs from the above-discussed problems significantly, the heuristics used in those problems cannot be applied directly and thus a novel heuristic, namely the elimination heuristic, is constructed to solve the second stage problem. Moreover, as shown in Table 1, both our proposed problem and solution method are novel to the literature while it can be applied in other network design problems.

Table 1. Comparisons of existing shortest path and minimum spanning tree problems.

Problem Type		Design Objective	Examples of Existing Heuristics
Source	**Sink**		
1	1	Minimize travel time between a node pair	Dijkstra's algorithm [13]
1	All	Minimize total edge cost that connects a single source to all other nodes	Prim's algorithm [14] Kruskal's algorithm [15]
All	All	Minimize total travel time between all node pairs	Floyd-Warshall algorithm [16,17]
1	Some	Minimize total edge cost from a single source to a subset of nodes	Kou, Markowsky, and Berman algorithm [18]
Some	Some	Minimize total travel time between all node pairs within a subset of nodes	This study

To summarize, the main contributions of this study include:

1. We propose a novel bikeway network design problem that covers all demand sources and minimizes the total travel times of all cyclists under a budget constraint;
2. We propose a two-stage solution method based on the genetic algorithm and an elimination heuristic which determines the selected node set that covers all demand sources and the bikeway layout respectively;
3. We investigate the effect of weights of the elimination factor, budget, and size of the selected node set on the final design. Two case studies in Hong Kong are provided to illustrate the effectiveness and applicability of the model.

The outline of this paper is listed below. Section 2 presents the problem descriptions and Section 3 describes the solution method. Section 4 provides the numerical results and Section 5 provides a real case example. A conclusion is given in Section 6.

2. Problem Descriptions

The design problem is described below. We firstly consider a potential bikeway network $G(V, E)$ with potential bike station location (i.e., node) set V, potential bikeway set E, and demand point set D. The total demand traveling from point k to point l is denoted as d_{kl}, where $k, l \in D$, and each bikeway that directly connects stations i and j, i.e., $e_{ij} \in E$, has the setup cost c_{ij} where $i, j \in V$. The travel time t_{ij} between the nodes can be different with respect to the travel direction to consider the case that the bikeway is built on a slope. In other words, the downhill travel time can be greatly reduced compared to the uphill travel time. Each demand point $k \in D$ is considered as "covered" when there is at least one proximate bike station for convenient bicycle pick up/ drop off (which should be less than 500 m, as revealed by reference [20]). This defines M_k as the set of potential bike stations which are proximate to demand point k, where $M_k \subset V$. It is also assumed that every demand point is covered by at least one potential bike station (i.e., $\nexists M_k = \varnothing, \forall k \in D$). The budget for the total bikeway construction is set to be B. For simplicity, all potential bike stations are assumed to have unlimited capacity.

This design problem involves two types of design decisions. The first type is to determine whether a node is picked as a selected node, denoted by a binary decision variable X_i which equals 1 if node i becomes a selected node and 0 otherwise. A selected node is defined as a node that covers at least one demand source and is connected by at least one bikeway. To separate the selected nodes from the unselected nodes, all selected nodes are put into a new set S, where $S \subset V$. The second type is the binary decision variable for opening a bikeway Y_{ij}, which equals 1 if the bikeway that connects nodes i and j is constructed and 0 otherwise. An additional note is that we have assumed that every built bikeway has sufficient capacity to accommodate all bike flows and thus the travel time can be a flow-independent constant. In other words, if the bikeway travel time is flow-dependent, a possible extension is to include the third type of decision variables, flow on each bikeway, in the design problem.

This problem involves determining the bikeway layout under four following conditions: (a) the total travel times of all cyclists is minimized, (b) all demand points are covered, (c) all selected node pairs are connected by bikeways, and (d) the total bikeway setup cost does not exceed the budget.

The design objective based on condition (a) can be expressed as

$$\min \sum_{i \in S} \sum_{j \in S} d_{ij} T_{ij}. \tag{1}$$

where T_{ij} denotes the lowest travel time between stations i and j based on the constructed bikeways, where $i, j \in S$. For the demand between points k and l, d_{kl}, the shortest path is determined among a maximum of $|M_k| \cdot |M_l|$ possible paths formed by the potential bike stations close to these two points. Among these paths, the shortest path is the one that has the minimum travel time and the origin and destination of the path are selected nodes. When both points share the same selected node, the travel time between the nodes becomes 0. When there are multiple shortest paths with respect to time between a pair of selected nodes, all demands are assigned to the path with the lowest setup cost. As a result, all demands of a pair of selected nodes are only assigned to a single shortest path.

Proposition 1. *For every feasible solution to the problem, there exists at least one path between every pair of demand sources.*

Proof. Condition (b) implies that at least one bike station proximate to a demand source is a selected node to cover all demand points. Thus, both the origin and the destination of every pair of demand sources have been covered by the selected node. From condition (c), every pair of selected nodes must be connected by at least one path. Therefore, there must be at least one path between every pair of demand sources and this completes the proof. □

Condition (b) (i.e., demand coverage constraint) is satisfied when all demand points are covered by at least one selected node, which can be expressed as

$$\sum\nolimits_{i \in M_k} X_i \geq 1, \forall k \in D. \tag{2}$$

Constraint (2) ensures that at least one potential bike station proximate to each demand point is a selected node. The connectivity between the selected nodes (i.e., conditions (c)) is hard to be expressed solely by X_i and Y_{ij} as the paths between the selected nodes can involve links that consist of unselected nodes. Unlike the minimum spanning tree, we can neither exclude cycles nor set exact bounds to the number of links because the final network layout is unlikely to be tree-like. The connectivity condition can, therefore, be described by the following set of necessary but not sufficient conditions: first, each selected node is connected by at least, but not exactly, one bikeway; and second, the two ends of an opened bikeway can be both unselected nodes.

Finally, condition (d) can be formulated as the total bikeway setup cost cannot exceed a given budget B, which can be expressed as

$$\sum_{i\in V}\sum_{j\in V} c_{ij}Y_{ij} \leq B. \quad (3)$$

As the problem contains the complicated connectivity constraint which cannot be expressed mathematically, metaheuristics should be adopted to solve the problem.

3. The Two-Stage Solution Method

This section presents a two-stage solution method to solve the proposed problem. A genetic algorithm (GA) is adopted firstly to determine the optimal selected node set, and a proposed elimination heuristic (EH) determines the optimal bikeway layout that minimizes the total travel cost without violating the budgetary constraint based on the selected node set.

3.1. The Genetic Algorithm

Figure 1 illustrates the general scheme of the GA for determining the optimal node set. At first, a population of solutions is initialized randomly, in which each individual (solution) in GA is represented by a chain of binary digits with length $|V|$ (number of potential stations). The number of '1's on each chain can be interpreted as the selected nodes of that solution. After evaluating the fitness of all initial solutions, the truncation selection method is adopted to select the x% of the fittest parents in the population and duplicate $1/x$% times to replace the less fit parents. For example, when the fittest 50% of parents are selected from a population of 100 individuals, the selected parents are duplicated 2 times so that a population of 100 individuals can be maintained. Then the parents would have a certain probability of undergoing crossover and mutation for generating new offspring. In the crossover, a crossover point along the chain is randomly selected, and all digits beyond that point are swapped between the two parents. In mutation, a random position in the solution is picked for flipping the bit at that position (as each position can be either 0 or 1). After the crossover and mutation processes, the solutions are compared to remove replicated solutions. Meanwhile, an elitist selection is carried out in GA by storing the best solution from the current generation to be carried over to the next iteration without alteration. The above process is repeated until the maximum number of iterations is reached.

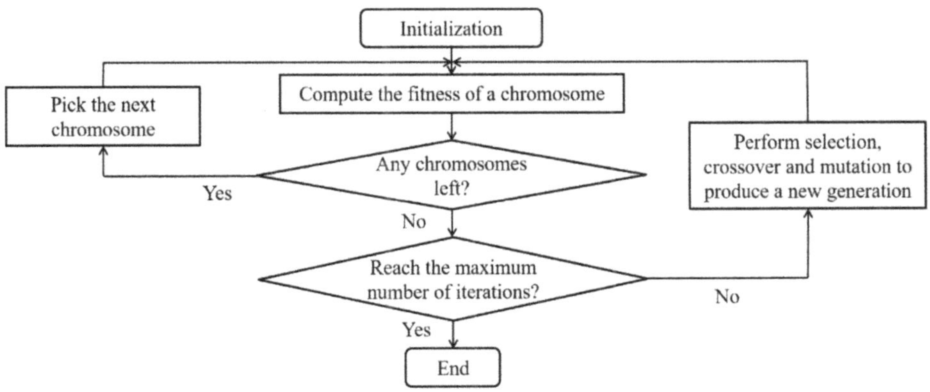

Figure 1. General scheme of the genetic algorithm.

Every generated solution needs to undergo a three-stage solution fitness evaluation. The first stage is a demand coverage check that verifies whether the selected nodes can cover all demand sources. The solution is regarded as infeasible if the selected nodes cannot cover all demand points: its fitness is not further evaluated and a large penalty is imposed. In contrast, a selected station set that can cover all demand points is regarded as feasible and its fitness is then evaluated by the elimination heuristic. The second stage is used to evaluate if the cost of the bikeway network falls below the budget. The elimination heuristic (introduced in Section 3.2) aims to determine a feasible bikeway layout for a given selected node set (from stage 1) which connects all selected nodes and does not exceed the budget. If all selected nodes cannot be *completely* connected under the given budget, the selected node set is also classified as infeasible, and a smaller penalty (compared with the penalty for demand coverage) is imposed on that infeasible node set. After screening out the infeasible solutions, the third stage is to determine the objective values (total system travel times) of the remaining feasible solutions, and the solution fitness can be calculated as the reciprocal of the objective value.

3.2. Elimination Heuristic

The EH is applied to every selected station set which covers all demand points (determined by GA) in order to determine a feasible bikeway layout that does not violate the budget constraint. The steps of the EH are as follows:

Step 1 Obtain the network G_F with the shortest paths between all pairs of selected nodes E_F. Terminate if budget B is not exceeded.

Step 2 Assign an elimination factor f_{ij} to each link e_{ij} in G_F according to its path count p_{ij} and cost c_{ij}. List and sort the links in descending order f_{ij}.

Step 3 Remove the link with the largest f_{ij} in the list and evaluate the network connectivity. Undo the link removal if the selected nodes become disconnected.

Step 4 Remove the evaluated link from the list.

Step 5 Repeat Step 3 until the cost is equal to or lower than B. If the cost is still higher than B after removing all possible links, a penalty is given to the selected node set to denote the solution infeasibility.

Step 1 is to construct a network $G_F(V_F, E_F)$ which consists of the node set V_F and path set E_F, in which V_F and E_F correspond to the selected node set S and the set of shortest paths between the selected node pairs respectively. To determine E_F based on the original network $G(V, E)$ and S, this study applies the minimum spanning tree algorithm (with respect to travel time) for $|S|$ times. The information of all shortest paths between all pairs of selected nodes (e.g., total cost, total travel time, and links included in the path) is then determined and stored. In other words, all selected nodes can reach other selected nodes with the lowest travel times. At this stage, the resultant bikeway network at this step has the lowest total system travel time and the highest total construction cost as most bikeways are constructed.

Step 2 is to determine the order of elimination of the built bikeways given in Step 1 by using the elimination factor f_{ij} to meet the budget constraint. The link-based factor is composed of two parts, the construction cost c_{ij}, and the path count p_{ij}, and expressed as $f_{ij} = a{\cdot}c_{ij} + b/p_{ij}$, where a and b are the non-negative weights for the cost and reciprocal of the path count, respectively. Path count describes the number of shortest paths using the link which can be determined based on the stored path information in Step 1. In other words, links that have higher costs or fewer shortest paths have a larger elimination factor, which implies higher chances to be eliminated. Using this factor is better than simply removing links with respect to cost because it avoids removing high-cost links used by a large number of shortest paths. For simplicity, the elimination factors of all links remain unchanged throughout the evaluation of each solution. In addition, the bikeways which perform as the only link that connects the selected node with other nodes are automatically removed from the list despite a low path count or a high construction cost.

Step 3 evaluates the connectivity between the selected nodes to determine whether any selected node pairs have been permanently disconnected after each bikeway removal. Instead of reconstructing all paths E_F, this step focuses on checking those paths that include the removed link. The heuristic determines whether the two ends of the removed link can be connected by a new path using other built bikeways. If they cannot be connected, the removed link is restored; if they can be connected, the travel times of those affected paths are updated.

The calculation of the updated travel time is not equivalent to the simple addition of the difference in travel time between the new path and the removed link, which can be explained by an example shown in Figure 2a,b. In these figures, nodes 1, 3, 4, and 6 are the selected nodes while nodes 2 and 5 are not. The solid lines denote the built bikeways, the black dotted lines with arrow denote the shortest paths between the selected nodes, and the italic numbers denote the link number. Figure 2a,b show that the initial shortest paths between the selected nodes and the case after link 4 is removed, respectively. When link 4 is removed (in Figure 2b), two shortest paths (between nodes 3-6 and nodes 1- 6) need to be re-routed and the updated shortest path between nodes 3 and 6 becomes 2-3-6. Meanwhile, if this updated shortest path sequence between nodes 3 and 6 is directly substituted into the original shortest path between nodes 1 and 6 (i.e., 1-2-4), the shortest path becomes 1-2-2-3-6 that link 2 is visited twice consecutively which shows a redundant visit on link 2. Therefore, the updated shortest path between nodes 1 and 6 should completely remove the duplicated links and thus becomes a path with link sequence 1-3-6.

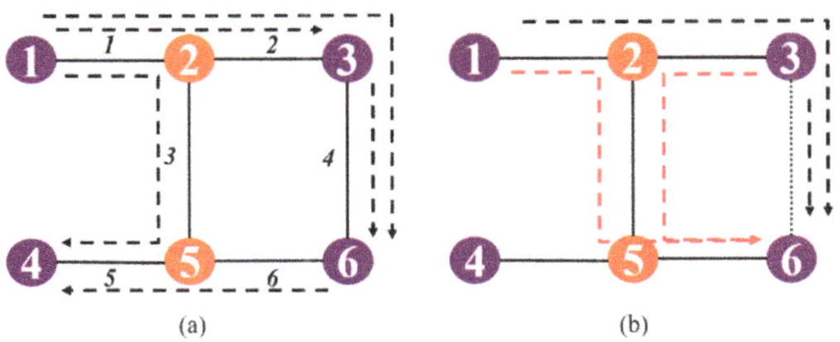

Figure 2. (**a**) Initial shortest routes between selected noes before link removal; (**b**) Updated shortest routes between selected nodes after link removal.

Based on these two cases, the network connectivity evaluation and path updates follow the following steps:

Step 3.1 Determine the new shortest path W that joins the two ends of the removed link V and store its link sequence.

Step 3.2 Update the shortest paths in E_F that contain V by directly substituting I with W.

Step 3.3 Determine the updated travel time T' and link sequence of each shortest path following the below rules:

(a) If it has no duplicated link, then T' is the sum of the original travel time and the travel time difference between V and W, and the updated link sequence is obtained by simply substituting V with W.

(b) If there are duplicated links, T' is calculated by firstly following Step 3.3(a) and then subtracting two times the sum of the duplicated links along the route, and the duplicated links are removed from the link sequence of the updated shortest path formed in Step 3.2.

After removing a link, the path counts of all built bikeways are updated and the built bikeways with zero path count are then immediately removed from the list.

Finally, Step 5 is the termination criteria of the EH in which the heuristic stops only if the total setup cost is lower than the given budget. To distinguish the infeasible solutions with the feasible ones, a large penalty is imposed when the network connectivity is not conserved. This gives preferences to the all-connected networks instead of disjoint networks.

4. Numerical Results

The numerical studies aim to (1) investigate the effect of the weights for elimination factor, (2) analyze the effect of the budget on the total system travel time and the solution fitness, and (3) examine the effect of the number of demand points on the construction cost. The proposed solution method was coded with Dev-C++ and ran on a computer with 3.40 GHz and 6.00 GB of RAM. The GA has a population size of 30 with a 50% truncation rate, a 40% crossover rate, and a 60% mutation rate.

4.1. Effect of the Weights in the Elimination Factor

The solution quality relies strongly on the elimination factor f_{ij}. The weights for the elimination factor, a (for the cost) and b (for the path count), should be calibrated to avoid overemphasizing either attribute. In the experiments, the genetic algorithm was run with the Sioux-Falls network (see Figure 3). To fit our test problem, nine demand sources are added to the network. The weight for the cost a was kept at 1, while the weight for the path count b varied from 0 to 10. Figure 4 shows the effect of the weight of path count on the solution fitness. The solution fitness improves significantly when b increases from 0 to 0.25, improves slightly from 0.25 to 3, and remains steady with negligible fluctuations when b is greater than 3. The values of a and b are thus set to 1 and 3, respectively, in the following sections.

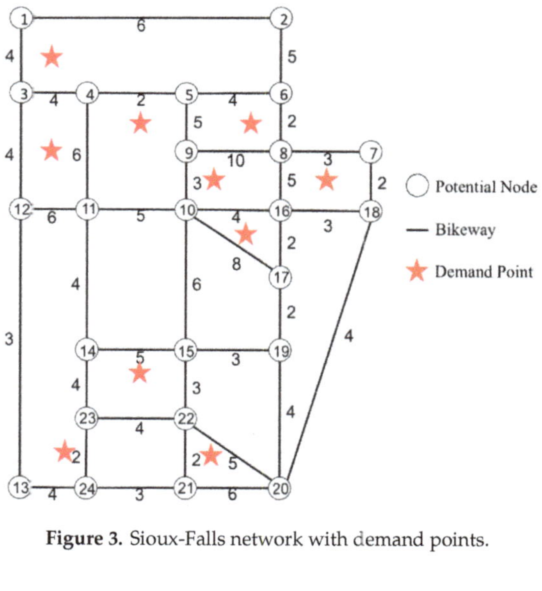

Figure 3. Sioux-Falls network with demand points.

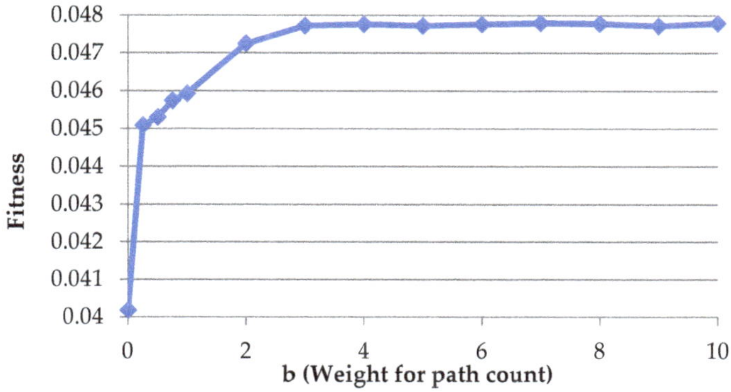

Figure 4. Effects of weight for path count b.

Figure 5 shows that the solution fitness changes with the value of *b* because the weights for the elimination factor change the order of link elimination and thereby affect the network layout. To study the sole effect of the value of b on the solution fitness, the two networks in Figure 5 use the same selected node set with different values of *b* (i.e., 0 and 3). The results show that the network in Figure 5b has a lower total system travel time than that in Figure 5a, given that both networks do not exceed the budget constraint, mainly because of the differences in the order of link elimination. For example, the high-cost link 4-11 is eliminated immediately when path count is not considered (i.e., $b = 0$). Nevertheless, it is kept when the path count is considered because its high path count lowers its order of link elimination. When this link is kept, the shortest paths which use link 4-11 do not have re-routing and thus a lower total system travel time can be achieved. This shows that the elimination factor can be a better approach than cost-based elimination in lowering the total travel time.

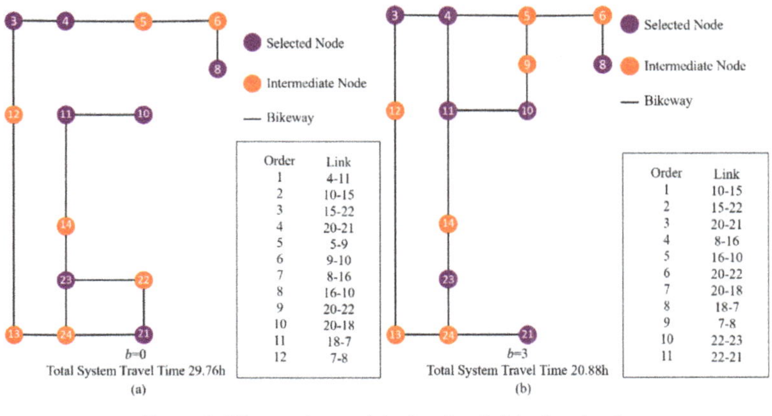

Order	Link
1	4-11
2	10-15
3	15-22
4	20-21
5	5-9
6	9-10
7	8-16
8	16-10
9	20-22
10	20-18
11	18-7
12	7-8

Order	Link
1	10-15
2	15-22
3	20-21
4	8-16
5	16-10
6	20-22
7	20-18
8	18-7
9	7-8
10	22-23
11	22-21

Figure 5. Bikeway layout (**a**) when $b = 0$; (**b**) when $b = 3$.

The difference in the value of *b* can result in different selected node set and thus the network layout. Table 2 lists the details of the optimal network layouts with different values of *b*. While all networks have 7 selected nodes, there are two different selected node sets and three different network layouts. For $b = 0$, its selected node set of $b = 0$ is identical to the sets of $b = 2$ and 3 but the resultant network layout has a higher total system travel time. The eliminated links when $b = 0$ are different from those when $b = 2$ and $b = 3$ due to the difference in the elimination factor of the links and thus due to the order of elimination. For $b = 1$, the network layout is different from the layouts when $b = 2$ and 3 due to the difference in the selected node set, so as the eliminated links. When $b = 2$ and 3, the selected node set and the network layout are identical and the total system travel time is the lowest possible. This shows that the path count can also influence the selected node set and thus the total system travel time.

Table 2. Solution fitness under different values of *b*.

Value of *b*	0	1	2	3
Size of selected node set	7	7	7	7
Solution fitness	0.04371	0.04664	0.04771	0.04771
Total travel time (hr)	22.88	21.44	20.96	20.96

4.2. Effect of the Budget on the Solution Fitness

This section demonstrates the trade-off between the budget and the solution fitness (as well as the total system travel time) via two cases. The first case adopts a fixed selected node-set consisting of 17 selected nodes from the network illustrated in Figure 3, and it is assumed that the selected node-set covers all the demands. With a defined selected node set, the elimination heuristic was solely used to evaluate the network layout and the total system travel time with different budgets. Figure 6 illustrates that the solution fitness increases when the budget increases, which is consistent with the intuition that a higher budget can construct more bikeways and thus reduces the total system travel time. When the budget is lower than 64, no infeasible solutions can be obtained as at least one pair of selected nodes is unconnected. When the budget equals 64, the bikeway layout is indifferent from the Steiner tree network in which the selected nodes are connected with minimal construction cost. This network has the highest total system travel time because most bikeways that contribute to some of the shortest paths are eliminated to minimize the cost, and therefore those selected node pairs have a longer travel time. On the other hand, the bikeway network has the lowest total system travel time when the budget is at or above 139. At this budget level, all shortest paths between all selected node pairs can be constructed. The minimal total travel time is achieved because all cyclists can travel on the corresponding shortest paths. Furthermore, when the budget increases from 64 to 139, there exist network layouts other than the Steiner tree and all shortest paths' network (as shown in Figure 7). While the budget is in between the upper and lower bounds, these networks have a lower total travel time than the Steiner Tree but a higher travel time than the all-to-all shortest path network. This shows the capability of the elimination heuristic in determining the budget-constrained solutions under the objective of travel time minimization. The existence of intermediate solutions between the solutions of the Steiner tree problem and the Floyd-Warshall algorithm also implies that the design of the bicycle network is sensitive to the budget. As the elimination heuristic can find these intermediate solutions, it allows the planners to use the budget as a parameter to design the bicycle network. The bicycle network obtained by the elimination heuristic can always meet the budget while minimizing the total system travel time.

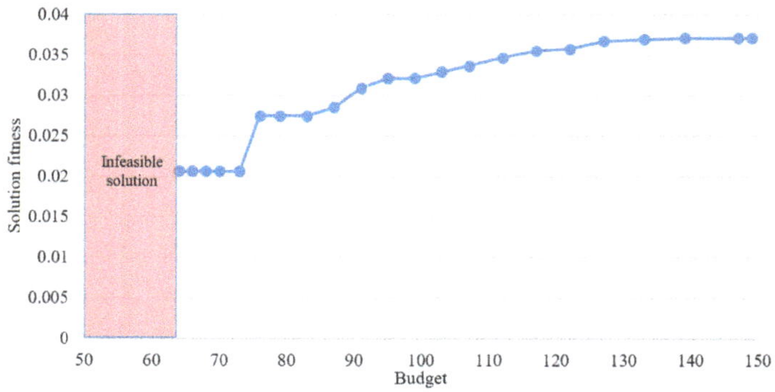

Figure 6. Effect of the budget on solution fitness with fixed selected node set.

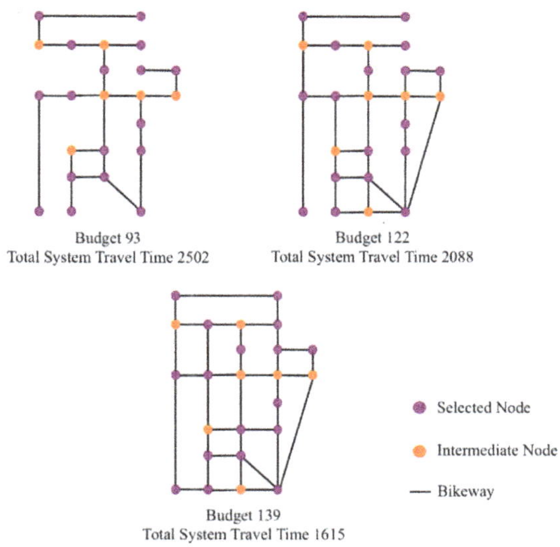

Figure 7. Bikeway network layout and total system travel time under different budgets.

In the second case, the selected node set is not predefined, but it needs to satisfy the demand coverage constraint. Unlike the first case, the budget level can alter the sets of selected nodes. Figure 8 shows that the solution fitness increases with the budget, which is consistent with the previous results. When the budget is lower than 42, all obtained solutions are infeasible because the built bikeways are insufficient to cover all the demand points. Figure 9 displays the network layout with a budget of 35. Some demand points are uncovered because the budget is not sufficient for building bikeways to connect all of their proximate stations. These results show that this model can determine whether the budget is enough for a feasible design and provide different designs according to the given budget.

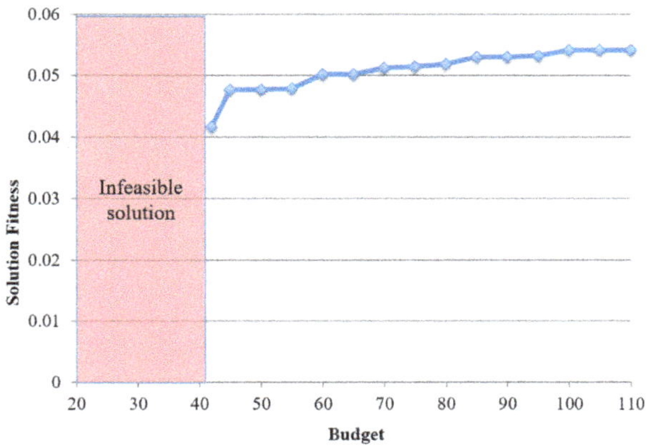

Figure 8. Effects of the budget on solution fitness without fixed selected node set.

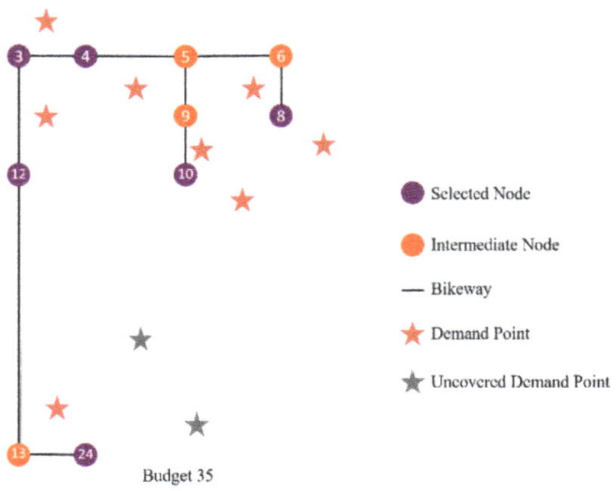

Figure 9. Bikeway layout when budget=35.

4.3. *Effect of Number of Demand Points on Construction Cost*

This section investigates the effect of the number of demand points on the construction cost. Different numbers of demand points were inputted into the model, and the minimum cost for covering all demand points was then obtained. The example network is identical to that in Figure 3. Figure 10 shows the relationships between the number of demand points and construction cost.

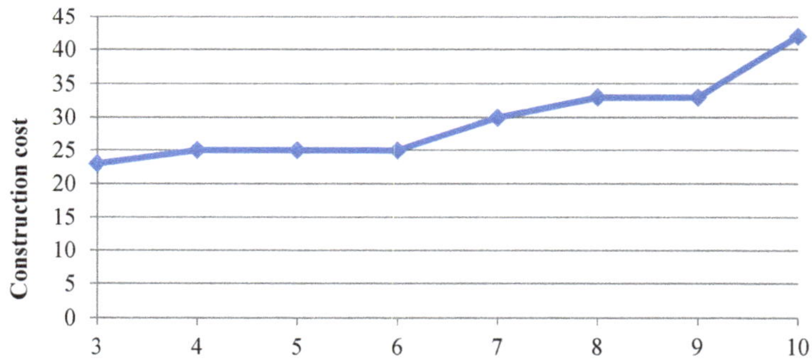

Figure 10. Effects of number of demand points on construction cost.

Figure 10 shows that the construction cost increases with the number of demand points. In general, more demand points require more selected nodes to cover all demands. As a result, more bikeways are constructed to connect these selected nodes, and the construction cost increases. Figure 11 illustrates how the number of demand points affects the number of selected nodes and thereby the construction cost. This figure also indicates that the budget amount affects the maximum number of covered demand points. When the budget is tight, only a few demand points can be covered. In this case, we should carefully investigate the relative importance of the demand points so that we can decide which demand points are worth to cover. In contrast, when the budget is higher, more demand points can be covered so as to increase the coverage of the bikeway network.

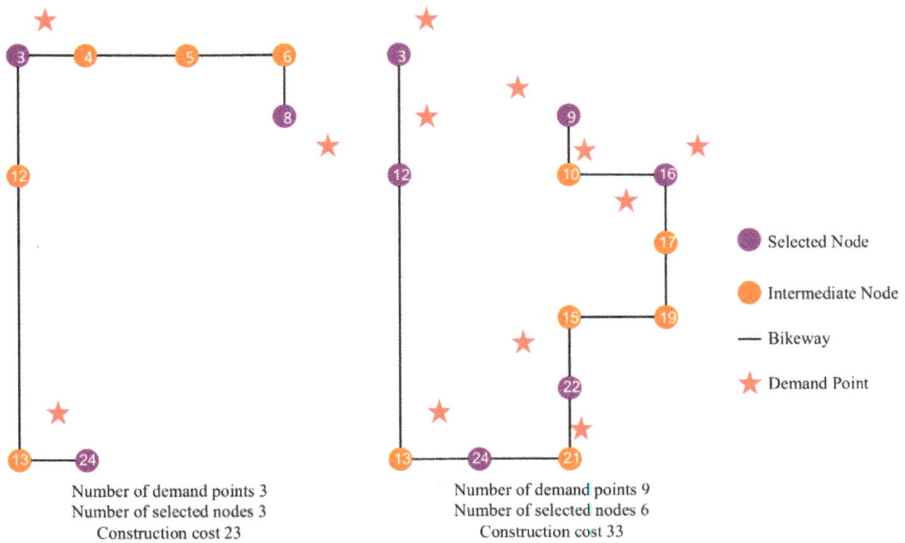

Figure 11. Effects of number of demand points on the number of selected nodes and construction cost.

5. Application in Hong Kong

This section aims to show the application of our model in Hong Kong. Two case studies were carried out in Kwu Tung North and Kam Tin (i.e., two proposed new towns in Hong Kong) to illustrate how our model can be applied to develop a new bicycle network and improve the existing network in Hong Kong. Developing a bikeway network coincides with the aim of the Hong Kong government in promoting cycling in new towns and development areas for short-distance commuting [21], in which the government has launched several projects that connect scattered cycle tracks in the New Territories and creates new bicycle networks in tourist spots [22]. Our proposed model can be useful to design the bikeway network in Hong Kong due to three reasons. First, it ensures popular destinations (e.g., public transport interchanges or tourist spots) to be covered by bikeways and interconnected. This is unlike the existing networks where the bikeways connect unpopular locations and thus many of them go unused [23]. Second, this model focuses on a network design with continuous and exclusive bikeways, which is favorable to the vast majority of inexperienced cyclists in Hong Kong who only cycle in a safe cycling environment. Thirdly, this model aims to minimize total travel time under the given budget, which is useful for the government that often has a fixed budget for bikeway construction.

5.1. Design Parameters

The unit cost of all bikeways is assumed to be the same, which is taken as c = HK$630 per meter, despite the construction materials, foundation, and technology applied [23]. The cost of building a bikeway between nodes i and j would be $C_{ij} = c \times l_{ij}$, where l_{ij} is the distance between nodes i and j. The distance l_{ij} would also be used for calculating the travel time between the bike stations. It is assumed that the cyclists travel at v = 20 km/h by bike [20]. The travel time between nodes i and j would be $t_{ij} = l_{ij}/v$.

5.2. Case Study in Kwu Tung North

Kwu Tung North (KTN) is one of the new development areas in the North East New Territories. It is planned to be 447 ha and to be developed as a mixed development node [24]. As the government aims to provide a quality living environment, a bicycle network will promote cycling within the area. Since there are no existing cycle tracks in KTN, we can demonstrate how our model designs a completely new bicycle network in this area. All the maps were obtained from OpenStreetMap [25], while the potential links and demand points have been identified from the Statutory Planning Portal of the Town Planning Board, Hong Kong [26]. We selected the residential area, the public transport interchange, and the business and technology park to be the demand points, as shown in Figure 12.

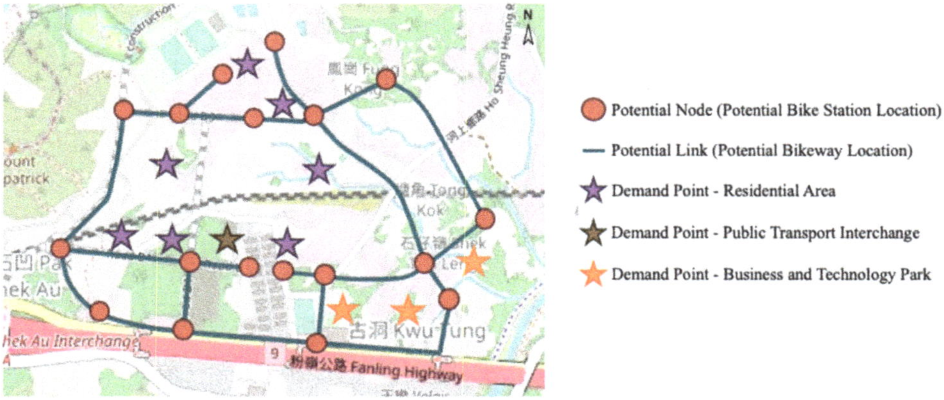

Figure 12. Demand points, potential nodes, and potential links in Kwu Tung North (KTN).

Figure 13a,b show the bikeway layouts under different budget levels. With a higher budget level (i.e., Figure 13b), the bikeways are denser and the total system travel time is reduced, which is consistent with the numerical results in Section 4. As shown in Figure 13a,b, a higher budget results in a larger number of selected nodes that are located closer to the demand sources and a longer continuous bikeway. From these results, we can conclude that our model can be applied to design a bikeway network in new development areas. The model is particularly useful when the government has a fixed budget, as it can guarantee a below budget design that minimizes the cyclists' total travel time.

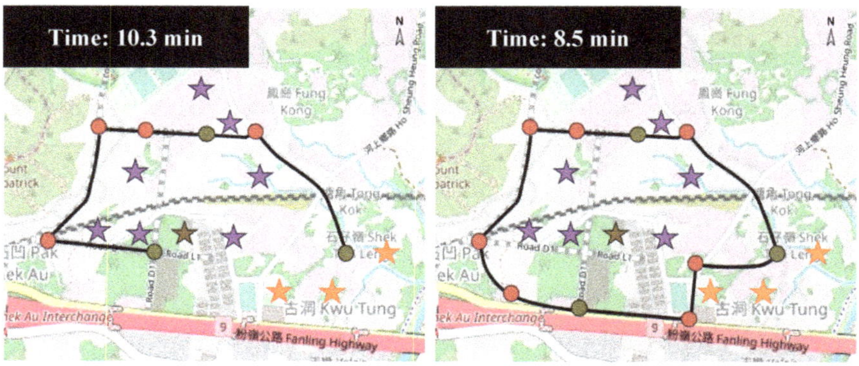

(**a**) (**b**)

Figure 13. Bikeway layout in KTN with a budget of (**a**) HK$1.08M; (**b**) HK$1.78M.

5.3. Case Study in Kam Tin

The government has planned a new residential development in Kam Tin (KT) South to increase the local housing supply [27]. The existing bicycle network should be expanded so that the new residential area can be connected to the Kam Sheung Road Station (KSRS). Meanwhile, there are only a small number of cycle tracks and parking sites in KT as shown in Figure 14. To facilitate cycling in KT, more cycle tracks and parking sites should be built. Figure 14 shows the demand points and the network. It is noted that some roads are not chosen as the potential bikeway locations because they are either too narrow or are owned by private owners.

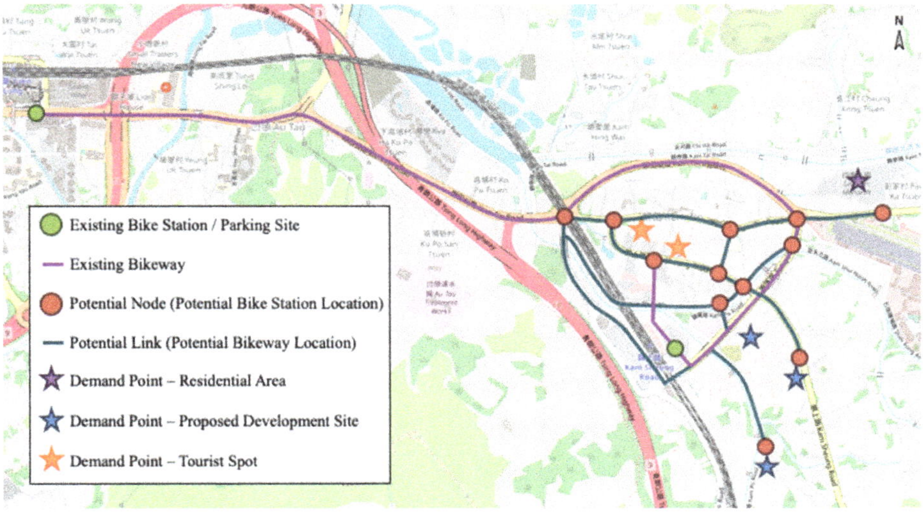

Figure 14. Demand points, potential nodes, and potential links in Kam Tin.

Figure 15a,b display the final bikeway network layouts under different budget levels. In this case, provided that there are existing bikeways and parking sites, the model was modified to ensure that these existing facilities are not removed in the elimination heuristics. The results of the bikeway designs are consistent with other sections that the total system travel time decreases when more budget is provided. Also, the travel time between the existing bike stations/ parking sites can be reduced because the stations can be connected by a shorter path when more budget is provided. These results show that our model handles the case of system expansion in which the new design can improve the system performance while taking the existing bicycle network and given budget level into consideration.

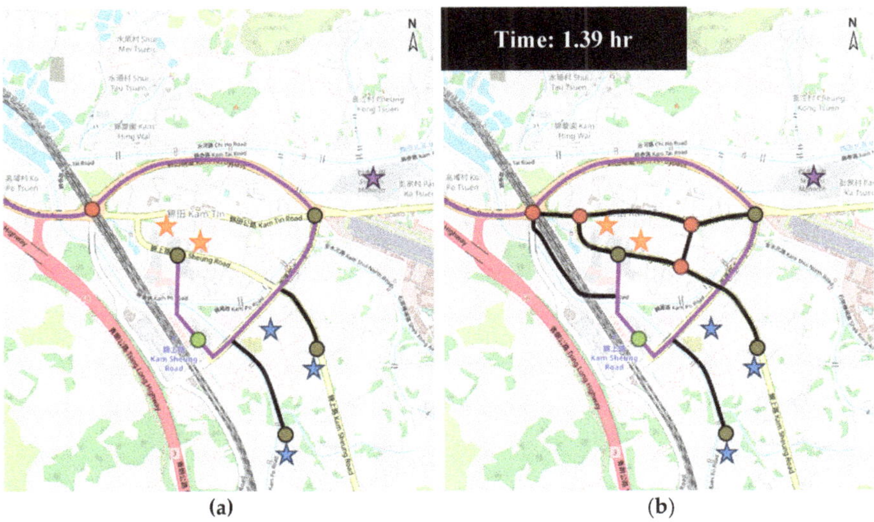

(**a**) (**b**)

Figure 15. Bikeway layout in KT with a budget (**a**) HK$0.61M; (**b**) HK$2.30M.

6. Conclusions and Future Directions

In spite of the significant role of bikeway in motivating cycling, there are very few existing bikeway network design models, and some design objectives and constraints have not been simultaneously taken into consideration. This study proposes a new bikeway network design problem that minimizes the total system travel time between a set of selected nodes under a budget constraint, in which all potential demands must be covered by the selected nodes and all selected nodes must be interconnected. An efficient two-step solution method is proposed to determine the selected node set and the bikeway layout separately. The genetic algorithm is employed to determine the selected node set that covers all demand points, while a novel elimination heuristic is employed to remove the bikeways between the shortest paths of the selected nodes (constructed by the shortest path algorithm) according to the elimination factor (which is a weighted sum of the setup cost and path count). Numerical studies showed that (1) the weights for elimination factors can significantly change the network layout and thus the solution quality; (2) a higher budget can change the network topology and improve the solution fitness whenever the selected node set is fixed or flexible; and (3) the construction cost increases with the number of demand points. The case studies in two Hong Kong new towns showed that the model can handle design problems with or without existing bikeways under various budget levels. This model can, therefore, be adapted to similar bike-sharing system designs because (1) it ensures that the resultant bike station locations have complete demand coverage, (2) all bike stations can be well-connected by the bikeway network, and (3) the designed network can satisfy the budget constraint. Furthermore, the model is not only applicable to new system designs but can also capture the existence of built bikeways and bike stations for system expansion.

Future research could be done in the following directions. First, other design constraints or measures, such as BLOS, could be included in the model. Second, the model assumes that the demand between every demand point pair is identical, while it can be relaxed in future studies by considering varying demand levels of all demand sources, elastic travel demand with respect to travel time, or uncertainties between the selected nodes. Third, the bikeway setup cost is assumed to be proportional to the length of links only, but it can also be influenced by other factors, such as the type of bikeway (e.g., exclusive bikeway, shared bikeway with vehicle roadway) and pavement type. Future studies can be extended to multi-type bikeway design problems. Fourth, the generalized travel cost of each bikeway can include other attributes in addition to travel time, including the number of turns and intersections, the percentage of greening, and the provision of cycling facilities along with a candidate link. Moreover, the route choice behavior of the cyclists can be remodeled by including other route choice attributes (such as traffic flow of automobile on the edges) or using other mathematical forms such as the bi-objective model [28]. Finally, the Genetic Algorithm used in the two-stage solution method can be replaced by other new meta-heuristics or hybrid meta-heuristics, such as the Artificial Bee Colony algorithm, Tabu Search, or Large Neighborhood Search. Comparative studies can be carried out to determine which of the solution methods has a higher computation efficiency.

Author Contributions: Conceptualization, methodology, writing—original draft preparation, writing—review and editing, supervision, project administration, C.S.S.; Validation, formal analysis, investigation, writing—original draft preparation, visualization, W.L.C.

Funding: This research received funding from the Ministry of Science and Technology, Taiwan ROC, with funding reference MOST 108 - 2218 - E - 009 - 062.

Acknowledgments: We thank the three anonymous reviewers for their constructive comments on the paper.

Conflicts of Interest: The authors declare no conflict of interest.

References

1. Pucher, J.; Dill, J.; Handy, S. Infrastructure, programs, and policies to increase bicycling: An international review. *Prev. Med.* **2010**, *50*, S106–S125. [CrossRef] [PubMed]

2. Smith, H. A Mathematical Optimization Model for a Bicycle Network Design Considering Bicycle Level of Service. Ph.D. Thesis, University of Maryland, College Park, MD, USA, 2011.
3. Fishman, E.; Washington, S.; Haworth, N. Bike share: A synthesis of the literature. *Transp. Rev.* **2013**, *33*, 148–165. [CrossRef]
4. Meddin, R.; DeMaio, P. The Bike-Sharing World Map. 2019. Available online: http://www.bikesharingworld.com/ (accessed on 1 March 2019).
5. Majumdar, B.B.; Mitra, S. Analysis of bicycle route-related improvement strategies for two Indian cities using a stated preference survey. *Transp. Policy* **2018**, *63*, 176–188. [CrossRef]
6. Buehler, R.; Dill, J. Bikeway networks: A review of effects on cycling. *Transp. Rev.* **2016**, *36*, 9–27. [CrossRef]
7. Duthie, J.; Unnikrishnan, A. Optimization framework for bicycle network design. *J. Transp. Eng.* **2014**, *140*, 04014028. [CrossRef]
8. Sohn, K. Multi-objective optimization of a road diet network design. *Transp. Res. Part A Policy Pract.* **2011**, *45*, 499–511. [CrossRef]
9. Mesbah, M.; Thompson, R.; Moridpour, S. Bilevel optimization approach to design of network of bike lanes. *Transp. Res. Rec. J. Transp. Res. Board* **2012**, *2284*, 21–28. [CrossRef]
10. Heinen, E.; Van Wee, B.; Maat, K. Commuting by bicycle: An overview of the literature. *Transp. Rev.* **2010**, *30*, 59–96. [CrossRef]
11. Frade, I.; Ribeiro, A. Bike-sharing stations: A maximal covering location approach. *Transp. Res. Part A Policy Pract.* **2015**, *82*, 216–227. [CrossRef]
12. Baker, B.M.; Ayechew, M. A genetic algorithm for the vehicle routing problem. *Comput. Oper. Res.* **2003**, *30*, 787–800. [CrossRef]
13. Dijkstra, E.W. A note on two problems in connexion with graphs. *Numer. Math.* **1959**, *1*, 269–271. [CrossRef]
14. Prim, R.C. Shortest connection networks and some generalizations. *Bell Syst. Tech. J.* **1957**, *36*, 1389–1401. [CrossRef]
15. Kruskal, J.B. On the shortest spanning subtree of a graph and the traveling salesman problem. *Proc. Am. Math. Soc.* **1956**, *7*, 48–50. [CrossRef]
16. Floyd, R.W. Algorithm 97: Shortest path. *Commun. ACM* **1962**, *5*, 345. [CrossRef]
17. Warshall, S. A theorem on boolean matrices. *JACM* **1962**, *9*, 11–12. [CrossRef]
18. Kou, L.; Markowsky, G.; Berman, L. A fast algorithm for Steiner trees. *Acta Inform.* **1981**, *15*, 141–145. [CrossRef]
19. Myung, Y.-S.; Lee, C.-H.; Tcha, D.-W. On the generalized minimum spanning tree problem. *Networks* **1995**, *26*, 231–241. [CrossRef]
20. Lin, J.-R.; Yang, T.-H. Strategic design of public bicycle sharing systems with service level constraints. *Transp. Res. Part E Logist. Transp. Rev.* **2011**, *47*, 284–294. [CrossRef]
21. Hong Kong Government. LCQ19: "Bicycle-Friendly" Environment. 2016. Available online: http://www.info.gov.hk/gia/general/201601/27/P201601260847.htm (accessed on 20 September 2017).
22. Legislative Council of Hong Kong. Meeting on 28 April- Background Brief on Cycle Tracks in Hong Kong. Hong Kong SAR: Legislative Council Panel on Transport. 2009. Available online: https://www.legco.gov.hk/yr08-09/english/panels/dev/papers/dev0428cb1-1347-5-e.pdf (accessed on 10 November 2019).
23. Coll, B.; Hefter, G.; Stokes, R.; Tabor, G. Cycling for Commuting in Hong Kong: An Observational Study. Bachelor's Thesis, Worcester Polytechnic Institute, Worcester, MA, USA, 2017.
24. Hong Kong Development Bureau. Kwu Tung North/Fanling North New Development Areas. 2014. Available online: http://ktnfln-ndas.gov.hk/en/2-1-3-about_project.php (accessed on 21 September 2017).
25. OpenStreetMap. The Maps are Made Available under the Open Data Commons Open Database License. Available online: http://opendatacommons.org/licenses/odbl/1.0/ (accessed on 21 September 2017).
26. Hong Kong Town Planning Board. Statutory Planning Portal. 2019. Available online: https://www2.ozp.tpb.gov.hk/gos/ (accessed on 11 November 2019).

27. Hong Kong Town Planning Board. Land Use Review of Kam Tin South and Pat Heung. 2014. Available online: http://www2.ozp.tpb.gov.hk/plan/ozp_plan_notes/en/S_YL-KTS_13_e.pdf (accessed on 21 September 2017).
28. Ehrgott, M.; Wang, J.Y.; Raith, A.; Houtte, C.V. A bi-objective cyclist route choice model. *Transp. Res. Part A* **2012**, *46*, 652–663. [CrossRef]

sustainability

MDPI

Article

The Difference in Night Visibility between Shared Bikes and Private Bikes during Night Cycling with Different Visibility Aids

Chengcheng Wu and Dawei Chen *

School of transportation, Southeast University, Nanjing 210096, China; 230179239@seu.edu.cn
* Correspondence: dw_chen@seu.edu.cn

Received: 24 October 2019; Accepted: 27 November 2019; Published: 9 December 2019

Abstract: In recent years, bike sharing has increasingly spread across the world. Compared with personal bikes, shared bikes are uniform and have bright surfaces to help the public to find them easily. At the same time, unfamiliarity is still a problem for some users of shared bikes. Therefore, these features should be understood to improve the night visibility of cyclists and improve traffic safety. Our study tested and compared differences in night visibility using five types of visibility aids. The results showed two cognitive differences between cyclists and drivers. First, cyclists believed that using flashing lights or static lights would provide better visibility than other visibility aids. However, using a static light and reflectors showed better results in our research. Secondly, compared to private bikes, cyclists showed more confidence in the nighttime visibility of shared bikes, especially with retroreflective strips. But the behavior of drivers in our study did not support such differences. A post-experiment survey was conducted to explore such cognitive differences, and showed that unfamiliarity with these strips was a possible reason for driver unawareness. This study will aid policy makers in incorporating suitable visibility aids within bike-sharing programs. Further, this study includes helpful advice for cyclists in terms of improving their night visibility.

Keywords: bike share; cycling safety; night-time visibility; cognitive difference between cyclists and drivers

1. Introduction

In recent years, bike sharing has increasingly spread across the world, with approximately 800 bike-sharing programs available in 2015 [1]. According to 2017 data from the local government in Nanjing (China), the annual growth rate of bicycle-vehicle collisions (BVCs) reached 1.5%. At the same time, there is emerging research on improving cycling safety, mostly focused on helmet use [2] and operational cycling speeds [3]. Compared with private bikes, shared bike users have lower operational speeds and are more reluctant to wear helmets [4]. While the former is likely to improve cycling safety, the latter reduces cycling safety. The safety influence of using shared bikes needs further research.

One of the explanations provided for vulnerable road user accidents is their poor visibility to other drivers [5–7]. Shared bike features show a possible difference in night visibility. Shared bikes commonly use a uniform and bright surface, so that the public can find them easily [8,9]. In China, the most popular shared bikes are Mobike and Hellobike. The former uses bright orange and the latter is blue and white. In Canada, Bike Share Toronto is green and black. Further, Bixi in Britain is green and gray. However, studies on the difference in visibility between shared bikes and private bikes are scarce.

On the other hand, using visibility aids to increase visibility is equally important, especially when cycling in low-light conditions [10,11]. Visibility aids can increase the distance at which car drivers detect cyclists at night [12,13]. Currently, front reflectors, static lights, and flashing lights are widely used to improve cyclist visibility at night [14–16]. Further, some countries and regions have started placing legal requirements on shared bikes to improve cycling visibility at night. For example, in Mainland China, night-time reflectors need to be added but there is no requirement for helmets and headlights. In the United States, Singapore, Australia [17], and some other locations, headlights are provided with shared bikes as mandatory requirement. Other promising visibility aids include retroreflective markers. Due to drivers' visual sensitivity to human motion patterns, when placing retroreflective markers on users' major moving joints (such as ankles, knees, wrists, etc.), drivers can recognize the presence of users more frequently and at much longer distances [18,19]. Much research has evaluated visibility during night cycling using different visibility aids [13,20], but there is little research on the difference between shared bikes and private bikes when using different visibility aids.

Therefore, the main objective of this study was to evaluate the difference between shared bikes and private bikes when using visibility aids under low light conditions. Specifically, the aim of this paper was to:

1. Evaluate the difference in visibility between shared bikes and private bikes with five types of visibility aids.
2. Explore the difference in behavior when using visibility aids. Cyclists' estimation of their own visibility is among the key influencing variables.

This study can contribute to achieving a better understanding of the visibility of shared bikes. Our results could help policy makers and cyclists to equip shared bikes with suitable visibility aids. Further, we can provide more reasonable advice on night cycling for cyclists.

2. Methodology

The experimental methods employed in this research are introduced in this section, including each step of this experiment and its related information.

2.1. Participants

In total, 50 participants completed the experiment, with a gender ratio of 1.27:1.00 (male: female).

Two subjects aged 25 and 40, respectively, with motor vehicle driving licenses issued by China's public security authority, were selected as the drivers of our test vehicle. Both of the subjects drive regularly—every week in the past two years—and have had no accidents in the last three years.

Recruitment

Offline leaflets and online advertising were both used to recruit participants (as bike riders) for our experiment. A total of 81 persons responded. However, due to logistics and test requirements, 53 persons initially participated, with 50 successfully completing the test.

Demographics

1. Data on gender, age, bike accidents, etc., for each participant were collected using a questionnaire before the test, and the statistical results are shown in Table 1. All participants came from Nanjing, China. Accidents were defined as incidents in which the participants were injured or caused an injury to anyone else while riding a bike during the past year.

Table 1. Sociological statistical information on participants.

Male		28
Female		22
Age	Max.	48
	Mean	33.4
	Min.	24
Number of accidents involved	Max.	2
	Mean	0.4
	Min	0
Bike preference (for typical use)	Private bike	34%
	Shared bike	66%

2.2. Driving Scenario

The layout of the test driving scenario is shown in Figure 1, where the driving course was a 400 m-long (1312 ft) section of straight road. The starting point was the departure point of bike riders and the terminal was the departure point of the test vehicle. The starting distance of these points between bike riders and the test vehicle was 400 m (1312 ft). The road width is 8 m, which is divided into two 4 m-wide (13.12 ft) mixed lanes for both motor vehicles and non-motorized vehicles. The driving road section was closed during the driving tests, in order to avoid any interference caused by other pedestrians or other motor vehicles.

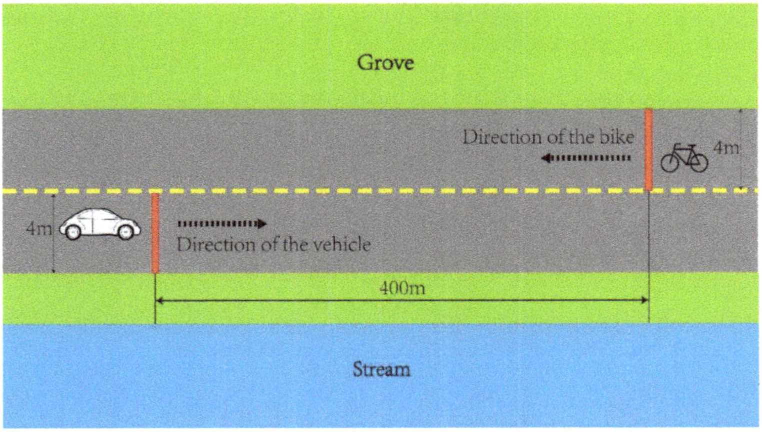

Figure 1. Test driving scenario.

Illumination of the Experimental Scene

All experiments were conducted at least one hour after sunset. A handheld light-measuring device was used to test the illumination conditions of the road section. We ensured that the illumination parameter for each test was not greater than 20 Lux at a height of 1.5 meters.

Weather in the Experimental Scene

All experiments were conducted in dry weather.

2.3. Test Vehicles and Bikes

A 20-inch HITO brand black bike was used as the private bike, with a net weight of 11 kg. This bike was suitable for riders with a height of between 140 cm and 180 cm.

Sharing-bikes from OFO company were chosen as our test shared bikes. The classical OFO 2.0 bike type was used, with the most commonly used type being a 26-inch family bike size. The color was golden yellow (CMYK color value C3, M26, Y96, K0).

The vehicle used in our experiment was a Volkswagen (POLO 1.4 L, 2007 version), with height of 1,465 mm, width of 1,650 mm, length of 3,916 mm, and wheelbase of 2,460 mm. It was a five-door, five-seat car, and the color was sky blue (CMYK color value C40, M0, Y0, K0).

2.4. Experiment Design

The experiments were designed to test the influence of visibility under low-light conditions using different visibility aids. At the same time, the difference between private bikes and shared bikes was analyzed to determine effects on night visibility.

Types of Visibility aids

The influence of the following five types of visibility aids was tested:

- Static light: an LED bike light was used with a maximum brightness of 250 LM and a direct radiation distance at night of 80 m.
- Flashing light: The flashing light used in this study has the same lighting capability as the static light and a flashing frequency of 0.5 times a second.
- Front reflectors: Yellow hard reflectors were mounted at the front of bikes and the front of the pedals.
- Retro-reflective strips on the major moveable joints: Red retro-reflective strips produced by China ONLINELOVE Company were used. The strip width was 1 cm (0.394 inches). Prism reflective films were attached to the strips. According to the test data provided by the manufacturer, the maximum visible reflection distance was 300 m (984.25 inch).
- No visibility factors.

Experiment Steps

The participants were asked to perform independent trials on the two kinds of bikes, which were equipped with one of five types of visibility aids. Each participant was required to complete 10 trials (five on shared bikes and five on private bikes), as listed in Table 2. For each participant, the order of these ten trials was random.

Table 2. Trials and independent variables.

No.	The Types of Bikes	Visibility Aids
1	Shared bike	Static light
2		Flashing light
3		Front reflectors
4		Retroreflective strips
5		No aids
6	Private bike	Static light
7		Flashing light
8		Front reflectors
9		Retroreflective strips
10		No aids

The participants were asked to ride to the end of the road section at a normal speed (around 15 km/h) in each trial. One experimenter assisted the participants to select and change different bikes and visibility aids and followed the participants to record two position coordinates of the participants:

1. The position coordinate of the bike rider at which the bike rider thought that the motor vehicle driver could see them was recorded as bike position 1 (BP1).

2. The position coordinate of the bike rider (participant) at which the motor vehicle driver said that he/she could see the bike rider was recorded as bike position 2 (BP2).

The other experimenter recorded two position coordinates of the test vehicle:

1. The position coordinate of the motor vehicle at which the bike rider thought that the motor vehicle driver could see them was recorded as vehicle position 1 (VP1).
2. The position coordinate of the motor vehicle at which the motor vehicle driver said that he/she could see the bike rider was recorded as vehicle position 2 (VP2).

Speed measuring devices and cameras (used to record the speed and position of bikes) were equipped on both the test vehicle and bikes for the test trials. The data recording process is shown in Figure 2.

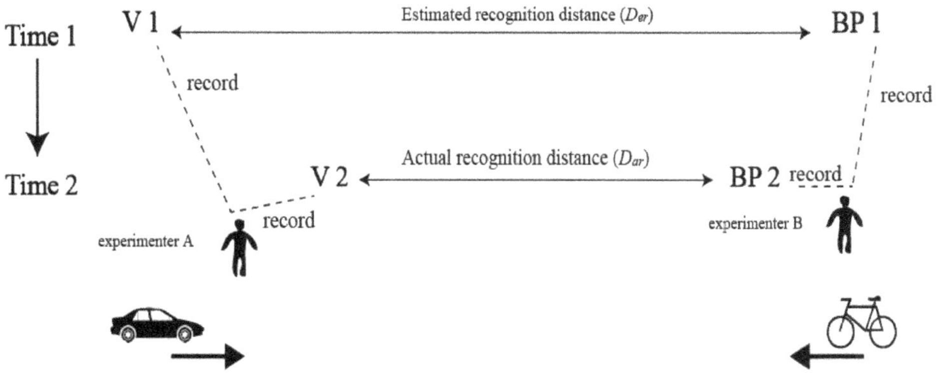

Figure 2. Data recording process.

Test Trial

One staff member introduced the test procedures without describing the test purpose to reduce potential interference caused by participants. The participants were asked to repeat the instructions to sure that they fully understand the experiment. A written description of the detailed procedures was provided to every participant before the test. We also required each participant to complete a random training trial to become familiar with the test procedure.

In addition, the following two criteria were used to evaluate the validity of a recorded result:

1. Speed requirement: with an average speed of 15 km/h, the speed variation of a bike rider should not be greater than 5 km/h. The speed variation of the motor vehicle should not be greater than 10 km/h, given the required driving speed of 40 km/h.
2. Track requirement: the cyclist is required to ride the bike in the right lane and the distance to the central line of the road should not be less than 2 m (6.56 feet). The driver should also drive the motor vehicle in the appropriate lane and not cross the central line.

If tests failed to meet one or all criteria, the tests were required to be performed again. If four consecutive tests completed by a participant did not qualify, this participant was removed from the study.

3. Results

Data analysis was performed using three dependent variables:

- The estimated recognition distance (D_{er}), or the distance between cyclists and drivers when the cyclists reported that they thought they were visible to drivers.

- The actual recognition distance (D_{ar}), or the distance between cyclists and drivers when the drivers reported that they could see the cyclists.
- The distance difference (*DD*), which means the difference between D_{er} and D_{ar}. A value of $\alpha = 0.05$ was used in all analyses to determine the significance of main effects.

The multiple-factor repetitive measurement and the analysis of variance were conducted on the estimated recognition distances, actual recognition distances, and distance differences as functions of the kind of bikes (2 levels: level 0 = private bikes and level 1 = shared bikes), visibility aids (5 levels: level 1 = static light; level 2 = flashing light; level 3 = reflector; level 4 = retro-reflective strips on the major moveable joints; level 5 = no visibility aids). Table 3 shows the least squares means of D_{er}, D_{ar}, and *DD* for private bikes and shared bikes across all visibility aids.

Table 3. Least squares means of estimated recognition distance, actual recognition distance, and distance difference in private bikes and shared bikes.

Types of Bikes	D_{er}	D_{ar}	DD
Shared bikes	78.9722(±3.0916) [a]	28.3100(±0.8167)	50.6622(±2.7873)
Private bikes	74.5960(±3.0916)	26.8667(±0.8167)	47.7293(±2.7873)

[a] Least squares means (standard error).

3.1. Estimated Recognition Distance (D_{er})

It can be seen from Table 4 that, across all kinds of visibility aids, the estimated recognition distance of shared bikes was significantly higher than that of private bikes, with $p = 0.0071$ and $F(1,49) = 8.39$.

Table 4. Differences of Least Squares Means in Pairwise Comparison of estimated recognition distance (shared bikes versus private bikes).

Group 1	Group 2	Estimate Differences	Standard Error	DF	t Value	Pr > |t|
shared bikes	private bikes	4.3762	1.5104	49	2.9	0.0071

For comparison of private bikes and shared bikes, as shown in Table 5, both static and flashing lights provide significantly higher D_{er} values than the use of reflectors, retro-reflective strips, and no aids. No aids were the least visible scenario based on the shortest distance of detection for all tested visibility aids. There was no significant difference in D_{er} between static light and flashing light, or between reflector and retro-reflective strips. Table 5 shows the differences of Least Squares Means in pairwise comparison.

Table 5. The Differences of Least Squares Means in Pairwise Comparison of estimated recognition distance (visibility aids).

Visibility Aids 1	Visibility Aids 2	Estimate Differences	Standard Error	DF	t Value	Pr > |t|
static light	reflector	14.6	2.3882	196	6.11	<0.001
static light	retroreflective strips	15.2658	2.3882	196	6.39	<0.001
static light	flashing light	3.6942	2.3882	196	1.55	0.1246
static light	no aids	24.5612	2.3882	196	10.28	<0.001
reflector	retroreflective strips	0.6658	2.3882	196	0.28	0.7809
reflector	flashing light	−10.9058	2.3882	196	−4.57	<0.001
reflector	no aids	9.9612	2.3882	196	4.17	<0.001
retroreflective strips	flashing light	−11.5717	2.3882	196	−4.85	<0.001
retroreflective strips	no aids	9.2954	2.3882	196	3.89	<0.001
flashing light	no aids	20.8671	2.3882	196	8.74	<0.001

The D_{er} of shared bikes with a static light was also significantly higher than that of private bikes, with $p = 0.0111$. Differently, there was no statistical difference between flashing light, reflector retro-reflective strips, or no aids (Table 6).

Table 6. Differences of Least Squares Means in Pairwise Comparison of estimated recognition distance (shared bikes with visibility aids versus personal bikes with visibility aids).

Shared Bikes' Visibility Aids	Personal Bikes' Visibility Aids	Estimate Differences	Standard Error	DF	t Value	Pr > |t|
Light lamp	light lamp	8.7167	3.3774	196	2.58	0.0111
reflector	reflector	0.4833	3.3774	196	0.14	0.8865
retroreflective strips	retroreflective strips	3.0883	3.3774	196	0.91	0.3624
flashing light	flashing light	4.165	3.3774	196	1.23	0.22
no aids	no aids	5.4275	3.3774	196	1.61	0.1108

3.2. Actual Recognition Distance (D_{ar})

In the tests of fixed effects, as indicated in Table 7, there was no statistical difference between private bikes and shared bikes for actual recognition distance, with $F(1,49) = 3.18$ and $p = 0.0849 > 0.05$.

Table 7. Differences of Least Squares Means in Pairwise Comparison of actual recognition distance (shared bikes versus private bikes).

Group 1	Group 2	Estimate Differences	Standard Error	DF	t Value	Pr > |t|
Shared bikes	Private bikes	1.4433	0.809	49	1.78	0.0849

As shown in Table 8, use of a bike with a reflector resulted in a higher actual distance than use of a bike with a flashing light, retro-reflective strips, or no aids, with positive estimate differences (6.1708 for retro-reflective strips, 7.6375 for flashing light, and 7.35 for no aids) and values of $p < 0.0001$. Interestingly, use of both a static light and a reflector had a larger positive estimate difference compared to use of a flashing light (6.0292 for static light with $p = 0.005$; 7.6375 for reflector with $p < 0.0001$), indicating that use of a flashing light reduced actual recognition distance compared to use of a static light and reflector.

Table 8. The Differences of Least Squares Means in Pairwise Comparison of actual recognition distance (visibility aids).

Visibility Aids 1	Visibility Aids 2	Estimate Differences	Standard Error	DF	t Value	Pr > |t|
static light	reflector	−1.6083	1.2792	196	−1.26	0.2112
static light	retroreflective strips	4.5625	1.2792	196	3.57	<0.001
static light	flashing light	6.0292	1.2792	196	4.71	<0.001
static light	no aids	5.7417	1.2792	196	4.49	<0.001
reflector	retroreflective strips	6.1708	1.2792	196	4.82	<0.001
reflector	flashing light	7.6375	1.2792	196	5.97	<0.001
reflector	no aids	7.35	1.2792	196	5.75	<0.001
retroreflective strips	flashing light	1.4667	1.2792	196	1.15	0.2539
retroreflective strips	no aids	1.1792	1.2792	196	0.92	0.3585
flashing light	no aids	−0.2875	1.2792	196	−0.22	0.8226

For use of the same retro-reflective strips, the actual distance was statistically higher with shared bikes than that for private bikes, with a positive estimate of differences of least squares means of 4.2417 (±1.809 standard error), $p = 0.0207$. This difference between shared and private bikes remained when there was no visibility aid, with a positive estimate of 4.25 (±1.809 standard error) and $p = 0.0205$. When equipped with a flashing light, private bikes were visible at a greater distance than shared bikes, with an estimate of 4.1583 (±1.809 standard error) and $p = 0.233$. Table 5 shows the differences of Least Squares Means in pairwise comparison. (Shown in Table 9)

Table 9. The Differences of Least Squares Means in Pairwise Comparison of actual recognition distance (Shared bikes' visibility aids versus Personal bikes' visibility aids).

Shared Bikes' Visibility Aids	Personal Bikes' Visibility Aids	Estimate Differences	Standard Error	DF	t Value	Pr > |t|
static light	static light	−0.2667	1.809	196	−0.15	0.8831
Reflector	reflector	3.15	1.809	196	1.74	0.0843
retroreflective strips	retroreflective strips	4.2417	1.809	196	2.34	0.0207
flashing light	flashing light	−4.1583	1.809	196	−2.3	0.0233
no aids	no aids	4.25	1.809	196	2.35	0.0205

3.3. The Distance Difference (DD) between Estimated Recognition Distance and Actual Recognition Distance

In the tests of fixed effects of *DD*, as shown in Table 10, there was no significant difference for different types of bikes ($F(1,49) = 3.31$ and $p = 0.0793 > 0.05$). Given those results, we only compared the *DD* under different visibility aids, as shown in Table 11.

Table 10. The Differences of Least Squares Means in Pairwise Comparison of distance differences (Shared bikes versus Private bikes).

Group 1	Group 2	Estimate Differences	Standard Error	DF	t Value	Pr > |t|
Shared bikes	Private bikes	2.9328	1.6125	29	1.82	0.0793

Table 11. The Differences of Least Squares Means in Pairwise Comparison of distance differences (visibility aids).

Visibility Aids 1	Visibility Aids 2	Estimate Differences	Standard Error	DF	t Value	Pr > |t|
static light	reflector	16.2083	2.5496	116	6.36	<0.001
static light	retroreflective strips	10.7033	2.5496	116	4.2	<0.001
static light	flashing light	−2.335	2.5496	116	−0.92	0.3617
static light	no aids	18.8196	2.5496	116	7.38	<0.001
reflector	retroreflective strips	−5.505	2.5496	116	−2.16	0.0329
reflector	flashing light	−18.5433	2.5496	116	−7.27	<0.001
reflector	no aids	2.6112	2.5496	116	1.02	0.3079
retroreflective strips	flashing light	−13.0383	2.5496	116	−5.11	<0.001
retroreflective strips	no aids	8.1162	2.5496	116	3.18	0.0019
flashing light	no aids	21.1546	2.5496	116	8.3	<0.001

As shown in Table 11, the use of static light gave a positive estimate of distance differences compared to the use of a reflector, retro-reflective strips, or no aids (16.2083 compared to a reflector, with $p < 0.001$; 10.7033 compared to retro-reflective strips, with $p < 0.001$; and 18.8196 compared to no aids, with $p < 0.001$). These results indicated that cyclists overestimate their visibility to a greater degree when using a static light than when using a reflector, retro-reflective strips, or no aids. At the same time, the results for night cycling without any visibility aids indicated less overestimation of cyclist visibility, with a lower *DD* value using no aid compared to that using a flashing light, retro-reflective strips, or static light.

Additionally, for both types of bikes and all kinds of visibility aids, the results showed a positive estimate value for *DD* (49.1958, with a standard error of 2.6682, $p < 0.001$) in the solution for fixed effects, which indicated a statistically higher estimated recognition distance relative to the actual recognition distance. In other words, night-cyclists might overestimate their visibility in low light conditions, as reported by Wood [18,19]. We also did a one-sample T-test for the hypothesis *H0* that the ensemble average of *DD* is zero, and the result showed that t (49) = 39.8296 and $p < 0.001$, which allowed the rejection of *H0*.

3.4. A Post-Experiment Survey

Analysis of the experiment data in Section 3.3 revealed that although night-cyclists have greater confidence when using a bike with a flashing light, our experiment indicated that static light and retro-reflective strips have better night visibility. Because few bikes in China are equipped with lights, especially flashing lights, it is necessary to assess the familiarity of drivers with flashing-light bikes.

We next did a survey to probe cyclists' views about private bikes and shared bikes in a night-cycling scene with different visibility aids. We then used a thirty-second video to assess drivers' familiarity with flashing-light bikes. The survey was completed by 187 respondents who ranged in age from 20 to 59. Of the respondents, 123 people were invited to answer questions as cyclists, and 65 people were invited to complete this survey as drivers.

In the survey of cyclists, 27.6% interviewees said that they felt a perceptible difference between using a shared bike or a personal bike. The rest (72.4%) reported no or only a slight difference between using a shared bike or a personal bike. Almost all interviewees considered shared bikes to be more visible for night cycling (93.5%). In response to a question on which type of visibility aid is the most visible for night cycling, 91.9% interviewees selected a flashing light, and the remaining respondents chose a static light.

The survey of drivers did not include written questions. Drivers watched a thirty-second video that was recorded from the driver's perspective about a bicyclist cycling at night from the opposite direction (nearly 200 meters away) on a bike with a flashing light. As they watched this video, we asked what they thought the flashing light was. No drivers thought it was a bike during the first 10 s of the video and 41.5% drivers reported that it might be a bike in the middle 10 s. After watching the whole video, 81.5% of viewers thought it was a bike, but only 61.5% were sure. This result indicated a low ability of drivers to recognize a bike with a flashing light, which may explain the difference between our result and that of Wood [18].

4. Discussion

We have evaluated the visibility difference between private bikes and shared bikes with five types of visibility aids. The results showed cyclists overestimate shared bikes' visibility in lowlight conditions. As a visibility aid, a flashing light does not provide as much visibility to others as cyclists expect it to. The difficulty of quick recognition one of the main the possible causes of this, as seen in the post-experiment survey.

Firstly, one result in this research is consistent with Wood [18,19], which indicated that cyclists overestimate their own night-time visibility. One possible explanation for this is that cyclists tend to stand on their own shoes to estimate recognition distance. A previous study showed that cyclist-motorists had fewer collisions with cyclists and detected them at a greater distance [5]. According to bike crashes research in New Zealand conducted by Tin et al. [21], cyclists' visibility may be improved if more people cycle and fewer people drive cars. In other words, cyclists or cyclist-motorists may have a longer recognition distance for other cyclists than motorists. So in our research, if our cyclist tends to estimate their recognition distance as a cyclist observing another cyclist, it would be a shorter estimate than a motorist would actually take to spot a cyclist.

One interesting result is that cyclists think shared bikes are more visible than private bikes while there is actually no significant difference. Different proportions of sensory visibility and cognitive visibility between cyclists and motorists are one possible reason for this. Sensory visibility shows the extent to which an object can be distinguished from its environment, due to its physical characteristics such as its shape, brightness, color, and angular size [22,23]. Obviously, shared bikes have a higher sensory visibility than private visibility due to its bright and uniform appearance. Cognitive visibility refers to the extent to which an object be noticed due to an observer's experience, expectations, and objectives [24–27]. In lowlight conditions, it can assume that shared bikes and private bikes may not have a significant cognitive difference. This means motorists may obtain the same observation times, no matter whether they encounter a sharing bike or private bike moving in the opposite direction. And they will extract similar information from traffic scenes based on their own expectations. Cyclists may estimate their own bikes' visibility mostly based on sensory visibility, while drivers mostly use cognitive visibility. However, for now this remains a hypothesis, which targeted research should explore further in the future.

Another cognitive difference is that flashing lights do not provide as good a visibility as cyclists estimate. A static light and flashing light showed a higher estimated recognition distance than reflector and retroreflective strips. But a static light and reflector had a significantly higher actual distance than flashing light, retroreflective strips, and no aids. A flashing light still provides higher sensory visibility, due to our visual sensitivity towards patterns of human motion [28,29]. However, it may be hard for drivers to recognize, which means it provides bad cognitive visibility, according to the post-experiment survey.

However, there is still a gender gap in using shared bikes. This could be a constraint for the validity of similar experiments if these gaps are not previously addressed. The gender distribution of cyclists and the gender distribution of drivers were not representative of the gender distribution of these two groups in the surveyed area. The experimental sample in our research included more males than females. It would be interesting to carry out the same experiment with a different sex ratio to study the effect of gender on cyclists' and drivers' estimation differences between shared bikes and private bikes.

We evaluated the visibility difference in shared bikes and private bikes using five types of visibility aids. The current study can advise policy makers in providing proper visibility aids and improving the safety of share bike programs. In addition, it suggested that there is a difference in sensory visibility and cognitive visibility between cyclists and motorists. Further research on cyclists and drivers' sensory visibility and cognitive visibility is necessary for improving cyclists' visibility in lowlight conditions.

Author Contributions: C.W. and D.C. conceived and designed this research. C.W. and other five graduate students completed these experiments and collected the data. C.W. wrote the paper. D.C. reviewed and edited the manuscript. All authors read and approved the manuscript.

Funding: This research received no external funding.

Conflicts of Interest: The authors have declared that no competing interest exist.

References

1. Frade, I.; Ribeiro, A. Bike-sharing stations: A maximal covering location approach. *Transp. Res. Part A* **2015**, *82*, 216–227. [CrossRef]
2. Ethan, D.; Basch, C.; Johnson, G.; Hammond, R.; Chow, C.; Varsos, V. An analysis of technology-related distracted biking behaviors and helmet use among cyclists in New York City. *J. Community Health* **2016**, *41*, 138–145. [CrossRef] [PubMed]
3. Schleinitz, K.; Petzoldt, T.; Gehlert, T. Risk compensation? —The relationship between helmet use and cycling speed under naturalistic conditions. *J. Saf. Res.* **2018**, *67*, 165–171. [CrossRef] [PubMed]
4. Fishman, E.; Schepers, P. Global bike share: What the data tells us about road safety. *J. Saf. Res.* **2016**, *56*, 41–45. [CrossRef]
5. Rogé, J.; Ndiaye, D.; Vienne, F. Useful visual field training: A way to improve elderly car drivers' ability to detect vulnerable road users. *Transp. Res. Part F Traffic Psychol. Behav.* **2014**, *26*, 246–257. [CrossRef]
6. Twisk, D.A.; Reurings, M. An epidemiological study of the risk of cycling in the dark: the role of visual perception, conspicuity and alcohol use. *Accid. Anal. Prev.* **2013**, *60*, 134–140. [CrossRef]
7. Noordzij, P.C. Cycling in the dark: An analysis of fatal bicycle accidents in The Netherlands. *J. Saf. Res.* **1976**, *8*, 73–76.
8. Demaio, P. Bike-sharing: History, impacts, models of provision, and future. *J. Public Transp.* **2014**, *12*, 3. [CrossRef]

9. Midgley, P. Bicycle-Sharing Schemes: Enhancing Sustainable Mobility in Urban Areas. *Dep. Econ. Soc. Aff.* **2011**, *8*, 1–12.
10. Thornley, S.J.; Woodward, A.; Langley, J.D.; Ameratunga, S.N.; Rodgers, A. Conspicuity and bicycle crashes: Preliminary findings of the taupo bicycle study. *Inj. Prev. J. Int. Soc. Child Adolesc. Inj. Prev.* **2008**, *14*, 11. [CrossRef]
11. Kwan, I.; Mapstone, J. Visibility aids for pedestrians and cyclists: A systematic review of randomised controlled trials. *Accid. Anal. Prev.* **2004**, *36*, 305–312. [CrossRef]
12. Madsen, J.C.O.; Andersen, T.; Lahrmann, H.S. Safety effects of permanent running lights for bicycles: A controlled experiment. *Accid. Anal. Prev.* **2013**, *50*, 820–829. [CrossRef] [PubMed]
13. Wood, J.M.; Tyrrell, R.A.; Marszalek, R.; Lacherez, P.; Carberry, T.; Chu, B.S. Using reflective clothing to enhance the conspicuity of bicyclists at night. *Accid. Anal. Prev.* **2012**, *45*, 726–730. [CrossRef] [PubMed]
14. Prati, G. The effect of an italian nationwide mandatory visibility aids law for cyclists. *J. Transp. Health* **2018**, *9*, 212–216. [CrossRef]
15. Lahrmann, H.; Madsen, T.K.O.; Olesen, A.V.; Madsen, J.C.O.; Hels, T. The effect of a yellow bicycle jacket on cyclist accidents. *Saf. Sci.* **2017**, *108*, 209–217. [CrossRef]
16. Carol, C.; Denise, K.; Miller, P.D.; Frank, C. The use of conspicuity aids by cyclists and risk of crashes involving other road users: A protocol for a population based case-control study. *BMC Public Health* **2010**, *10*, 1–6.
17. National Transport Commission. National Transport Commission Australian Road Rules, Regulation 259. 2009. Available online: http://www.ntc.gov.au/ViewPage.aspx?documentid=00794 (accessed on 20 May 2010).
18. Wood, J.M.; Tyrrell, R.A.; Marszalek, R.; Lacherez, P.; Carberry, T. Bicyclists overestimate their own night-time conspicuity and underestimate the benefits of retroreflective markers on the moveable joints. *Accid. Anal. Prev.* **2013**, *55*, 48–53. [CrossRef]
19. Wood, J.M.; Lacherez, P.F.; Marszalek, R.P.; King, M.J. Drivers' and cyclists' experiences of sharing the road: Incidents, attitudes and perceptions of visibility. *Accid. Anal. Prev.* **2009**, *41*, 772–776. [CrossRef]
20. Rogé, J.; Ndiaye, D.; Aillerie, I.; Aillerie, S.; Navarro, J.; Vienne, F. Mechanisms underlying cognitive conspicuity in the detection of cyclists by car drivers. *Accid. Anal. Prev.* **2017**, *104*, 88–95. [CrossRef]
21. Tin, S.; Woodward, A.; Ameratunga, S. Incidence, risk, and protective factors of bicycle crashes: Findings from a prospective cohort study in New Zealand. *Prev. Med.* **2013**, *57*, 152–161. [CrossRef]
22. Hancock, P.A.; Wulf, G.; Thom, D.; Fassnacht, P. Driver workload during differing driving maneuvers. *Accid. Anal. Prev.* **1990**, *22*, 281–290. [CrossRef]
23. Wulf, G.; Hancock, P.A.; Rahimi, M. Motorcycle conspicuity: An evaluation and synthesis of influential factors. *J. Saf. Res.* **1989**, *20*, 153–176. [CrossRef]
24. Horswill, M.S.; Helman, S.; Ardiles, P.; Wann, J.P. Motorcycle accident risk could be inflated by a time to arrival illusion. *Optom. Vis. Sci.* **2005**, *82*, 740–746. [CrossRef] [PubMed]
25. Mitsopoulos-Rubens, E.; Lenné, M.G. Issues in motorcycle sensory and cognitive conspicuity: The impact of motorcycle low-beam headlights and riding experience on drivers' decisions to turn across the path of a motorcycle. *Accid. Anal. Prev.* **2012**, *49*, 86–95. [CrossRef] [PubMed]
26. Prijs, K. Problems of Sensory-and Cognitive Conspicuity of Motorcyclists at Junctions: A Car to Motorcycle Comparison of Visual Search and Give-Way Intentions by Car Drivers, from two Angles of Approach. Master's Thesis, Utrecht University, Utrecht, The Netherlands, 2015.
27. Clabaux, N.; Brenac, T.; Perrin, C.; Magnin, J.; Canu, B.; Van Elslande, P. Motorcyclists' speed and "looked-but-failed-to-see" accidents. *Accid. Anal. Prev.* **2012**, *49*, 73–77. [CrossRef] [PubMed]
28. Balk, S.A.; Tyrrell, R.A.; Brooks, J.O.; Carpenter, T.L. Highlighting human form and motion information enhances the conspicuity of pedestrians at night. *Perception* **2008**, *37*, 1276–1284. [CrossRef] [PubMed]
29. Wood, J.M.; Tyrrell, R.A.; Carberry, T.P. Limitations in drivers' ability to recognize pedestrians at night. *Hum. Factors J. Hum. Factors Ergon. Soc.* **2005**, *47*, 644–653. [CrossRef]

sustainability

MDPI

Article

Research on the Psychological Model of Free-Floating Bike-Sharing Using Behavior: A Case Study of Beijing

Dandan Xu [1], Yang Bian [1,*] and Shinan Shu [2]

1. Beijing Key Laboratory of Traffic Engineering, College of Metropolitan Transportation, Beijing University of Technology, Beijing 100124, China; superdan@emails.bjut.edu.cn
2. Beijing Municipal Institute of City Planning and Design, Beijing 100045, China; shushinan@emails.bjut.edu.cn

* Correspondence: bianyang@bjut.edu.cn; Tel.: +86-10-67390918

Received: 6 March 2020; Accepted: 3 April 2020; Published: 8 April 2020

check for updates

Abstract: As a clean, sustainable transport tool, bicycles have significant advantages in short-distance travel. Despite many efforts assumed in Beijing to improve the cycling environment, the popularity of cycling remains relatively low. However, the advent of the free-floating bike-sharing (FFBS) system has engendered an unexpected cycling enthusiasm in Beijing. Therefore, it is of great importance to delve into why travelers prefer FFBS as a transportation form from a psychological perspective. In this paper, 352 valid questionnaires were collected from an online survey, and an extended theory of planned behavior (TPB) was adopted to examine the psychological determinants of intention and actual behavior to use FFBS. The results showed that men and car-owners prefer vehicles and show a lower willingness to use FFBS. In contrast, residents under the age of 60, residents with FFBS riding experience, and residents skilled in cycling are inclined to use FFBS; the economic convenience of FFBS is the most important attractant for FFBS, while bad weather is the biggest hindrance factor for residents to use FFBS; however, imperfection in infrastructure has no significant impact on reducing residents' willingness to use FFBS. These results have important implications for planners to better understand the FFBS use behavior, and several suggestions are proposed to support the policymaking about FFBS.

Keywords: free-floating bike-sharing system; influence factor; social-psychological variables; intention; use frequency

1. Introduction

Cycling is widely acknowledged as a sustainable and clean mode of travel with significant advantages in short-distance travel [1]. It is commonly considered a healthy lifestyle to reduce deaths caused by the urban sedentary lifestyle and an effective way to alleviate air pollution caused by automobile exhaust [2–4]. During the 1980s, China was called the "kingdom of bicycles", and bicycles were the most common urban transportation. However, with the rapid development of motorization and urbanization, the proportion of cycling in Beijing had declined from 62.7% in 1986 to 12.4% in 2015 (see Figure 1) [5], due to a variety of factors. The increasing travel distance with the fast growth at the urban scale is partly responsible for the decrease of bicycle usage. Fifty-five percent of journeys in Beijing are still within five kilometers according to a 2014 survey, thanks to its reasonable urban layout and extensive living facilities. However, twelve percent of these short-distance trips were still made by car, which were fundamentally suitable for cycling [6]. Therefore, it is very important to promote the prevalence of bicycles, particularly for short-distance travel, for the sake of the sustainable development of urban traffic.

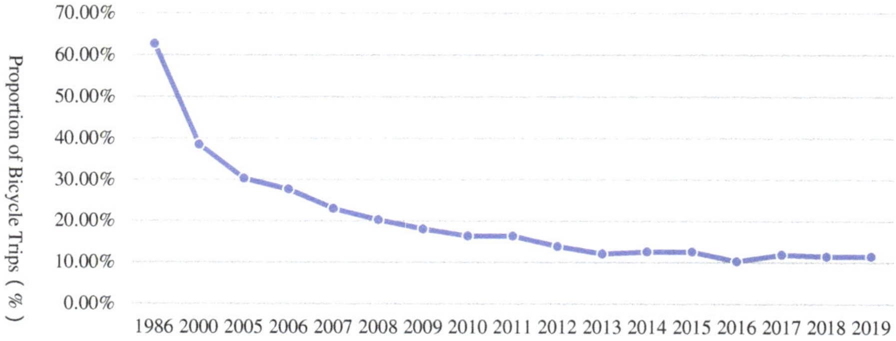

Figure 1. Proportion of bicycle trips in Beijing.

According to the success of Copenhagen and other cities, many measures have been taken in Beijing to improve the cycling environment. From 2013 to 2016, more than 700 kilometers of bike lanes were constructed in Beijing; 328,000 square meters of colored intersections were paved; 1702 meters of bike lanes were broadened. However, the popularity of the usage of bicycles as a transportation form is yet dissatisfactory. To promote sustainable transportation, the bike-sharing system has been valued. There are mainly three types of bike-sharing programs, including the public bike-sharing system (PBS), the closed campus bike-sharing system (CBS), and free-floating (FFBS). CBS (Figure 2a) is a kind of bike-sharing system that is used only inside the campus. PBS (Figure 2b) is often run or subsidized by the government, involving massive docking station constructions, whilst FFBS is a newer form, which has been developing rapidly in China since 2016. It is found that FFBS (Figure 2c) not only provides a new kind of sharing mode, but also gives rise to an unexpected change for sustainable transport [7].

FFBS, which is represented by OFO and MOBIKE, has emerged in China at the right moment as a result of the mobile Internet. FFBS are completely sponsored and operated by enterprises. They have overcome the limit of fixed sites, that is people can rent or return bikes anywhere. Each bike can be located by a Global Positioning System (GPS) module in its smart lock. Users can rent bikes by scanning a QR code (a kind of two-dimensional code) with the help of a smartphone. [7] By June 2018, over 1.9 million bikes were available, and the average amount of daily usage reached approximately 2,250,000. The declining trend of the share rate of cycling for the past few years has been altered, or even slightly increased by 1.1%, with the emergence of FFBS [5]. Such progress is beyond the other approaches. Therefore, it is of great importance to understand the motivations and barriers underlying the usage of shared bikes, so that further methods can be proposed to enhance the sustainability of FFBS and, more importantly, to promote the development of cycling.

(**a**) CBS (**b**) PBS (**c**) FFBS

Figure 2. Bike-sharing program. (**a**) CBS: campus bike-sharing; PBS: (**b**) public bike-sharing; (**c**) FFBS: free-floating bike-sharing.

2. Literature Review

In order to understand the cycling prevalence of FFBS, numerous studies have been conducted. The present research is mainly concerned with the impact of various factors such as social demography, travel details, infrastructure, meteorological environment, and FFBS operating system. Factors influencing the use of bicycles and PBS, which constitute a part of the basis for the current research, were also included in the literature review.

FFBS usage is related to social demographic factors. Most studies demonstrated that male cyclists outnumber female cyclists. Pucher et al. [8] believed that there was a negative correlation between cycling frequency and age. The results of Guo showed that the usage of FFBS was affected by household bicycle/car ownership [9]. It is generally believed that PBS users have high average incomes [10], high levels of education [11], and full-time or part-time employment [12]. In addition to social demographic factors, travel details also affect FFBS usage: travel distance has a greater influence than travel time and purpose [9,13,14]. Likewise, there is a relationship between infrastructure and FFBS usage. According to the study of Gamez-Perez [15], a safe and adequate infrastructure is a guarantee for the potential usage of FFBS. Barnes [16] found that setting up a dedicated bike road could increase the amount of cyclists by 1% to 2%. Stinson and Bhat [17] made a survey and found that cyclists preferred continuous roads in cycling. Environmental factors are often a barrier to the use of FFBS. The study results of Li [18] showed that reduced air pollution exhibited a favorable influence on non-motorized transport usage. It was also found that rainy weather discouraged the use of public bikes [19]. Demand for FFBS is evidently reduced by temperature and poor air quality [20]. Additionally, the exclusive advantages of FFBS are particularly important for its popularity. First of all, convenience is the major motivator for FFBS use [21]. Other merits such as favorable ease of access with a smart phone, convenience of pickup and parking, low expense [22], and extensive presence of docking stations within 250 m of their workplace were found to be statistically significant contributions to the preference for shared bikes [10], while primary obstacles for FFBS usage came from malfunction and limited regulations [23]. Additionally, FFBS is confronted with competition from the others transportation in terms of cost and travel experience [24].

However, the interrelationship between these factors is complex, and the governing mechanism is unclear. Scholars try to explore the unobservable potential variables based on the perspective of social psychology. Some new aspects of social psychology have been explored with the widespread application of planned behavior theory (TPB) [25], among which attitude, social norms, and perceived behavior control are the three most important potential variables.

In the early stage, the cycling behavior was studied based on TPB, and some general laws were found. Dill [26] argued that a positive attitude toward cycling led to the increase of the likelihood of cycling. Similarly, Abraham et al. [27] argued that negative views of driving also encouraged people to use bicycles. The studies of De [28] showed that cyclists were more likely to get help and support from cycling groups. At the same time, Bamber [24] found that people who had ridden a bicycle had less anxieties about and more willingness toward cycling. TPB can be also used to explain the intentions and behaviors towards FFBS. Scholars studied the social effects of shared bikes: the membership increase of shared bikes seemed to be slightly due to contagion from neighboring membership [29]. Social influence has a positive effect on users' trust attitude and hence subjective well-being [30]. According to a structural equation model (SEM) of intentions, the significance of factors is a sequence of subjective norms > attitude > perceived pleasure > effects of flexibility and convenience [31].

Compared with bicycles, except for some common socio-demographic factors, the use of PBS and FFBS is more susceptible to the system factors and perceived potential variables, while the infrastructure displays an insignificant influence. More details are shown in Table 1.

Table 1. Influence factors for bicycles, PBS and FFBS.

	Bicycle	PBS	FFBS
Social demographic factor	Age, Gender, Bicycle ownership	Age, Income, Education, Full-/part-time occupation	Bicycle/car ownership
Travel factor	Travel distance, Travel purpose		Travel time, Travel purpose
Infrastructure factor	Bicycle road		
Environmental factor	Air pollution	Rainy weather	Poor air, Temperature
System factors		Convenience, Member or not	Convenience of pickup and parking, Low expense, Malfunction, Limited regulations
Latent variable	Attitude toward cycling, Attitude toward driving, Experience	Experience, Subjective norm, Attitude, Perceptual behavior control	Experience, Subjective norm, Attitude, Perceptual behavior control, Preference, External restrictions

2.1. Theory of Planned Behavior

The following variables are involved according to the theoretical framework of standard TPB: ATT (attitude), SN (subjective norm), PBC (perceived behavior control), BI (intention), and BV (behavior) [25]. Attitude indicates negative or positive assessment of persons towards some behavior; a subjective norm refers to the detailed feelings of other people regarding decision making; and PBC is similar to the concept of self-control over one's own behavior. On the basis of TPB, behavior is mainly predicted by intentions, which inversely suffer from the influences of ATT towards the behavior, PBC over the behavior, and SN pertaining to the behavior.

TPB is a kind of primary social-cognitive theory developed by Ajzen [25], who implied that personal behavior is mainly predicted by behavioral intentions, which are further determined by three salient motivational factors: attitude, subjective norms, and perceived behavioral control (PBC). Despite that the basic model could provide an interpretation for a majority of the alterations in intention, several scholars have proposed that it would lead to an evident improvement in its explanatory power with consideration of attitudes toward bad weather, attitudes toward cars [32], and external restrictions (infrastructure). On the basis of TPB, Han et al. [33] predicted the intention of bike traveling, using the attractiveness of unsustainable alternatives as a moderator and personal norm and past behavior as predictors. The results offered a thorough understanding of the role of volitional and non-volitional processes, personal norm, past behavior, and the attractiveness of unsustainable alternatives in explaining the intention formation of bike traveling. The predicted intention of car usage [24] indicated that role beliefs increased the explanatory power of the Ajzen model, and car use habit significantly contributed to the predictive power of the Ajzen model, while the personal norm exerted an insignificant effect either on intention or on behavior.

However, the explanatory power of these variables was involved in a few related explorations for FFBS usage. As a consequence, the present study employed an extended the TPB model that accounted for the basic TPB model, as well as attitudes toward bad weather, attitudes toward cars, and external restrictions regarding infrastructure, to explain FFBS usage intentions and behaviors.

2.1.1. Original PBC Variables

Perceived behavior control (PBC): Perceived behavior control means self-control over one's own FFBS usage behavior.

Subjective norm (SN): Subjective norms mean the social pressure received from several important referents to perform a particular behavior.

Attitude towards FFBS (BATT): Attitude towards FFBS is the original attitude, explained as the extent of one's favor or praise towards FFBS usage behavior (BATT is used to represent attitude towards FFBS here to distinguish it from other attitudes).

2.1.2. Additional Variables

Attitude towards cars (CATT): Attitude towards cars represents the inclination toward FFBS usage on the condition of the preference for cars.

Attitude towards bad weather (WATT): Attitudes towards bad weather, exhaust fumes, and haze represent the inclination towards FFBS usage in the case of bad weather.

External restrictions regarding infrastructure (IER): External restrictions regarding infrastructure reflect restrictions for FFBS use that are beyond the control of the user.

In order to explain the prevalence in the choice of FFBS as a mode of transport, three main concerns are involved in the present paper, including: (1) identification of the psychosocial factors that play a role in FFBS use intention and actual behaviors; (2) the complex interrelation among those factors; (3) advice for promoting FFBS use. It should be noted that this paper mainly discusses the influence of cars on the use of FFBS, considering that travelling by bus belongs to a green mode of transportation.

3. Research Hypotheses

For the sake of an interpretation of the FFBS usage behavior in Beijing, the TPB model was incorporated in the current study. Besides, a special concern of the current study refers to the direct influence of attitudes toward bad weather, attitudes toward cars, and external restrictions regarding infrastructure, together with their effects as mediators that account for FFBS usage behaviors. A hypothesis framework was proposed based on the summary of the above discussion. The hypotheses and the proposed model (Figure 3) are shown as follows.

Hypothesis 1 (H1). *Intention shows a positive relation toward FFBS usage behaviors.*

Hypothesis 2 (H2). *PBC (H2a), SN (H2b), and BATT (H2c) will positively predict FFBS usage intentions.*

Hypothesis 3 (H3). *WATT (H3b) and IER (H3c) will have negative, direct effects on users' intentions to use FFBS.*

Hypothesis 4 (H4). *PBC will positively predict users' behavior toward the use of FFBS.*

Hypothesis 5 (H5). *IER has an influence on intention by reducing BATT.*

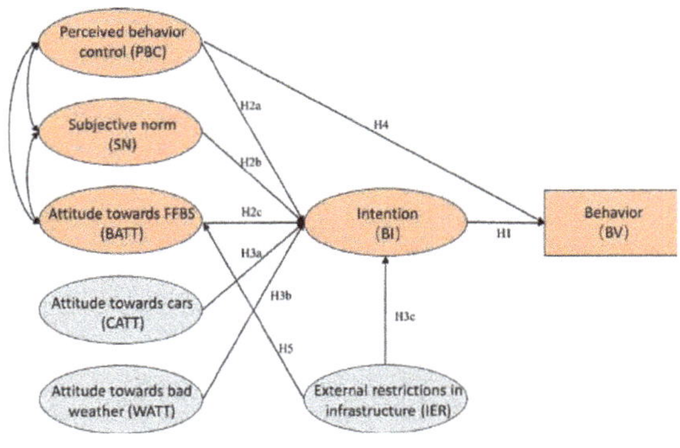

Figure 3. The proposed path model.

4. Methods

4.1. Participants and Procedure

In this study, data were collected within five days in July 2018 from an online survey in Beijing. The questionnaire link was forwarded to 12 WeChat groups covering people with differences in age, education, and cycling experience. To ensure that all the respondents were residents living in Beijing, the questionnaire link was set with IP restrictions accessible only for respondents in Beijing. Three-hundred-eighty-five group members responded, and each were paid CNY 4 (approximately $0.44). After the filtration of incomplete data and extreme data (i.e., those respondents whose answering time was 6 min less than the minimum time limit), there were in all 352 valid questionnaires from 197 women and 155 men, with an acceptable proportion of 91.43%.

4.2. Measure

The questionnaire included 3 parts: introduction, demographic information, and latent variable items. The introduction part helped participants complete the questionnaire; the demographic information part mainly included gender, age, income, education, occupation, vehicle ownership, cycling ability, frequency of FFBS use, and working location; and the latent variable items were derived from the hypothetical framework of this paper and previous studies. According to the actual usage of shared bikes, a six-point scale was designed to measure the variable BV, and the other variables were evaluated according to a five-point Likert scale, ranging from 1 = "strongly unlikely" to 5 = "strong likely". A further revision was employed for the questionnaire to enhance its clarity and reliability based on feedback, after a pretest of 100 people.

- Perceived behavior control (PBC): Three items were used to assess perceived behavior control. Respondents demonstrated the following three behavioral beliefs for FFBS: PBC 1: I have the ability to pay the deposit; PBC 2: I can afford a membership card; PBC 3: I can afford to use it;
- Subjective norm (SN): Five items were used to assess subjective norm. SN1: My friends would support me to use shared bikes; SN2: My family would support me to use shared bikes; SN3: Most people in society would support me to use shared bikes; SN4: News medium would support me to use shared bikes; SN5: Social media would support me to use shared bikes;
- Attitude towards FFBS (BATT): Five items were used to assess residents' attitude towards FFBS. BATT1: It is economical because I don't have to buy a bicycle; BATT2: It is economical because it's cheap to use; BATT3: It is convenient because it can be rented anywhere; BATT4: It is convenient because I can stop to do other things at any time in travel; BATT5: It is convenient for I need not to worry about maintenance and theft;
- Attitude towards cars (CATT): Five items were used to assess residents' attitude towards cars. CATT1: It is economical; CATT2: It is convenient; CATT3: It is efficient; CATT4: It is comfortable; CATT5: It is safe;
- Attitude towards bad weather (WATT): Two items were used to assess residents' attitude towards bad weather. WATT1: It is uncomfortable as it exposes me to exhaust fumes and haze directly; WATT2: It is uncomfortable because I might suffer from bad weather;
- External restrictions regarding infrastructure (IER): Three items were used to assess residents' attitude towards bad weather. IER 1: Incomplete bike lanes would limit my use of FFBS; IER 2: Insufficient parking space would limit my use of FFBS; IER 3: Poor cycling environment (shade) would limit my use of FFBS;
- FFBS usage intentions (BI): Three items were used to measure residents' FFBS usage intentions. IN1: I intend to travel by shared bikes frequently in the next three months; IN2: I try to travel by shared bikes frequently in the next three months; IN3: I plan to travel by shared bikes frequently in the next three months;
- FFBS usage behavior (BV): The FFBS usage behavior was measured by one item: BV1: How often have you used BSS in the past three months? (1 never 2 once a month 3 once a week 4 2–4 times a week 5 almost every weekday 6 almost every day)

5. Results

5.1. Demographic Results

After deleting responses that did not meet the minimum response time requirement, 352 valid samples with 155 males and 197 females were collected from the survey. This distribution was basically consistent with the permanent resident population structure in Beijing in 2019 with 50.8% male and 49.2% female [34]. The average age was 30.77 (SD = 9.58). The proportion of the surveyed people under 12 or over 65 was relatively small compared to the age distribution of the permanent resident population in Beijing. The reason was mainly due to the survey being conducted by means of WeChat. However, considering the prohibition of children under 12 to use a shared bike and the inconvenience of elderly people over 65, who seldom ride a bike, such a discrepancy did not have a sensible influence on the following discussion. Different income levels, occupations, education levels, and working locations were covered. Thirty-one-point-two-five percent of participants owned cars, and 41.48% of participants owned private bicycles. With regard to FFBS usage frequency, only 10 participants had never used shared bikes in the past three months (2.84%), and more than half of participants used bicycles regularly (63.07%). More details are shown in Table 2.

Table 2. Summary of respondents' demographic information (n = 352).

Gender	**Frequency**	**Percentage**	**Occupation**	**Frequency**	**Percentage**
Male	155	44.03%	Government/public institution	41	11.65%
Female	197	55.97%	Enterprise employers	122	34.66%
Age	**Frequency**	**Percentage**	Business/service personnel	10	2.84%
<18	2	0.56%	Medical personnel	4	1.14%
19–24	101	28.69%	Teacher	16	4.55%
25–34	152	43.21%	Student	111	31.53%
35–44	58	16.49%	Worker	5	1.42%
>45	39	11.07%	Community worker	3	0.85%
Monthly Income (US$)	**Frequency**	**Percentage**	Individual owner or operator	5	1.42%
<221	76	21.59%	Retirement	11	3.13%
221–442	47	13.35%	Free employment	8	2.27%
443–738	56	15.91%	Unemployed	5	1.42%
739–1180	64	18.18%	Others	11	3.13%
1181–2214	82	23.3%	**Education**	**Frequency**	**Percentage**
>2214	27	7.67%	Primary school	1	0.28%
Car Ownership	**Frequency**	**Percentage**	Junior high school	7	1.99%
Have	110	31.25%	Senior high school	22	6.25%
Have no	242	68.75%	College	165	46.87%
Private Bicycle Ownership	**Frequency**	**Percentage**	Master or above	157	44.6%
Have	146	41.48%	**Working Location**	**Frequency**	**Percentage**
Have no	206	58.52%	inside the second ring road	25	7.10%
FFBS Usage Frequency	**Frequency**	**Percentage**	the second ring road-the third ring road	68	19.32%
Never	10	2.84%	the third ring road-the fourth ring road	140	39.77%
Once a month	64	18.18%	the fourth ring road-the fifth ring road	60	17.05%
Once a week	56	15.91%	outside the second ring road	49	13.92%
Times a week	142	40.34%	**Cycling Ability**	**Frequency**	**Percentage**
Almost every workday	33	9.38%	Have	343	97.44%
Almost every day	47	13.35%	Have no	9	2.56%

Independent *t*-tests proved to be suitable for the analysis of the difference between two independent groups of small samples [35] and were conducted to analyze the differences between two independent groups in terms of seven extended TPB variables (e.g., male/female, old/young, have cycling experience/not, owning a car or a bike/not, riding ability/not), respectively. Men usually had a more friendly attitude towards cars than women; residents under 60 years old had stronger perceived behavioral control, subjective norms, and external restrictions regarding infrastructure; they were also more likely to use FFBS. Attitudes toward cars showed significant differences in car ownership: residents who owned cars had better attitudes toward cars. However, the difference was absent between the group of bicycle owners and that without a bicycle. Residents with FFBS cycling experience showed stronger subjective norms and less preference for cars than those without FFBS cycling experience. In addition, people with good cycling skills had stronger subjective norms (see Table 3).

For the sake of further exploring the significant difference of demographics in FFBS usage frequency, a variance analysis [36] was conducted after a normality test. The results showed that gender, age, cycling ability, car ownership, occupation, and education level contributed to the variation in the use frequency of FFBS (see Table 4). Men, young people, and students were more likely to use shared bikes. Residents with a car and cycling ability showed higher enthusiasm toward shared cycling. A high education level also showed higher bike-sharing use frequency.

Table 3. Independent *t*-tests between demographics and theory of planned behavior (TPB) variables.

Variable	Gender M (SD)		*t*-Tests		Age M (SD)		*t*-Tests	
	Male (n = 155)	Female (n = 197)	*t* Value	Effect Size	Over 60 (n = 8)	Under 60 (n = 344)	*t* Value	Effect Size
PBC	0.36 ± 0.26	0.35 ± 0.27	0.24	0.81	0.21 ± 0.13	0.36 ± 0.26	−3.049	0.015 *
SN	4.20 ± 0.74	4.24 ± 0.82	−0.516	0.606	3.63 ± 0.90	4.24 ± 0.78	−2.191	0.029 *
BATT	4.21 ± 0.69	4.27 ± 0.73	−0.714	0.475	4.30 ± 1.06	4.24 ± 0.70	0.158	0.879
CATT	2.67 ± 0.78	2.50 ± 0.79	2.114	0.035 *	2.15 ± 0.89	2.58 ± 0.79	−1.542	0.124
WATT	1.93 ± 0.78	1.84 ± 0.80	1.071	0.285	2.00 ± 0.85	1.87 ± 0.79	0.452	0.651
IER	4.17 ± 0.86	4.26 ± 0.86	−0.935	0.351	3.54 ± 0.89	4.24 ± 0.85	−2.277	0.023 *
BI	4.15 ± 0.72	4.01 ± 0.94	1.585	0.114	3.46 ± 0.94	4.09 ± 0.85	−2.071	0.039 *
Variable	**Car Ownership M (SD)**		***t*-Tests**		**Bike Ownership M (SD)**		***t*-Tests**	
	Have (n = 242)	Have No (n = 110)	*t* Value	Effect Size	0.0 (n = 206)	1.0 (n = 146)	*t* Value	Effect Size
PBC	0.37 ± 0.26	0.33 ± 0.27	1.394	0.164	0.38 ± 0.26	0.32 ± 0.26	1.85	0.065
SN	4.23 ± 0.73	4.21 ± 0.91	0.2	0.842	4.18 ± 0.82	4.29 ± 0.73	−1.381	0.168
ATT	4.26 ± 0.67	4.21 ± 0.80	0.588	0.557	4.22 ± 0.78	4.28 ± 0.59	−0.875	0.382
CATT	2.67 ± 0.76	2.37 ± 0.81	3.314	0.001 **	2.59 ± 0.79	2.55 ± 0.80	0.486	0.628
WATT	1.87 ± 0.76	1.88 ± 0.86	−0.036	0.971	1.92 ± 0.81	1.81 ± 0.76	1.337	0.182
IER	4.26 ± 0.80	4.15 ± 0.97	1.013	0.312	4.18 ± 0.84	4.28 ± 0.88	−1.112	0.267
BI	4.11 ± 0.80	3.99 ± 0.95	1.25	0.212	4.04 ± 0.85	4.12 ± 0.85	−0.909	0.364
Variable	**Experience M (SD)**		***t*-Tests**		**Riding Ability M (SD)**		***t*-Tests**	
	≥1 (n = 342)	0 (n = 10)	*t* Value	Effect Size	Have (n = 343)	Have No (n = 9)	*t* Value	Effect Size
PBC	0.36 ± 0.26	0.26 ± 0.24	1.189	0.235	0.35 ± 0.26	0.38 ± 0.33	−0.281	0.779
SN	4.25 ± 0.77	3.46 ± 1.04	3.155	0.002 **	4.26 ± 0.75	3.00 ± 1.21	3.104	0.014 *
BATT	4.26 ± 0.68	3.56 ± 1.18	1.877	0.093	4.25 ± 0.70	3.84 ± 1.01	1.71	0.088
CATT	2.60 ± 0.78	1.80 ± 0.65	3.188	0.002 **	2.59 ± 0.79	2.16 ± 0.68	1.617	0.107
WATT	1.86 ± 0.78	2.30 ± 1.11	−1.731	0.084	1.86 ± 0.78	2.28 ± 1.03	−1.553	0.121
IER	4.22 ± 0.87	4.30 ± 0.67	−0.288	0.773	4.22 ± 0.86	4.19 ± 1.04	0.132	0.895
BI	4.12 ± 0.80	2.43 ± 1.03	6.528	0.000 **	4.10 ± 0.83	2.93 ± 0.72	4.184	0.000 **

* $p < 0.05$ ** $p < 0.01$.

Table 4. Analysis of variance between demographics and FFBS usage frequency.

Gender	M (SD)	F	p
Male	3.99 ± 1.27	5.132	0.024 *
Female	3.69 ± 1.20		
Age	**M (SD)**	**F**	**p**
<18	4.00 ± 1.22	3.299	0.011 *
19–24	3.91 ± 1.26		
25–34	3.99 ± 1.21		
35–44	3.64 ± 1.32		
>45	3.26 ± 0.99		
Income	**M (SD)**	**F**	**p**
<221	3.79 ± 1.28	0.813	0.541
221–442	4.04 ± 1.30		
443–738	3.61 ± 1.15		
739–1180	3.78 ± 1.19		
1181–2214	3.94 ± 1.26		
>2214	3.77 ± 1.24		
Cycling Ability	**M (SD)**	**F**	**p**
Have	3.85 ± 1.23	6.844	0.009 **
Have no	2.40 ± 0.55		
Car Ownership	**M (SD)**	**F**	**p**
Have	3.59 ± 1.14	5.979	0.015 *
Have no	3.93 ± 1.27		
Private Bicycle Ownership	M (SD)	F	p
Have	3.87 ± 1.17	0.361	0.548
Have no	3.79 ± 1.29		

Occupation	M (SD)	F	p
Government/public institution	3.77 ± 1.23	1.892	0.034 *
Enterprise employers	3.88 ± 1.30		
Business/service personnel	3.70 ± 0.95		
Medical personnel	3.60 ± 0.55		
Teacher	3.88 ± 1.26		
Student	4.03 ± 1.23		
Worker	2.60 ± 0.89		
Community worker	2.67 ± 1.15		
Individual owner or operator	3.60 ± 1.14		
Retirement	3.33 ± 0.89		
Free employment	4.00 ± 1.12		
Unemployed	2.20 ± 0.45		
Others	3.70 ± 1.16		
Education	**M (SD)**	**F**	**p**
Primary school	3.00 ± null	6.386	0.000 **
Junior high school	2.71 ± 0.95		
Senior high school	3.41 ± 0.91		
College	3.62 ± 1.21		
Master or above	4.15 ± 1.24		
Working Location	**M (SD)**	**F**	**p**
inside the second ring road	3.65 ± 1.06	1.253	0.288
the second ring road-the third ring road	4.03 ± 1.27		
the third ring road-the fourth ring road	3.89 ± 1.20		
the fourth ring road-the fifth ring road	3.70 ± 1.23		
outside the fifth ring road	3.60 ± 1.39		

* $p < 0.05$ ** $p < 0.01$ **M**: mean, **SD**: standard deviation.

5.2. Reliability and Validity Analysis

The scale in this study was redesigned based on TPB. Therefore, the quality of the scale needed to be re-evaluated to test the validity and reliability of the questionnaire and ensure that the questionnaire could reflect the objective reality.

5.2.1. Reliability Analysis

In order to examine the reliability of the collected data, a corresponding analysis was carried out for the extended TPB scale. Calculation of Cronbach's alpha provided a method for the estimation of the internal consistency of the questionnaire and of each factor of the remaining items according to the exploratory factor analysis. As shown in Table 2, the reliability for each of the remaining factors was larger than 0.60, and the total scale $\alpha = 0.792$, indicating the acceptable reliability of the TPB questionnaire.

5.2.2. Validity Analysis

Exploratory factor analysis [37] provides assistance in the determination of the contribution from indicators to the evaluation of each latent variable. Before the factor analysis, the KMO test and the Bartlett test were required to analyze the correlation between variables and the independence between variables, respectively. KMO = 0.804, and the Bartlett test showed a significance level of $p < 0.01$, indicating the suitability for the factor analysis. The approach to derive principal components by means of varimax rotation was subsequently employed for the identification of seven factors. Items should be assigned to factors based on their factor loadings. In the case of a factor loading less than 0.4 or the presence of cross-loading in more than one factor, the items should be removed [38]. The eigenvalues for the seven factors—PBC, SN, BATT, CATT, WATT, IER, and BI—were 2.251, 3.353, 3.396, 2.745 1.570, 2.389, and 2.333, respectively. The seven factors could explain the variations of 8.66%, 12.90%, 13.06%, 10.56%, 6.04%, 9.19%, and 8.97%, respectively. The accumulative variance for the contribution rate was 69.38%, and each of the seven factors had a value above one (see Table 5). Consequently, the validity and reliability of the extended TPB questionnaire were confirmed, and thus, the extended TPB questionnaire was a feasible measurement for the use behavior of FFBS.

Table 5. Factor analysis (n = 352).

Items	Mean Score	Eigen-Values	Cronbach's Alpha	Factors Loading	Variance Explained (%)	Cumulative Variance Explained (%)
Factor 1: PBC	3.579					
PBC1	3.46	2.251	0.823	0.866	8.66%	8.66%
PBC2	3.543			0.879		
PBC3	3.733			0.8		
Factor 2: SN	4.224					
SN1	4.54			0.695		
SN2	4.199	3.353	0.854	0.619	12.90%	21.56%
SN3	4.128			0.843		
SN4	4.116			0.855		
SN5	4.139			0.868		
Factor 3: BATT	4.242					
BATT1	3.801			0.71		
BATT2	4.043	3.396	0.83	0.76	13.06%	34.62%
BATT3	4.455			0.771		
BATT4	4.423			0.774		
BATT5	4.489			0.702		
Factor 4: CATT	3.425					
CATT1	2.727			0.651		
CATT2	3.586	2.745	0.774	0.792	10.56%	45.18%
CATT3	3.094			0.675		
CATT4	3.955			0.731		
CATT5	3.781			0.772		
Factor 5: WATT	1.875	1.57	0.678		6.04%	51.22%
WATT1	1.966			0.82		
WATT2	1.784			0.822		
Factor 6: IER	1.777					
IER1	1.767	2.389	0.859	0.879	9.19%	60.41%
IER2	1.96			0.824		
IER3	1.605			0.831		
Factor 7: BI	4.072					
BI1	4.165	2.333	0.931	0.779	8.97%	69.38%
BI2	4.054			0.825		
BI3	3.997			0.805		

5.3. Correlation Analysis

As Table 6 shows, SN, BATT, CATT, WATT, and IER were highly correlated with the intention to use FFBS; PBC, SN, BATT, CATT, and IER were highly correlated with the FFBS use behavior.

Table 6. The correlation matrix between the factors of the scale and the total score.

	PBC	SN	BATT	CATT	WATT	IER	BI	BV
PBC	1							
SN	0.004	1						
BATT	−0.043	0.295 **	1					
CATT	0.019	−0.046	−0.087	1				
WATT	−0.085	−0.058	−0.275 **	−0.139 **	1			
IER	−0.265 **	−0.226 **	−0.335 **	0.049	0.250 **	1		
BI	−0.045	0.401 **	0.569 **	−0.167 **	−0.244 **	−0.293 **	1	
BV	−0.107 *	0.163 **	0.179 **	−0.117 *	−0.064	−0.126 *	0.468 **	1

* $p < 0.05$ ** $p < 0.01$.

5.4. Structural Equation Modeling

Based on an unbiased method via Amos 17.0, a structural equation model was established to explore the relationship among potential variables and confirm the results of the exploratory analysis through a confirmatory factor analysis (see Figure 4). Such an analysis approach is able to estimate latent variables with the indicators and to further relate them with an observed measure of the studied object, which is the frequency of FFBS use for the present case.

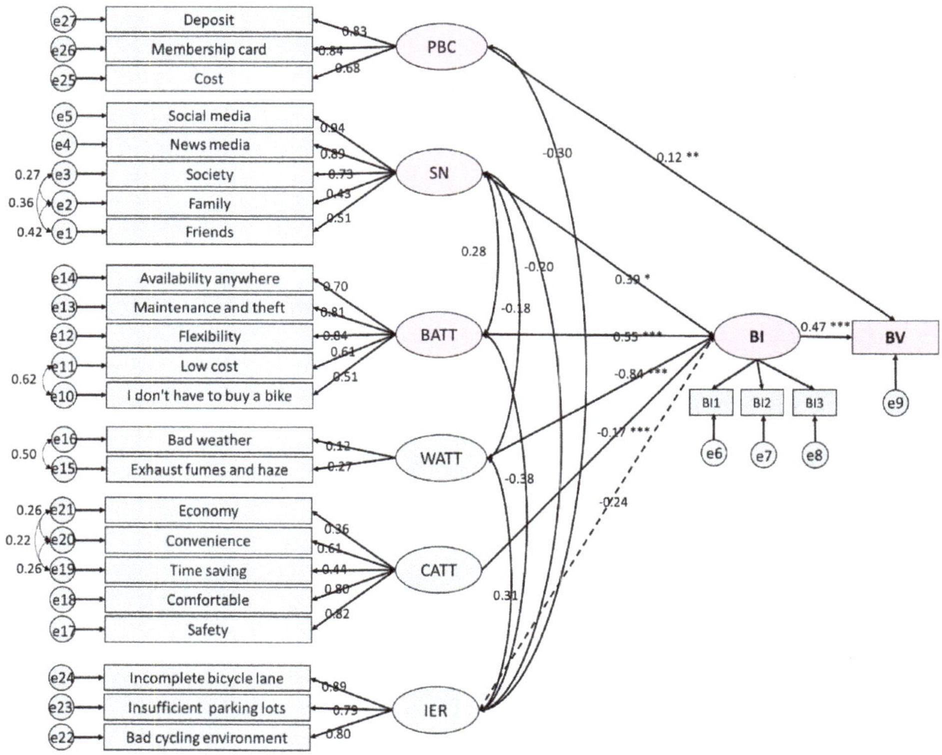

(*p <0.05, **p <0.01, ***p <0.001)

Figure 4. Path diagram of the proposed model to explain the use of FFBS.

Results of Hypotheses Testing

Attitudes toward FFBS (β = 0.55, $p < 0.001$) produced the foremost predictive contribution to intention, followed by subjective norms (β = 0.39, $p < 0.05$), supporting H2b and H2c, respectively. Although perceptual behavior control had no direct effect on intention, it positively predicted users' behavior to use FFBS (β = 0.12, $p < 0.01$), rejecting H2a and supporting H4. Attitudes toward bad weather (β = −0.84, $p < 0.001$) produced the largest direct and negative predictive influence on intention, with the second foremost contribution from attitudes toward cars (β = −0.17, $p < 0.001$), supporting H3b and H3a, respectively. However, the direct effects of external constraints regarding infrastructure (β = −0.24, $p > 0.05$) on intention were of little significance, thereby rejecting H3c and H5. Lastly, intention gave a significant prediction of users' behavior to use FFBS (β = 0.47, $p < 0.001$), completely supporting H1.

The χ^2 test and its fit indices (root mean squared error of approximation (RMSEA), efficiency of fit index (GFI), incremental fit index (IFI), non-normed fit index (NNFI), Tucker–Lewis coefficient (TLI), and comparative fit index (CFI)) were employed to determine whether the actual data were in general agreement with the proposed model. According to the results of the fit indexes from Table 7, the values of χ^2/df ranged from one to three, indicating the general consistency between the model and data. Additionally, the RMSEA < 0.08, and only the NFI was slightly below the standard value. Meanwhile, the remaining fit indexes were greater than 0.9. The results indicated a favorable fit ability; thus, the model met our expectations regarding statistical adequacy.

Table 7. Results of the goodness of fit of the revised model: root mean squared error of approximation (RMSEA), efficiency of fit index (GFI), incremental fit index (IFI), non-normed fit index (NNFI), Tucker–Lewis coefficient (TLI), and comparative fit index (CFI).

Fit Index	χ^2/df	RMSEA	GFI	IFI	NFI	TLI	CFI
Measured value	1.732	0.046	0.901	0.953	0.896	0.945	0.953
Standard value	1–3	<0.08	>0.90	>0.90	>0.90	>0.90	>0.90
Adaptation judgment	Yes	Yes	Yes	Yes	No	Yes	Yes

6. Discussion

The current study was primarily aimed at identifying the psychosocial variables that play a role in FFBS use intention and actual behaviors, as well as to further explore the complex interrelation among those factors and to provide a useful effective method to promote FFBS use.

As expected, the extended planned behavior theory in this paper could reasonably explain FFBS usage intention and behavior. The structure involved herein showed that the positive indicators associated with FFBS usage could be revealed through three latent variables: perceived behavioral control, subjective norms, and residents' attitudes toward FFBS, among which residents' attitudes toward FFBS were related to indicators that made cycling a competitive mode of transport. On the other hand, the negative indicators associated with FFBS usage could be revealed through three latent variables: attitudes toward cars, attitudes toward bad weather, and external constraints regarding infrastructure, among which bad weather was the biggest obstacle for residents to use FFBS.

6.1. Influence of Demographic Variables

The current study investigated the correlation between extended TPB variables and demographic factors. According to the results of independent t-tests, residents' gender, age, car ownership, FFBS riding experience, and riding ability had significant influences on the psychological variables. In the current study, men showed a better attitude toward cars; one reasonable explanation might be that cars exhibit more attraction for men than for women and that women account for a relatively small proportion of licensed drivers. [39]. This implied that motivation for car usage was quite different for men and women. Thus, men could be affected by factors that might make little sense for women.

People under 60 years old were more willing to use FFBS, which might be due to the higher ability of young people to accept new things, and studies showed that people who were more receptive to new things were more likely to perceive the ease of use and usefulness of shared bikes [40]. There was a dissimilarity in attitudes towards cars between people with cars and those without. This was related to car owners being used to driving for travel, and thus, they were less susceptible to new ways of travel. In the current study, there was a clear difference between the perceptions of users that had FFBS experience and those that never used a shared bike [41], which was consistent with previous studies. Adequate advice for the promotion in FFBS is to encourage people to experience FFBS riding [42]. In addition, it was reasonable that FFBS was more attractive for skilled cyclists with stronger subjective norms. On the other hand, according to the results of the analysis of variance, the significant influence of different demographic variables on the frequency of shared bikes was identified through this study. Men, younger people, and students were more likely to use shared bikes. Car owners and skilled cyclists showed higher enthusiasm toward shared cycling. Well-educated cyclists also showed higher inclination to becoming bike-sharing users.

6.2. Influence of Psychosocial Variables

6.2.1. Perceptual Behavior Control

The perceptual behavior control explained by resources and ability (deposit, membership card, cost) had a direct and predictive effect on FFBS use behavior: the stronger the perceived behavioral control, the more likely residents were to use FFBS, consistent with previous work [43]. Therefore, it is adequate advice to lower the cost for FFBS usage.

6.2.2. Subjective Norms

A general consensus has been achieved by most researches that subjective norms have predictive effects on intention [44], also confirmed in this study. In addition, the results from the current study revealed the crucial role that social media, news media, and society play in FFBS prevalence. With the development of society, people are more and more dependent on social media and news media for information. Moreover, shared bikes have obvious social attributes, which leads to the significant influence of media on residents' behaviors. Policies could encourage the recommendation of this environmentally-friendly travel mode to neighboring friends, relatives, and social contacts [7,45]. On the other hand, urban governments should dispel residents' anxieties through news media, social atmosphere, and word-of-mouth among groups [3] to improve residents' willingness to use shared bikes.

6.2.3. Attitudes toward FFBS

The results of this paper showed that a positive attitude towards shared bikes was the key to promote residents to use shared bikes. One possible explanation might be the unique advantages of FFBS that it can be rented and returned anywhere and anytime with a low cost and that in some cases, it could replace one's own bicycle, as well as the users are relieved from the maintenance work and security anxieties. These are beyond the capability of other vehicles. This result was in accordance with the finding of Yan [31]: the more these characteristics are perceived by the cyclist with use, the more important convenience becomes to explain their decisions. This paper also explored the success of FFBS over the PBS launched by the Beijing government. Although there are seemingly plenty of PBS stations in Beijing, people often have to walk more than 200 m to a fixed site to rent a bike, and they are not sure about the location or the existence of a PBS station near the destination. In addition, the procedure to register membership and get a card is time consuming, and thus, it is impractical for non-members to use a bike right away. However, all these inconveniences are eliminated with FFBS: people can even get an available shared bike at the gate of the community by scanning a QR code with a smartphone and leave the bike anywhere at the destination. Therefore, the convenience of user experience belongs to an important user experience that FFBS operators should further improve, and some potential approaches for the improvement of the service level should also be considered.

6.2.4. Attitudes toward Cars

The current study confirmed the competitive relationship between cars and FFBS. The more friendly residents are toward cars, the less likely they are to choose shared bikes. Koller found that car usage is often accompanied by functional, economic, emotional, and social values [32]. Interestingly, contrary to expectations, the results of this paper indicated that the "comfort and safety of cars" was the main reason for people to choose cars over "time saving". Such a conclusion was established based on the transportation mode itself and the perception of users, without considering other potential factors such as the daily intention to ride a bike. Therefore, in an area suitable for bicycle travel (such as the transit-oriented development area), targeted measures and designs to limit cars, such as higher parking fees, less parking convenience, and building a dense road network in small blocks, can effectively improve residents' intention to choose FFBS.

6.2.5. Attitudes toward Bad Weather

Previous studies had consistently reported the negative effects of bad weather, such as rainy weather [19], temperature, and poor air quality [18] as observation variables. In the present study, attitudes toward bad weather represented by variables of rainy weather, exhaust fumes, and haze were also a significant factor for intention, consistent with previous work. This may be due to the limitation of FFBS. Residents' intention to travel by shared bikes will be greatly reduced under harsh travel environment such as rain, snow, exhaust fumes, and haze. Therefore, it is a key consideration for FFBS operators to address the restrictions of bicycles by corresponding optimization, such as the incorporation of fenders and adding rain shelters.

6.2.6. External Restrictions Regarding Infrastructure

Previous studies have emphasized the importance of infrastructure for bike or shared bike trips; however, this paper showed a different conclusion: limitations regarding infrastructure such as inadequate bike lanes had no significant effect on the change in the willingness to use FFBS. One possible explanation is that in terms of infrastructure in Beijing, which is far from satisfactory even with the improvement in recent years, the convenience brought by FFBS outweighs the obstacles of infrastructure. This conclusion explains why Beijing's efforts in infrastructure are not as significant an effect on shared bike usage.

7. Conclusions

This study conducted an online survey in Beijing and then adopted an extended TPB to examine the psychological determinants of intention and actual behavior to use FFBS. Both reliability and validity analysis confirmed the reasonability of the extended TPB with regard to interpreting Beijing residents' intentions to adopt FFBS. The main findings in this study were as follows.

1. The SEM model outputs showed that H1, H2b, H2c, H3a, H3b, and H4 were supported, while H2a and H3c were rejected. Psychological variables influencing residents' use of FFBS were divided into motivation factors and hindering factors. Among them, perceived behavior control, subjective norms, and attitude toward FFBS were the motivating factors, while friendly attitude toward cars, attitude toward bad weather, and external restrictions of infrastructure were the hindering factors.
2. The economic convenience of FFBS was the most important attraction factor for residents to use FFBS, while bad weather was the biggest hindrance factor for residents to use FFBS. However, barriers of imperfect infrastructure had no significant impact on reducing residents' willingness to use FFBS.

3. The overall process simplification and high convenience were the main reasons for FFBS's success in promoting sustainable transport development over PBS and efforts in infrastructure.
4. For FFBS operating enterprises, the use threshold (reducing costs, etc.) and system convenience (scheduling, etc.) are key to improve the utilization of FFBS. For governments, the car restrictions (raising parking costs, etc.) and the bicycle travel restrictions improving (adding shelters, etc.) play an important role in the development of sustainable transport.

Limitations and Future Research Needs

This study also had limitations. First, both cars and buses form a competitive relationship with FFBS in short-distance travel. Considering the green travel mode of buses, this paper mainly discussed the influence of car attitude on the use of FFBS. The influence of buses will be studied in future research. In addition, publicity and experience of using FFBS have been found influential in the use of FFBS, yet the effects of the intervention are not clear; thus, the effect assessment of psychological intervention measures could be carried out in future work, so as to evaluate the effects.

Author Contributions: Conceptualization, D.X.; data curation, S.S.; methodology, D.X.; resources, Y.B.; writing, original draft, D.X.; writing, review and editing, Y.B. and S.S. All authors have read and agree to the published version of the manuscript.

Funding: This research received no external funding.

Conflicts of Interest: The authors declare no conflict of interest.

References

1. Heredia, A.; Monzón, A.; Jara-Díazlt, S. Understanding cyclists' perceptions, keys for a successful bicycle promotion. *Transp. Res. Part A Policy Pract.* **2014**, *63*, 1–11. [CrossRef]
2. Heinen, E.; Maat, K.; Wee, B.V. The role of attitudes toward characteristics of bicycle commuting on the choice to cycle to work over various distances. *Transp. Res. Part D Transp. Environ.* **2011**, *16*, 102–109. [CrossRef]
3. Cervero, R.; Duncan, M. Walking, bicycling, and urban landscapes: Evidence from the San Francisco Bay Area. *Am. J. Public Health* **2003**, *93*, 1478–1483. [CrossRef] [PubMed]
4. Yan, X.; Handy, S.L.; Mokhtarian, P.L. Factors associated with proportions and miles of bicycling for transportation and recreation in six small US cities. *Transp. Res. Part D Transp. Environ.* **2010**, *15*, 73–81.
5. Beijing Transport Annual Report 2017. Available online: http://www.bjtrc.org.cn/JGJS.aspx?id=5.2&Menu=GZCG (accessed on 26 August 2019).
6. The Fourth Comprehensive Traffic Survey in Beijing. Available online: http://www.bjtrc.org.cn/JGJS.aspx?Id=INN20120227005&Menu=YJXMNR&sort=JCYJ (accessed on 26 August 2019).
7. Xu, D.D.; Bian, Y.; Rong, J.; Wang, J.C.; Yin, B.C. Study on Clustering of Free-Floating Bike-Sharing Parking Time Series in Beijing Subway Stations. *Sustainability* **2019**, *11*, 5439. [CrossRef]
8. Pucher, J.; Buehler, R. Why Canadians cycle more than Americans: A comparative analysis of bicycling trends and policies. *Transp. Policy* **2006**, *13*, 265–279. [CrossRef]
9. Guo, Y.Y.; Zhou, J.B.; Wu, Y. Identifying the factors affecting bike-sharing usage and degree of satisfaction in Ningbo, China. *PLoS ONE* **2017**, *12*, e0185100. [CrossRef]
10. Fishman, E.; Washington, S.; Haworth, N.; Watson, A. Factors influencing bike share membership: Analysis of Melbourne and Brisbane. *Transp. Res. Part A Policy Pract.* **2015**, *71*, 17–30. [CrossRef]
11. Shaheen, S.; Cohen, A.P.; Martin, E.W. Public bikesharing in north America: Early operator understanding and emerging trends. *Transp. Res. Rec. J. Transp. Res. Board* **2013**, *1568*, 83–92. [CrossRef]
12. Woodcock, J.; Tainio, M.; Cheshire, J.; O'Brien, O.; Goodman, A. Health effects of the London bicycle sharing system: Health impact modelling study. *Br. Med. J.* **2014**, *348*, 1–14. [CrossRef]
13. Pan, C.; Guo, Z. A study on commuting behavior selection of commuters under social interaction effect. *Transp. Syst. Eng. Inf.* **2017**, *17*, 101–106.
14. Zhao, P.; Li, S. Bicycle-Metro integration in a growing city: The determinants of cycling as a transfer mode in metro station areas in Beijing. *Transp. Res. Part A Policy Pract.* **2017**, *99*, 46–60. [CrossRef]

15. Gámez-Pérez, K.M.; Arroyo-López, P.; Cherry, C.R. Defining a Primary Market for Bikesharing Programs. *Transp. Res. Rec. J. Transp. Res. Board* **2017**, *2634*, 50–56. [CrossRef]
16. Barnes, G.; Thompson, K. A longitudinal analysis of the effect of bicycle facilities on commute mode share. In Proceedings of the 85th Annual Meeting of Transportation Research Board, Washington, DC, USA, 22–26 January 2006.
17. Stinson, M.; Bhat, C. Frequency of bicycle commuting: Internet-based survey analysis. *Transp. Res. Rec. J. Transp. Res. Board* **2004**, *1878*, 122–130. [CrossRef]
18. Li, W.; Kamargianni, M. Air pollution and seasonality effects on mode choice in China. *Transp. Res. Rec. J. Transp. Res. Board* **2017**, *2634*, 101–109. [CrossRef]
19. Lu, C.; Xu, F.; Dong, S. Observations of Public Bikesharing. *Transp. Res. Board* **2017**, *2662*, 93–101. [CrossRef]
20. Campbell, A.A.; Cherry, C.R.; Ryerson, M.S. Factors influencing the choice of shared bicycles and shared electric bikes in Beijing. *Transport. Res. Part C Emerg. Technol.* **2016**, *67*, 399–414. [CrossRef]
21. Fishman, E. Bikeshare: A Review of Recent Literature. *Transp. Rev.* **2015**, *36*, 92–113. [CrossRef]
22. Gu, T.; Kim, I.; Currie, G. To be or not to be dockless: Empirical analysis of dockless bikeshare development in China. *Transp. Res. Part A Policy Pract.* **2019**, *119*, 122–147. [CrossRef]
23. Li, X.; Zhang, Y.; Du, M. Social factors influencing the choice of bicycle: Difference analysis among private bike, public bike sharing and free-Floating bike sharing in Kunming, China. *KSCE J. Civ. Eng.* **2019**, *23*, 2339–2348. [CrossRef]
24. Bamberg, S.; Schmidt, P. Incentives, morality, or habit? Predicting students' car use for university routes with the models of Ajzen, Schwartz and Triandis. *Environ. Behav.* **2003**, *35*, 264–285. [CrossRef]
25. Ajzen, I. The theory of planned behavior. *Organ. Behav. Hum. Decis. Process.* **1991**, *50*, 179–211. [CrossRef]
26. Dill, J.; Voros, K. Factors affecting bicycling demand: Initial survey findings from the Portland region In Proceedings of the 86th Annual Meeting of Transportation Research Board, Washington, DC, USA, 21–25 January 2007.
27. Abraham, J.; Mcmillan, S.; Brownlee, A.; Hunt, J.; Mcmillan, A. Investigation of cycling sensitivities. In Proceedings of the 81th Annual Meeting of Transportation Research Board, Washington, DC, USA, 13–17 January 2002.
28. de Geus, B.; De Bourdeaudhuij, I.; Jannes, C.; Meeusen, R. Psychosocial and environmental factors associated with cycling for transport among a working population. *Health Educ. Res.* **2008**, *23*, 697–708. [CrossRef] [PubMed]
29. Schoner, J.; Lindsey, G.; Levinson, D. Is bike sharing contagious? Modeling its effects on system membership and general population cycling. *Transp. Res. Rec. J. Transp. Res. Board* **2016**, *2587*, 125–132. [CrossRef]
30. Ma, L.; Zhang, X.; Ding, X.Y. Bike sharing and users' subjective well-Being: An empirical study in China. *Transp. Res. Part A Policy Pract.* **2018**, *118*, 14–24. [CrossRef]
31. Yan, R.; Chu, J.L.; Yang, X.G. Intention of urban residents to use public bike-sharing system: A case study in Hefei. *J. Transp. Inf. Saf.* **2017**, *35*, 101–107.
32. Koller, M.; Floh, A.; Zauner, A. Further insights into perceived value and consumer loyalty: A "green" perspective. *Psychol. Market.* **2011**, *28*, 1154–1176. [CrossRef]
33. Han, H.; Meng, B.; Kim, W. Emerging bicycle tourism and the theory of planned behavior. *J. Sustain. Tour.* **2017**, *25*, 292–309. [CrossRef]
34. Beijing statistical yearbook 2019. Available online: http://202.96.40.155/nj/main/2019-tjnj/zk/indexch.htm (accessed on 8 March 2020).
35. Cressie, N.A.; Whitford, H. How to use the two sample t-Test. *Biom. J.* **1986**, *28*, 131–148. [CrossRef]
36. Larson, M.G. Analysis of Variance. *Circulation* **2008**, *117*, 115–121. [CrossRef]
37. Spearman, C. General intelligence, objectively determined and measured. *Am. J. Psychol.* **1904**, *15*, 201–293. [CrossRef]
38. Badland, H.; Schofield, G. Understanding the relationships between private automobile availability, overall physical activity, and travel behavior in adults. *Transportation* **2008**, *35*, 363–374. [CrossRef]
39. Kronsell, A.; Rosqvist, L.S.; Hiselius, L.W. Achieving climate objectives in transport policy by including women and challenging gender norms: The Swedish case. *Int. J. Sustain. Transp.* **2016**, *10*, 703–711. [CrossRef]
40. Chen, C.H.; Li, X.Y. Research on the effect factors of customer use intention of shared bicycles. *Chin. J. Manag.* **2018**, *15*, 1601–1610.

41. Rondinella, G.; Fernandez-Heredia, A.; Monzón, A. Analysis of perceptions of utilitarian cycling by level of user experience. In Proceedings of the 91th Annual Meeting of Transportation Research Board, Washington, DC, USA, 12–19 January 2012.
42. Broach, J.; Dill, J.; Gliebe, J. Where do cyclists ride? A route choice model developed with revealed preference GPS data. *Transp. Res. Part A Policy Pract.* **2012**, *46*, 1730–1740. [CrossRef]
43. Chen, S.Y. Using the sustainable modified TAM and TPB to analyze the effects of perceived green value on loyalty to a public bike system. *Transp. Res. Part A Policy Pract.* **2016**, *88*, 58–72. [CrossRef]
44. Pelling, E.L.; White, K.M. The theory of planned behavior applied to young people's use of social networking web sites. *Cyberpsychol. Behav.* **2009**, *12*, 755–759. [CrossRef]
45. Gong, X.S.; Xu, L.; Li, Z.L. Research on the relationship between different types of reference group influence and residents' consumption intentions. *Chin. J. Manag.* **2017**, *14*, 1829–1839.

sustainability

MDPI

Article

Empirical Study on Bikesharing Brand Selection in China in the Post-Sharing Era

Guangnian Xiao * and Zihao Wang

School of Economics and Management, Shanghai Maritime University, Shanghai 201306, China; 201930710100@stu.shmtu.edu.cn

* Correspondence: gnxiao@shmtu.edu.cn

Received: 7 March 2020; Accepted: 11 April 2020; Published: 13 April 2020

check for updates

Abstract: With the rapid popularization of mobile Internet technology and smart terminal equipment in recent years, the volume and usage of dockless bikesharing (hereafter referred to as bikesharing), which is green, environmentally friendly and convenient, have grown rapidly, making it one of the China's "new four major inventions." The development of the bikesharing in China consists of a pre-sharing era and a post-sharing era. In the pre-sharing era, capital-driven vicious market competition and lack of precise control have led to the abuse of urban space. Since the post-sharing era, the industry structure has returned to rationality, and many participants have been forced out of the market. The bikesharing has formed an oligopoly market consisting of head players such as Hellobike, Mobike, and Ofo. Therefore, how to improve the level of refined operations, promote sustainable development, improve cyclist satisfaction, and contribute to China's strength in transportation have become urgent problems for bikesharing companies and traffic management departments. From the perspective of the cyclist experience, the brand choice of the bikesharing is taken as the research object. An online revealed preference survey is used to collect data on cyclists' socio-economic attributes and subjective evaluations on the bikesharing. The conditional Logit model is used to explore the important factors that influence cyclists on the choice of bikesharing brands. Research results include: (1) age, occupation type, after-tax monthly income of the faculty group, riding comfort, rent, picking up/returning convenience, word of mouth, and volume have a significant impact on cyclists' bikesharing brand choices; (2) gender, educational background, monthly living expenses of the student group, appearance, deposit, deposit returning speed, rate of broken bikes, ease of use of software, and rent discount have no significant impact on cyclists' bikesharing brand choices. The research results are of great significance for improving the service quality of bikesharing companies and promoting the healthy development of the shared economy in China. Based on the results of the study, policy recommendations are made on the improvement for riding comfort, human-centered design, and word of mouth, and the construction of shared facilities.

Keywords: post-sharing era; bikesharing; brand choice; conditional Logit model; sustainable development

---

1. Introduction

With the continued growth of the Internet economy in recent years, dockless bikesharing (hereafter referred to as bikesharing) has emerged as a solution to the "last mile" of travel. In comparison with traditional docked shared bikes (hereafter referred to as public bikes), the bikesharing is a new mode of transportation extensively welcomed by travelers, wherein dockless shared bikes (hereafter referred to as shared bikes) can be picked up and returned whenever and wherever possible with convenient use and large volume. The bikesharing is the pathfinder and pioneer of shared economy. The utilization rate of the bikesharing is elevated by separating the ownership and use right of bikes. Risk capitals entered the bikesharing industry in a large scale since 2016, facilitating dozens of bikesharing brands to enter the market and contributing to the exponential growth of volume and usage of the bikesharing. Since then, the bikesharing entered a pre-sharing era. However, after 2018, the excessive volume of the bikesharing and "severe winter of capitals" brought most bikesharing brands into a business distress. What bikesharing companies focus on is turned from increasing market share into improving cyclist satisfaction and promoting sustainable development. The pattern of the bikesharing industry tends to stabilize and rationalize, and bikesharing companies gradually enter the profitable phase. Most problems encountered in the pre-sharing era, including management disorder and illegal parking, are solved step by step. According to iiMedia Research [1], growth rates of Chinese bikesharing presented a declining trend in 2018 (Figure 1), but the market scales continued to grow steadily as the rent was elevated by a large margin. The market scale is predicted to reach 30 billion yuan (Figure 2) in 2020.

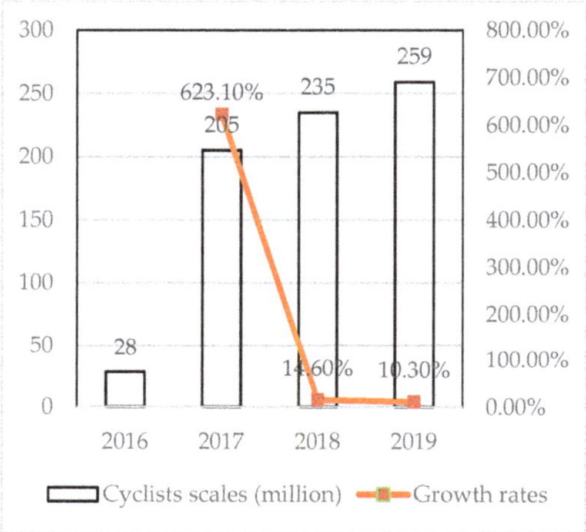

Figure 1. Cyclist scales and growth rates of Chinese bikesharing during 2016 to 2019

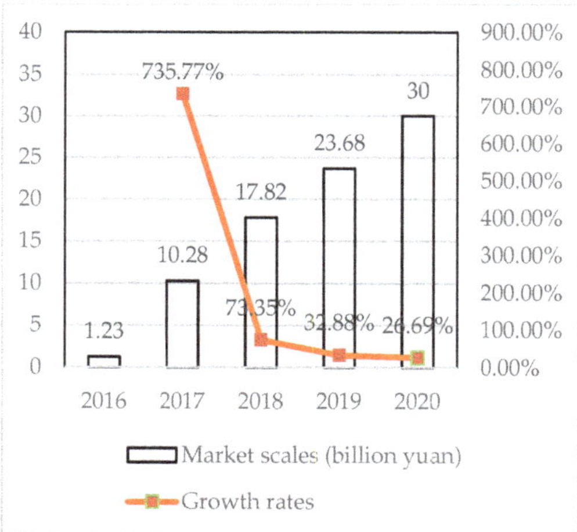

Figure 2. Market scales and growth rates of Chinese bikesharing during 2016 to 2020

In the post-sharing age, the cost used to gain customers is increasing day by day and the demographic dividend is gradually disappearing, so that bikesharing companies are laying a greater emphasis on the refinement of market operation and the precision of customer service. The Deposit-free mode is a novel connector between bikesharing companies and the general public. The Deposit-free mode achieves quick success because this mode targets at extensive consumer groups, creating a low-riding threshold, high-riding cost performance, and improved travel experience. In addition, an oligopoly market consisting of Hellobike, Mobike, and Ofo was formed gradually. To compare the basic conditions of three bikesharing companies, deposit, rent, bike design, online picking up/returning operation, membership cards, and rent discount are summarized as shown in Table 1. To compare the relative competitiveness and promote the management of the three brands, we explore bikesharing brand choice with an empirical study.

Table 1. Basic conditions of three bikesharing brands.

Brand	Deposit	Rent	Bike Design	Online Picking up/Returning Operation	Membership Cards and Rent Discount
Hellobike	Deposit-free for over 650 Zhima points and 199 yuan otherwise.	1 yuan/15 min.	Strong braking force, large seat area, simple wheel tread pattern, and aluminum-alloy wheel frame.	A dedicated application and Alipay interface.	20 and 16.9 yuan for a month card and continuous monthly payment, temporary cards (1 yuan/twice, 2 yuan/five times).
Mobike	Deposit-free.	1.5 yuan/first 15 min and 0.5 yuan/every additional 15 min.	Wide solid tire, short axle distance, good ground gripping force, and aluminum-alloy wheel frame.	A dedicated application, WeChat small program and Meituan interface.	20 yuan/month card.
Ofo	199 yuan for non-student groups.	0.8 yuan/min and 0.5 yuan/km, but at most 2 yuan/h.	Large tire radius, heavy bike, and simple frame design.	A dedicated application and WeChat small program interface.	20 yuan/month card.

2. Literature Review

The review of existing studies focuses on the bikesharing evolution, the cyclist profile and trip characteristics, and the influential factors for determining cyclists' bikesharing brand choice.

2.1. Bikesharing Evolution

The sharing economy improves people's work efficiency, reduces living cost, and monetizes underused resources. With the development of information technologies and the improvement of institutional policies, the shared mobility (including carsharing and bikesharing) gradually has a revolutionary impact on the choice of travel modes [2]. The shared mobility provides users short-term access and modes of transportation on demand. In the urban environment, the shared mobility is affected by the transportation infrastructure, zoning, land use, urban design, housing, and economic development.

Bikesharing provides users with on-demand access for one-way (point-to-point) or round trips from a variety of locations. The abundance of bicycles in dense neighborhoods often creates a "network-effect" that further encourages cycling and other forms of commuting. In addition, bikesharing also contributes to increased mobility, reduced carbon emission, decreased automobile use, economic development, and health benefits [3].

Bikesharing emerged in Europe as a transportation mode in 1965. The bikesharing evolution is categorized into three generations: the first generation (white bikes) that began in Amsterdam in 1965, the second generation (coin-deposit systems) that started in Copenhagen, Denmark in 1995, and the third generation (IT-based systems) that emerged in the Rennes, France city-based system in 1998. The third generation is the prototype of today's bikesharing. In the future, the fourth-generation mode will be characterized by: (1) flexible, clean and dockless characteristics, (2) redistribution innovations; (3) smartcard integration with other transportation modes, and (4) technological advances including GPS tracking, touchscreen kiosks, and electric bikes [4].

In addition to the public's use of bikesharing systems, closed campus systems are increasingly deployed in most universities at present. These closed campus systems are available only to the particular campus community they serve and the main service modes of bikesharing in universities are membership-based and non-membership-based self-service models [5]. In the future, bikesharing may receive more attention as a sustainable transportation alternative as a result of rising fuel prices, public health concerns, smart growth initiatives, and climate-change concerns [6].

2.2. Cyclist Profile and Trip Characteristics

The cyclist profile is described from socio-economic attributes (including gender, age, educational background, and income) and psychological latent variables. Ran et al. developed a binary Logit model to identify the cyclist profile based on 621 valid questionnaires collected by an online revealed preference (RP) survey from multiple cities, including Shanghai, Beijing, Nanjing, and Hefei [7]. The results indicate that most cyclists are women, young, and highly educated. To compare cyclist profile between shared bikes and private bikes, Buck et al. extracted the profile for cyclists of private bikes from the household travel survey of Washington, DC area in 2007–2008, cyclists of Capital Bikeshare (CaBi) with short-term membership by an intercept survey, and cyclists of CaBi with annual membership by an online survey [8]. The online survey is conducted by sending an email to CaBi's approximately 18 000 annual members and asking them to participate in the survey through a survey website. A total of 5 464 members completed the survey during a one-month period, with a response rate of 31%. A descriptive statistical analysis shows that CaBi cyclists with short-term and annual memberships are more likely to be female, young, underpaid, and own few private bikes and cars, compared with cyclists with private bikes. Raux et al. achieved one-day travel diaries and socio-economic attributes of 3 161 respondents in Lyon, France with an online survey and analyzed the cyclist profile of the Velo'v bikesharing program [9]. The majority of cyclists with Velo'v annual membership are the young, highly paid, and those who live near bikesharing stations and own private cars. Additionally, cyclists whose bikes were stolen are more willing to take shared bikes [10,11]. In terms of psychological latent variables, environmental awareness and subjective norm are positive factors to promote the bikesharing use [12].

The trip characteristics may be reflected by departure time, trip duration, travel distance, trip purpose, travel route, and travel destination, etc. Deng et al. explored the spatio-temporal characteristics of Mobike based on picking up/returning records of 485.5 thousand bikes between 10 to 25 May 2017, and the geographical data extracted from application programming interfaces of Baidu [13]. They concluded that a huge bikesharing demand appears in morning/evening peaks of working days and that bikesharing trips have various characteristics, including tide type, one-way type, loosely connected type, closely connected type, and distance-proof type. In addition, cyclists are more likely to take shared bikes in the weather of no rainfall, light wind, moderate temperature, and clean air quality [14–16]. The top three purposes of bikesharing trips are commuting (35.9%), entertainment (22.5%), and going home (18.1%). Bikesharing trips are characterized by short travel distance and trip duration. Specifically, over 73.0% of trips have a trip duration of less than 15 min, and only 5.5% of trips have a trip duration of over 30 min [17]. The main reasons for short bikesharing trips are twofold: one is a high proportion of trips for transferring to bus/subway modes, and the other is the marketing activity of "red-packet bikes," which attracts cyclists to ride shared bikes for at least 10 min, rewarded with a red packet of a random amount [18]. Most cyclists prefer travel routes with bike lanes and travel destinations with diversified land use, high transit accessibility, and sufficient parking space [19–22].

2.3. Influential Factors

In general, the bikesharing choice is influenced by various factors, including travel distance, picking up convenience, travel time reliability, credit supervision mechanism, social interaction, and service quality. Travel distance is one of the most important factors to influence the travel choice. The bikesharing is preferred for trips with no more than 3 km [23]. Picking up convenience depends on an adequate supply and a reasonable bikesharing distribution. The adequate supply may be ensured by a low supply cost and a huge subsidy on the condition of increasing the company's profit [24]. The reasonable bikesharing distribution lies in an accurate prediction of picking up/returning demand and an efficient redistribution. Dong et al. proposed a model called "DestiFlow" to explore the flow characteristics based on points of interest clustering [25]. DestiFlow is further applied with the time-series location data of Ofo and Mobike. Caggiani et al. suggested a new comprehensive dynamic bike redistribution methodology that comprises a prediction of the picking up/returning demand and a relocation Decision Support System [26]. Zhang et al. further introduced a dynamic pricing scheme with negative prices to achieve a more balanced bikesharing distribution by guiding cyclists to ride from oversupplied areas to undersupplied areas [27]. Travel time reliability for the bikesharing is often believed to be high for trips from and to stations/electrical fences that have a small number of shops within a walking distance [28]. The credit supervision mechanism may work with the condition that the parking behavior of cyclists can be effectively monitored and sanctioned [29]. Yao et al. further demonstrated that the credit supervision mechanism was more appealing to attract cyclists to ride shared bikes when a negative credit is introduced [30]. The social interaction influences cyclists' bikesharing choice by inducing family members or close friends to make the same choices [31]. The service quality is a relatively comprehensive factor and is proved to have a positively significant effect on the intention to ride shared bikes [32].

Another hot topic on the concerning bikesharing is to determine the key factors influencing the bikesharing choice in a specific context. Barbour et al. designed a web-based survey to collect the data on bikesharing usage between February and April, 2018 in the USA and constructed a random parameter Logit model to assess how much a bikesharing trip was displacing an auto trip [33]. They found that age, gender, income, household size, commuting type and length, vehicle ownership, and respondents' body mass index are significant factors for modal substitution decisions. Zhou et al. explored the spatio-temporal patterns of taxi and bikesharing trips in Chicago from 2014 to 2016 [34]. They applied random forests to model the choice between taxies and bikesharing and indicated that travel distance, the number of temporary stops, and recreational facilities played a significant role in determining the choice. Martín et al. developed a multinomial Logit model to investigate individual and contextual factors on the choice of the bikesharing based on data from a household travel survey conducted in 2014 in Vitoria-Gasteiz, Spain [35]. They indicated that age, gender, population density, and mixed land use significantly influenced the bikesharing choice. Li et al. developed a multinomial Logit model to investigate the mode choice among private bikes, public bikes, and shared bikes based on 522 questionnaires conducted in Kunming, China in 2018 [36]. They demonstrated that trips with shared bikes were more likely to be long-distance and for transfer. Moreover, most cyclists of shared bikes are young, underpaid, students, and have no registered permanent residence. Ma et al. constructed a combined model of a factor analysis and a two-layer nested Logit model to explore the factors influencing the riding behavior of the bikesharing for college students [37]. They found that the service quality, rent, mixed traffic, and the number of signal lights at intersection have a significant effect on the bikesharing choice. Du et al. discussed a case study based on the usage data from Mobike in Shanghai and demonstrated that the top three factors influencing the riding frequency were residential areas, parks & green areas, and population size [38]. Gu et al. compared public bikes and shared bikes from various aspects and concluded that shared bikes may be a better choice in mega cities where local governments had strong control and supervision abilities [39].

Researches on the bikesharing brand selection mainly focuses on analyzing the critical influencing factors of the bikesharing choice, but studies investigating the selection of mainstream bikesharing brands are limited. Therefore, bikesharing brands with top three market shares in China were taken as the study object. We analyze the main factors influencing brand selection with the conditional Logit model. Policy suggestions were proposed for bikesharing companies and related governmental sectors to promote sustainable development of the bikesharing industry.

3. Methods

In the brand choice analyzed in this study, the variables not only include socio-economic attributes of cyclists (cyclist-specific attributes) but also those associated with bikesharing brands (alternative-specific attributes), so that the multinomial Logit model does not apply. Hence, the conditional Logit model [40] is selected to model the choice of bikesharing brands. The conditional Logit model allows us to take into account multiple alternative-specific attributes simultaneously and thus evaluate the effect of these attributes on the choice of bikesharing brands. Additionally, the conditional Logit model has been applied in multiple transportation fields, including carsharing use [41] and travel mode choice [42]. As a member of disaggregate probability models, the conditional Logit model is a discrete choice and analysis method in microeconometrics, and its theoretical basis is that cyclists pursue "utility" maximization when selecting shared bike brands. The random utility function for individual i to select brand j is as follows:

$$U_{ij} = c_j + \sum_{k=1}^{M_1} z_{ik}\alpha_{jk} + \sum_{k=1}^{M_2} x_{ijk}\beta_k + \varepsilon_{ij}(i = 1, \ldots, N, j = 1, 2, 3) \tag{1}$$

where c_j is the inherent constant of the shared bike brand j, and this constant is zero for Ofo because it is taken as the reference brand; z_{ik} denotes the kth specific variable of the ith cyclist and α_{jk} is the corresponding parameter of the jth brand; x_{ijk} is the kth specific variable of cyclist i and the jth bikesharing brand and β_k is the corresponding parameter; M_1 and M_2 are the numbers of cyclist-specific and alternative-specific attributes; ε_{ij} is a random error; and N denotes the sample size.

Based on the above random utility function and the theory of random utility maximization, the probability for individual i to select brand j is as follows:

$$P_{ij} = \frac{\exp\left(c_j + \sum_{k=1}^{M_1} z_{ik}\alpha_{jk} + \sum_{k=1}^{M_2} x_{ijk}\beta_k\right)}{\sum_{s=1}^{3} \exp\left(c_s + \sum_{k=1}^{M_1} z_{ik}\alpha_{sk} + \sum_{k=1}^{M_2} x_{isk}\beta_k\right)}(i = 1, \ldots, N, j = 1, 2, 3) \tag{2}$$

4. Data Collection and Statistical Analysis

4.1. Questionnaire Design and Respondent Recruitment

The questionnaires were designed to analyze the choice of bikesharing brands based on the principles of conciseness, clearness, non-repetition, comprehensiveness, and reasonability. The questionnaire may be divided into the following three parts. The first part involves socio-economic attributes of cyclists, including gender, age, educational background, occupation type, and after-tax monthly income/monthly living expense. The second part consists of subjective evaluations of the three bikesharing brands, including riding comfort, appearance, rent, deposit, deposit returning speed, picking up/returning convenience, word of mouth, rate of broken bikes, ease of use of software (dedicated application and embedded interfaces of other applications), volume, and rent discount. For riding comfort, the question is "Which bikesharing branch do you think provides the greatest riding comfort?" and the options are "A. Hellobike, B. Mobike, C. Ofo, D. No difference." For other subjective evaluations, similar questions are raised. The third part involves the choice of a bikesharing brand that is most commonly used by the respondents. To ensure the quality of the questionnaire, a small-scale pre-survey was performed before the formal survey. The questionnaire and survey plans were adjusted in accordance with the feedback of cyclists.

The target group of this questionnaire survey was college students and faculties. The bikesharing was born to solve the "last mile" travel difficulty of college faculties and students, and offered convenience for their school life. By far, colleges remain an important market for bikesharing companies. College faculties and students are familiar with the bikesharing due to high use frequency.

Therefore, analyzing the brand choice behaviors for cyclists from faculties and students in colleges may provide a beneficial reference for the sustainable bikesharing development.

4.2. Descriptive Statistical Results

An online RP survey was carried out via Wenjuanxing (www.wjx.cn) [43], which is the biggest online Chinese survey platform focusing on questionnaire establishment, distribution, management, and analysis services. The platform is named as "SurveyStar" in English and operated by Changsha Ranxing Science and Technology Ltd. In comparison with the traditional paper-based questionnaire survey, the online survey has many advantages, including online data verification, fast collection, and avoidance of incorrect manual type-in [44,45]. From 29 March 2019 to 1 April 2019, the cyclists were solicited through WeChat Moments and direct invitation. A total of 190 valid questionnaires from students and faculties in colleges were recovered, and most of the cyclists lived in Shanghai. The number of cyclists who filled in and submitted the questionnaires through mobile browser and WeChat are 79 and 111, respectively. The average answering time was 95 s. The numbers of cyclists who choose Hellobike, Mobike, and Ofo are 33 (17.37%), 63 (33.16%), and 94 (49.47%) in this study. According to "China's Shared Bicycle Industry Monitoring Report for the Second Quarter of 2017" [46], the percentages of active cyclists who choose Hellobike, Mobike, and Ofo are 12.10%, 34.00%, and 53.90%. According to the Chi-square test, the brand choice distributions of the sample in this study and the population have no significant difference. The descriptive statistical results of data in the first and the second part of the questionnaire are shown in Tables 2 and 3. Due to the restriction of data availability, we take gender as example to evaluate the sample representativeness in terms of respondents' socio-economic attributes. According to the survey made in Nanjing University of Information Science & Technology in 2017 [47], the percentages of male and female cyclists are 37.84% and 62.16%. Based on the Chi-square test, the gender distributions of the sample in this study and the population have no significant difference.

Table 2. Descriptive statistical results of respondents' socio-economic attributes.

Socio-Economic Attributes		Quantity	Percentage/%
Gender	Male	85	44.7
	Female	105	55.3
Age	≤ 18	10	5.3
	(18,25]	176	92.6
	(25,30]	0	0.0
	(30,40]	1	0.5
	(40,50]	1	0.5
	> 50	2	1.1
Educational background	High school or below	3	1.6
	University and junior college	177	93.2
	Master	9	4.7
	Ph.D.	1	0.5
Occupation type	Student	183	96.3
	Faculty	7	3.7
Monthly living expense of the student group (yuan)	≤1000	11	6.0
	(1000,2000]	120	65.6
	(2000,3000]	46	25.2
	>3000	6	3.2
After-tax monthly income of the faculty group (yuan)	≤3000	0	0.0
	(3000,5000]	1	14.3
	(5000,7000]	6	85.7

Table 3. Descriptive statistical results of respondents' subjective evaluations.

Subjective Evaluations	Mobike (%) [1]	Hellobike (%)	Ofo (%)	No Difference (%)
Riding comfort	34.74	17.89	18.42	28.95
Appearance	36.84	13.68	20.53	28.95
Rent	11.58	12.63	39.47	36.32
Deposit	15.26	18.42	37.37	28.95
Deposit returning speed	35.26	15.26	18.42	31.05
Picking up/returning convenience	25.79	15.79	28.95	29.47
Word of mouth	46.32	15.26	17.89	20.53
Rate of broken bikes	44.21	16.84	14.74	24.21
Ease of use of software	26.84	21.05	26.32	25.79
Volume	24.74	15.79	43.68	15.79
Rent discount	19.47	15.26	35.26	30.00

[1] The second column refers to the percentage of cyclists who think that Mobike is the best, and the following two columns have the same meanings. The last column refers to the percentage of cyclists who think that the three brands make no difference.

5. Variable Definition and Result Analysis

5.1. Variable Definition

Ofo is taken as the reference brand, so that the corresponding variables and parameters of Hellobike and Mobike but not those of Ofo are defined for individual socio-economic attributes. The socio-economic and subjective evaluation variables are defined in Table 4. For the sake of brevity, h, m, and o in the suffixes of variable and parameter names represent Hellobike, Mobike, and Ofo, respectively.

Table 4. Variable definition.

Variable Attribute	Variable Meaning	Variable Names (Parameter Names)	Variable Value
Socio-economic variables	Gender	*male_h* (c_male_h), *male_m* (c_male_m)	1: Male; 0: Female
	Young	*young_h* (c_young_h), *young_m* (c_young_m)	1: ≤18; 0: otherwise
	Middle aged	*midele_age_h* (c_middle_age_h), *middle_age_m* (c_middle_age_m)	1: (18, 40]; 0: otherwise
	Educational background	*education_h* (c_education_h), *education_m* (c_education_m)	1: High school and below; 2: University and junior college; 3: Master; 4: Ph.D.
	Occupation type	*student_h* (c_student_h), *student_m* (c_student_m)	1: Student; 0: Non-student
	Low monthly living expense of the student group	*low_expense_h* (c_low_expense_h), *low_expense_m* (c_low_expense_m)	1: ≤1000 yuan; 0: otherwise
	High monthly living expense of the student group	*high_expense_h* (c_high_expense_h), *high_expense_m* (c_high_expense_m)	1: >3000 yuan; 0: otherwise
	High after-tax monthly income of the faculty group	*high_income_h* (c_high_income_h), *high_income_m* (c_high_income_m)	1: >5000 yuan; 0: otherwise
Subjective evaluation variables	Riding comfort	*comfort* (c_comfort)	The cyclist thinks that riding comfort of Hellobike is the greatest: *comfort* = 1 for Hellobike, *comfort* = 0 for other brands. The cyclist thinks that riding comfort of Mobike is the greatest: *comfort* = 1 for Mobike, *comfort* = 0 for other brands. The cyclist thinks that riding comfort of Ofo is the greatest: *comfort* = 1 for Ofo, *comfort* = 0 for other brands. The cyclist thinks that riding comfort of three brands make no difference: *comfort* = 1 for all brands. The valuation method of other variables is similar.
	Appearance	*appearance* (c_appearance)	
	Rent	*rent* (c_rent)	
	Deposit	*deposit* (c_deposit)	
	Deposit returning speed	*speed* (c_speed)	
	Picking up/returning convenience	*convenience* (c_convenience)	
	Word of mouth	*mouth* (c_mouth)	
	Rate of broken bikes	*rate* (c_rate)	
	Ease of use of software	*software* (c_software)	
	Volume	*volume* (c_volume)	
	Rent discount	*discount* (c_discount)	

5.2. Result Analysis

The parameters of the conditional Logit model were estimated by programming in Stata14.0. In general, a conditional Logit model is believed to be acceptable if the goodness of fit exceeds 0.2. The goodness of fit in this study is 0.467, indicating that the model is preferable. The predicted accuracy is 78.54%, which demonstrates that the model had good prediction effect on the choice of bikesharing brands. For the reference brand Ofo, the corresponding coefficients of individual socio-economic variables are all zeros, and those corresponding to other variables are shown in Table 5. At the significance level of 10%, an asterisk is marked at the upper right of the corresponding coefficient of variables with significant influences, and p value is shown in bold.

Table 5. Calibration results.

Parameter Names	Parameter Estimates	z Statistics	p Values	Odds Ratios
c_male_h	0.086	0.19	0.852	1.090
c_male_m	0.097	0.16	0.870	1.102
c_young_h *	−0.340	−1.68	**0.092**	0.712
c_young_m	−0.155	−0.76	0.445	0.856
c_middle_age_h	−0.207	−0.85	0.394	0.813
c_middle_age_m	−0.192	−0.68	0.495	0.825
c_education_h	−0.538	−0.49	0.623	0.584
c_education_m	−0.222	−0.16	0.876	0.801
c_student_h	7.332	0.72	0.4684	1528.436
c_student_m *	10,452.311	2.64	**0.008**	1.000×10^{30}
c_low_expense_h	1.385	1.32	0.186	3.994
c_low_expense_m	−0.383	−0.22	0.824	0.682
c_high_expense_h	−0.506	−0.34	0.732	0.603
c_high_expense_m	−0.114	−0.07	0.948	0.892
c_high_income_h	2.387	1.19	0.233	10.881
c_high_income_m *	30.233	2.65	**0.008**	1.349×10^{13}
c_comfort *	1.168	3.30	**0.001**	3.214
c_appearance	0.144	0.48	0.634	1.155
c_rent *	0.623	1.80	**0.072**	1.865
c_deposit	0.520	1.62	0.105	1.683
c_speed	−0.197	−0.58	0.561	0.821
c_convenience *	1.325	4.60	**0.000**	3.762
c_mouth *	0.670	1.97	**0.049**	1.955
c_rate	−0.481	−1.22	0.223	0.618
c_software	0.330	1.13	0.260	1.391
c_volume *	0.457	1.79	**0.073**	1.579
c_discount	−0.054	−0.18	0.860	0.947

Note: * is marked at the upper right of the corresponding coefficient of variables with significant influences, and p value is shown in bold.

The sample size is relatively small in this study, so that the significance level is set as 0.1. In other words, when other variables remain unchanged and p value is smaller than 0.1, the influence of this variable on the choice of other brands is significant relative to the reference brand Ofo; when p value is greater than 0.1, the influence of this variable on the choice of other brands is insignificant relative to the reference brand Ofo.

The influence of gender is not significant, which is related to the fact that male and female cyclists have no obvious difference in the recognition for the same bikesharing brand. This is consistent with the conclusion of the study by Gu et al. [39]

The age has no significant influence on the choice of Mobike, but it has significant influence on the choice of Hellobike. The young are reluctant to use Hellobike, possibly because Hellobike charges a relatively high rent. In addition, young students are usually freshman, and may participate in less activities than elder students. This contributes to less demand for taking shared bikes.

The influence of educational background is not significant. The brand choice is closely related to the individual recognition for brands and external variables influencing cyclist experience, rather than relies on excessive professional knowledge.

The occupation type has no significant influence on the brand choice of Hellobike, but it has a significant influence on the brand choice of Mobike. This may be associated with the fact that the rent of Mobike is low, and it can effectively attract cyclists of the student group.

The monthly living expense of the student group has insignificant influence. Overall, the proportion of the bikesharing rent in their living expense is small, and students who usually ride shared bikes generally buy memberships to enjoy the discount of monthly payment, effectively lowering riding costs. Hence, increasing or reducing the living expense has a minor influence on the decision making in the bikesharing brand choice.

The after-tax monthly income of the faculty group has a significant influence. In the post-sharing era, the bikesharing rent has increased by a large margin and even exceeds the expense of subway or bus lines. Because travelers of the faculty group need to complete commuting every day, the rent of shared bikes imposes a burden. In addition, cyclists who take shared bikes are not qualified to enjoy Shanghai's bus transfer discount (1-yuan transportation fee is exempted once a traveler transfers between two bus lines or between a bus line and subway).

The riding comfort has a significant influence. Cycling is an active mode and consumes a lot of physical strength, especially when travel routes include road segments with a certain slope. Although Shanghai is located in a flat region, many signal lamps exist in the city area, greatly impacting the riding speed of shared bikes and increasing the cycling time. Therefore, cyclists tend to choose bikesharing brands with great riding comfort. If a cyclist deems that the riding comfort of Hellobike or Mobike is the greatest, his/her probability of selecting this shared bike brand will increase by 2.214 times relative to the reference brand Ofo.

The appearance has insignificant influence. Like the pathfinder Ofo, companies of other brands usually design their shared bikes with a single color, so that colors of shared bikes were not enough in China in the pre-sharing era. Due to this phenomenon, cyclists generate aesthetic fatigue over the appearance of shared bikes and care less about the appearance. In addition, the main goal for cyclists to use bikesharing is to complete their trips. Hence, the appearance does not generate significant effect on the trip experience and the brand choice.

The rent exerts a significant influence. In the post-sharing era, the main goal of bikesharing companies turns from grabbing market share into ensuring sustainable development. As such, they have lifted up "the tide of rise in price" and reduced the granting frequency and quantity of coupons to explore new profit models. In addition, mileage and rent calculation errors caused by incorrect locating and forgetting to lock the bike occur from time to time. Consequently, cyclists become increasingly sensitive to the rent. If a cyclist thinks that the rent of Hellobike or Mobike is the most reasonable, his/her probability of selecting this bikesharing brand will increase by 0.865 times relative to the reference brand Ofo.

The deposit has an insignificant influence. In the post-sharing era, most bikesharing brands have offered deposit-free ride (or conditional deposit-free ride) across China, relieving consumers from the worry about the "escape" of bikesharing companies and attracting a large batch of new cyclists. Therefore, the deposit has no significant influence on the choice of bikesharing brands.

The influence of deposit returning speed is not remarkable. Except that Ofo only provides students with deposit-free ride, most bikesharing companies have offered deposit-free rides (or conditional deposit-free rides) for all groups. Hence, the deposit insignificantly influences cyclists to select bikesharing brands.

The influence of the picking up/returning convenience is significant. In the pre-sharing era, the volume of shared bikes was enormous, so that cyclists may pick up and return shared bikes almost whenever and wherever. This situation brought great convenience to cyclists. Since the post-sharing era, most cities exercised total volume control, regional restriction, and electronic fences. Hence, the convenience for cyclists to pick up and return shared bikes is reduced significantly. Under this circumstance, cyclists may tend to select bikesharing brands with high convenience to pick up and return shared bikes to decrease the trip duration. If a cyclist thinks that the picking up/returning convenience of Hellobike or Mobike is the greatest, his/her probability of choosing this brand will increase by 2.762 times relative to the reference brand Ofo.

The word of mouth has a significant influence. The bikesharing industry is developing rapidly with emerging marketing and management measures in the post-sharing era, including coupons and electronic fences. However, a high time cost is required to acquire the information. Therefore, most cyclists tend to obtain word-of-mouth information of bikesharing brands from relatives and friends or via the Internet to lower the information collection cost when selecting bikesharing brands, which puts the word of mouth into an important factor. If a cyclist believes that the word of mouth of Hellobike or Mobike is the best, his/her probability of choosing this brand will increase by 0.955 times relative to the reference brand Ofo.

The rate of broken bikes has an insignificant influence. With the improvement of production technologies and maintenance level, the quality of shared bikes gradually improves, and the rate of broken bikes gradually declines. In addition, the rate of broken bikes is further decreased by smart locks mounted on shared bikes. The smart lock is embedded with a GPS module and a SIM card, enabling the wireless transmission of positioning and healthy data from shared bikes to a server. The server automatically allocates maintenance and care tasks to proper staff to realize timely detection and processing according to the workload and position of all staff. Hence, the rate of broken bikes has no significant influence on the selection of bikesharing brands.

The influence of the ease of use of software is significant. Besides developing independent applications, the three bikesharing brands have embedded interfaces in common applications (Hellobike-Alipay, Mobike-WeChat small program and Meituan, and Ofo-WeChat small program), improving the accessibility of the software used to take shared bikes. Hence, the ease of use of software has no significant influence of brand selection among cyclists.

The influence of the volume is significant. In the post-sharing era, most cities have implemented total volume control policies and allocated quotas for each bikesharing company to maintain traffic order of non-motor vehicles. After obtaining the quota, each bikesharing company needs to determine the volume in each region according to the predicted demand. In the region with large volume, it is easy for cyclists to find shared bikes, improving cyclists' satisfaction. In addition, cyclists tend to select bikesharing brands with large volume. If a cyclist thinks that the volume of Hellobike or Mobike is the largest, the probability for him/her to choose this brand will increase by 0.904 times relative to the reference brand Ofo.

The rent discount has insignificant influence. In the pre-sharing era, bikesharing companies attracted cyclists with red packets, coupons, and other marketing means, elevating their market shares. However, bikesharing companies rarely granted red packets and coupons within a large scope in the post-sharing era for sustainable development. Hence, the rent discount has no significant influence on the choice of bikesharing brands.

6. Conclusions

6.1. Policy Suggestions for Shared Bike Companies

(1) Make shared bikes "easier to ride." In the post-sharing era, the riding comfort is one of important variables for cyclists to choose bikesharing brands. Hence, on the premise that the design and manufacturing of shared bikes reach related standards, bikesharing companies should reduce the weight of the frame and tire to the greatest extent and improve the integrating degree between the chain and tires. For example, the light ride series launched by Mobike optimizes the transmission ratio to reduce riding resistance.

(2) Introduce human-centered design. In the post-sharing era, bikesharing companies should shift the emphasis of the design from cost-effectiveness to humanistic care, and develop more auxiliary functions to serve cyclists. Some bikesharing companies have already made attempts. The rear fender of the fourth-generation shared bikes launched by Hellobike is upgraded into a full-wrapping type, effectively preventing the splash of muddy water. The mode of adjusting seats is changed from the original rotation into a novel wrench to improve the convenience of the adjustment. In addition, a reflection block is installed behind the seat to enhance nighttime riding safety. Besides the above design, bikesharing companies may increase more useful functions, e.g., installing a mobile phone stand at a proper position of the handlebar, so that cyclists can use the navigation of mobile phones safely.

(3) Optimize the rent design. In the post-sharing era, the bikesharing rent is obviously elevated, which reduces the enthusiasm of taking shared bikes. To increase the utilization rate and promote sustainable development, bikesharing companies may take a variety of measures. First, they should take full consideration of the affordability of low-income groups and launch exclusive month cards or coupons dedicated for students and special groups (households enjoying the minimum living guarantee and the unemployed). Second, they may release new membership cards like single-day cards, three-day cards, and week cards to meet diversified demands of more cyclist groups. Third, they should try their best to enhance the business cooperation with businesses, including superstores, commercial complex, and entertainment, to exempt/reduce the rent of cyclists who go to and back from these businesses. The exempted/reduced rent is paid by bikesharing companies and businesses together or businesses alone.

(4) Adjust operation areas and volume dynamically. Unreasonable operation areas and volume of shared bikes lead to difficulties in picking up/returning shared bikes in certain areas. To alleviate the problem of "cold and warm unevenness," bikesharing companies should make accurate demand prediction and redistribute shared bikes in advance to meet the picking up/returning demand. For example, during winter and summer vacations of colleges, shared bikes on camps should be transferred to nearby subway stations, bus stations and residential areas. In addition, bikesharing companies may predict short-time picking up/returning demands with deep learning models considering spatio-temporal dependency based on historical data. The predicted demands contribute to a reasonable determination of the dynamic operation areas and volume.

(5) Plan electronic fences via big data of cycling trajectories. In the post-sharing era, bikesharing companies have erected electronic fences according to governmental requirements and charged a certain dispatching fee for returning shared bikes beyond electronic fences. Although electronic fences have effectively relieved illegal parking of shared bikes, cyclist experience is negatively influenced by unreasonable site selection of electronic fences. Therefore, bikesharing companies should strengthen the analysis of cycling trajectories and conduct systematic planning of site selection of electronic fences. For instance, the Hubble big data platform cooperating with Hellobike connects human–vehicle–life scene through data interaction and drives intelligent full-chain operation like intelligent supply and demand prediction, and intelligent planning of electronic fences, improving the efficiency of Hellobike in managing parking regions of shared bikes.

(6) Improve the word of mouth of cyclists. The word of mouth of cyclists has a significant effect on the brand choice. Besides providing basic services like deposit returning and cyclist service, bikesharing companies should pay high attention to value-added services. For example, they may push the cycling data of shared bikes to social platforms like WeChat Movement, so that cyclists can check the cycling mileage, calorie consumption, ranking and other real-time data. This enhances the loyalty of viscosity of cyclists. They may also establish a donation mechanism of the cycling mileage (mileages accumulated to a certain degree can be donated) to enhance sense of social responsibility of cyclists.

6.2. Policy Suggestions for Related Governmental Sectors

(1) Provide strong support to promote sustainable development of bikesharing companies. Bikesharing companies are usually punished by traffic management departments due to the illegal parking of cyclists. To relieve the negative effect of the illegal parking, traffic management departments should arrange enough staff with law enforcement power to supervise and manage cyclists. Related governmental sectors with right to allocate public resources like fiscal subsidies should provide a certain support to bikesharing companies, and put the bikesharing industry into a new model of mutual benefit and win–win result of the shared economy.

(2) Construct a credit system containing illegal use of shared bikes. Mobike and Hellobike have launched online credit systems, where cyclists will receive credit rewards when using shared bikes legally, but their credits will be deducted if using shared bikes illegally. When the credit of a cyclist is too low, he will fail to enjoy the discount of month cards or other discounts and the rent will be improved. However, this credit system remains at the company level, and illegal cyclists may escape punishments by taking shared bikes of other brands. Therefore, the illegal use behaviors of cyclists should be contained in local public credit information platform, and severe violators should be included into the blacklist of violating social credits.

(3) Construct shared maintenance centers and sheds. The maintenance cost is high due to a large transportation distance of shared bikes between stations and maintenance centers. Hence, traffic management departments may unite bikesharing companies to construct a batch of shared maintenance centers in areas with high volume of shared bikes. Hence, their staff may maintain shared bikes nearby, shortening the maintenance time and increasing the utilization rate of maintenance centers. In addition, severe weather conditions (gale, heavy rain, heavy fog, heavy snow, etc.), may negatively affect the demand of picking up shared bikes sharply and damage shared bikes. Therefore, traffic management departments may unite shared bike companies to construct shared bike sheds in areas with high volume of shared bikes, so that staff store shared bikes temporarily nearby when encountering harsh weather conditions, reducing the rate of broken bikes.

6.3. Summary and Future Research Direction

Under the market pattern of oligopoly in the post-sharing era, three main bikesharing brands in China—Hellobike, Mobike, and Ofo—were selected as the study objects. The influences of socio-economic attributes of cyclists and their subjective evaluations for shared bikes on the brand choice was analyzed using the conditional Logit model. The model is calibrated based on the data from an online questionnaire survey. Results indicate that most socio-economic attributes of cyclists (including educational background, occupation type, and after-tax monthly income of the faculty group) and subjective evaluation variables (including riding comfort, rent, picking up/returning convenience, word of mouth, and volume) are important factors deciding the choice of shared bike brands.

Results cannot only be used to make policy suggestions for bikesharing companies and related governmental sectors and promote the sustainable development of shared bikes in China in the post-sharing era, but also provide a basis for traffic management departments to quantitatively evaluate performances of bikesharing companies and determine the total volume and the distribution of the total volume among bikesharing companies. From the angle of cyclists, results reveal the main factors concerned by cyclists who choose bikesharing brands among a minority of competitors in China. From the angle of bikesharing companies, results indicate that bikesharing companies may gradually increase their market shares and support sustainable development by optimizing important factors.

The study may be extended from the following two aspects. On the one hand, GPS data and swiping data of bus IC cards may be used to acquire trip attributes of cyclists, including departure time, trip duration, trip purpose and travel destination. Furthermore, it is possible to analyze the relationship between the choice of bikesharing brands and these trip attributes. On the other hand, it is interesting to collect the choice data of bikesharing brands among cyclists in typical second-tier and third-tier cities in China for building the conditional Logit model to compare the key factors influencing the brand choice of cyclists in cities at different tiers.

Author Contributions: Conceptualization, G.X. and Z.W.; methodology, G.X.; software, Z.W.; formal analysis, G.X.; data curation, Z.W.; writing—original draft preparation, G.X.; writing—review and editing, G.X. and Z.W. All authors have read and agreed to the published version of the manuscript.

Funding: This research was funded by National Natural Science Foundation of China, grant number 71701125, 71901196; Shanghai Commission of Science and Technology, grant number 17692109300, 17DZ1204000.

Conflicts of Interest: The authors declare no conflict of interest.

References

1. iiMedia Research. *2018 China Shared Bikes Development Status Rese arch*; iiMedia Research: Hongkong, China, 2018.
2. Cohen, A.; Shaheen, S. *Planning for Shared Mobility*; American Planning Association: Chicago, IL, USA, 2018.
3. Shaheen, S.; Cohen, A. *Shared Micromoblity Policy Toolkit: Docked and Dockless Bike and Scooter Sharing*; eScholorship: Berkeley, CA, USA, 2019.
4. Shaheen, S.; Guzman, S.; Zhang, H. Bikesharing in Europe, the Americas, and Asia: Past, present, and future. *Trans. Res. Rec.* **2010**, *2143*, 159–167. [CrossRef]
5. Shaheen, S.; Cohen, A.; Zohdy, I. *Shared Mobility: Current Practices and Guiding Principles*; Federal Highway Administration: Washington, DC, USA, 2016.
6. Shaheen, S.; Martin, E.; Cohen, A.; Finson, R.S. *Public Bikesharing in North America: Early Operator and User Understanding*; Mineta Transportation Institute: San Jose, CA, USA, 2012.
7. Ran, L.; Li, F. An analysis on characteristics and behaviors of traveling by bike-sharing. *J. Transp. Inform. Saf.* **2017**, *6*, 93–100.
8. Buck, D.; Buehler, R.; Happ, P.; Rawls, B.; Chung, P.; Borecki, N. Are bikeshare users different from regular cyclists? A first look at short-term users, annual members, and area cyclists in the Washington, D.C., region. *Transport. Res. Rec.* **2013**, *2387*, 112–119. [CrossRef]
9. Raux, C.; Zoubir, A.; Geyik, M. Who are bike sharing schemes members and do they travel differently? The case of Lyon's "Velo'v" scheme. *Transport. Res. Part A Policy* **2017**, *106*, 350–363. [CrossRef]
10. Ji, Y.; Fan, Y.; Ermagun, A.; Cao, X.; Wang, W.; Das, K. Public bicycle as a feeder mode to rail transit in China: The role of gender, age, income, trip purpose, and bicycle theft experience. *Int. J. Sustain. Transp.* **2017**, *11*, 308–317. [CrossRef]
11. Castillo-Manzano, J.I.; Castro-Nuño, M.; López-Valpuesta, L. Analyzing the transition from a public bicycle system to bicycle ownership: A complex relationship. *Transp. Res. Part D Transp. Environ.* **2015**, *38*, 15–26. [CrossRef]
12. Xiao, G.; Wang, R.; Zhang, C.; Ni, A. Demand prediction for a public bike sharing program based on spatio-temporal graph convolutional networks. *Multimed. Tools Appl.* **2020**. [CrossRef]
13. Deng, L.; Xie, Y.; Huang, D. Bicycle-sharing facility planning based on riding spatio-temporal data. *Planners* **2017**, *33*, 82–88.
14. Campbell, A.A.; Cherry, C.R.; Ryerson, M.S.; Yang, X. Factors influencing the choice of shared bicycles and shared electric bikes in Beijing. *Transp. Res. Part C Emerg. Technol.* **2016**, *67*, 399–414. [CrossRef]
15. Li, W.; Kamargianni, M. Providing quantified evidence to policy makers for promoting bike-sharing in heavily air-polluted cities: A mode choice model and policy simulation for Taiyuan-China. *Transp. Res. Part A Policy Pract.* **2018**, *111*, 277–291. [CrossRef]
16. Kim, K. Investigation on the effects of weather and calendar events on bike-sharing according to the trip patterns of bike rentals of stations. *J. Transp. Geogr.* **2018**, *66*, 309–320. [CrossRef]
17. Caulfield, B.; O'mahony, M.; Brazil, W.; Weldon, P. Examining usage patterns of a bike-sharing scheme in a medium sized city. *Transp. Res. Part A Policy Pract.* **2017**, *100*, 152–161. [CrossRef]
18. Zhang, L. The satisfactory survey for shared bikes in Shijiazhuang urban area. *Mod. Bus. Transp. Ind.* **2017**, *38*, 67–68.
19. Faghih-Imani, A.; Eluru, N. Analysing bicycle-sharing system user destination choice preferences: Chicago's divvy system. *J. Transp. Geogr.* **2015**, *44*, 53–64. [CrossRef]

20. Kou, Z.; Cai, H. Understanding bike sharing travel patterns: An analysis of trip data from eight cities. *Phys. A Stat. Mech. Appl.* **2019**, *515*, 785–797. [CrossRef]
21. Shen, Y.; Zhang, X.; Zhao, J. Understanding the usage of dockless bike sharing in Singapore. *Int. J. Sustain. Transp.* **2018**, *12*, 686–700. [CrossRef]
22. Sun, F.; Chen, P.; Jiao, J. Promoting public bike-sharing: A lesson from the unsuccessful Pronto system. *Transp. Res. Part D Transp. Environ.* **2018**, *63*, 533–547. [CrossRef]
23. Zhao, R.; Yang, L.; Liang, X.; Guo, Y.; Lu, Y.; Zhang, Y.; Ren, X. Last-mile travel mode choice: Data-mining hybrid with multiple attribute decision making. *Sustainability* **2019**, *11*, 6733. [CrossRef]
24. Chen, Y.; Wang, D.; Chen, K.; Zha, Y.; Bi, G. Optimal pricing and availability strategy of a bike-sharing firm with time-sensitive customers. *J. Clean. Prod.* **2019**, *228*, 208–221. [CrossRef]
25. Dong, J.; Chen, B.; He, L.; Ai, C.; Zhang, F.; Guo, D.; Qiu, X. A spatio-temporal flow model of urban dockless shared bikes based on points of interest clustering. *ISPRS Int. J. Geo-Inf.* **2019**, *8*, 345. [CrossRef]
26. Caggiani, L.; Camporeale, R.; Ottomanelli, M.; Szeto, W.Y. A modeling framework for the dynamic management of free-floating bike-sharing systems. *Transp. Res. Part C Emerg.* **2018**, *87*, 159–182. [CrossRef]
27. Zhang, J.; Meng, M.; David, Z. A dynamic pricing scheme with negative prices in dockless bike sharing systems. *Transport. Res. Part B Meth.* **2019**, *127*, 201–224. [CrossRef]
28. Mcbain, C.; Caulfield, B. An analysis of the factors influencing journey time variation in the cork public bike system. *Sustain. Cities Soc.* **2018**, *42*, 641–649. [CrossRef]
29. Chen, R. "Bike litter" and obligations of the platform operators: Lessons from China's dockless sharing bikes. *Comput. Law Secur. Rev.* **2019**, *35*, 105317. [CrossRef]
30. Yao, Y.; Liu, L.; Guo, Z.; Liu, Z.; Zhou, H. Experimental study on shared bike use behavior under bounded rational theory and credit supervision mechanism. *Sustainability* **2019**, *11*, 127. [CrossRef]
31. Manca, F.; Sivakumar, A.; Polak, J.W. The effect of social influence and social interactions on the adoption of a new technology: The use of bike sharing in a student population. *Transp. Res. Part C Emerg.* **2019**, *105*, 611–625. [CrossRef]
32. Hsu, C.; Liou, J.; Lo, H.; Wang, Y. Using a hybrid method for evaluating and improving the service quality of public bike-sharing systems. *J. Clean. Prod.* **2018**, *202*, 1131–1144. [CrossRef]
33. Barbour, N.; Zhang, Y.; Mannering, F. A statistical analysis of bike sharing usage and its potential as an auto-trip substitute. *J. Transp. Health* **2019**, *12*, 253–262. [CrossRef]
34. Zhou, X.; Wang, M.; Li, D. Bike-sharing or taxi? Modeling the choices of travel mode in Chicago using machine learning. *J. Transp. Geogr.* **2019**, *79*, 102479. [CrossRef]
35. Martín, B.; Páez, A. Individual and geographic variations in the propensity to travel by active modes in Vitoria-Gasteiz, Spain. *J. Transp. Geogr.* **2019**, *76*, 103–113. [CrossRef]
36. Li, X.; Zhang, Y.; Du, M.; Yang, J. Social factors influencing the choice of bicycle: Difference analysis among private bike, public bike sharing and free-floating bike sharing in Kunming, China. *KSCE J. Civ. Eng.* **2019**, *23*, 2339–2348. [CrossRef]
37. Ma, S.; Zhou, Y.; Yu, Z.; Zhang, Y. College students' shared bicycle use behavior based on the NL model and factor analysis. *Sustainability* **2019**, *11*, 4538. [CrossRef]
38. Du, Y.; Deng, F.; Liao, F. A model framework for discovering the spatio-temporal usage patterns of public free-floating bike-sharing system. *Transp. Res. Part C Emerg.* **2019**, *103*, 39–55. [CrossRef]
39. Gu, T.; Kim, I.; Currie, G. To be or not to be dockless: Empirical analysis of dockless bikeshare development in China. *Transp. Res. Part A Policy* **2019**, *119*, 122–147. [CrossRef]
40. Mcfadden, D. Conditional logit analysis of qualitative choice behavior. *Front. Econom.* **1973**, 105–142. Available online: http://elsa.berkeley.edu/pub/reprints/mcfadden/zarembka.pdf. (accessed on 29 March 2020).
41. Becker, H.; Ciari, F.; Axhausen, K. Modeling free-floating car-sharing use in Switzerland: A spatial regression and conditional logit approach. *Transp. Res. Part C Emerg.* **2017**, *81*, 286–299. [CrossRef]
42. Duy Quy, N.; Amoh-Gyimah, R.; Anh Thi Phuong, T.; Cao Tho, P. Mode choice among university students to school in Danang, Vietnam. *Travel Behav. Soc.* **2018**, *13*, 1–10.
43. Wenjuanxing Homepage. Available online: www.wjx.cn (accessed on 29 March 2020).
44. Xiao, G.; Cheng, Q.; Zhang, C. Detecting travel modes using rule-based classification system and Gaussian process classifier. *IEEE Access* **2019**, *7*, 116741–116752. [CrossRef]

45. Xiao, G.; Cheng, Q.; Zhang, C. Detecting travel modes from smartphone-based travel surveys with continuous hidden Markov models. *Int. J. Distrib. Sens. Netw.* **2019**, *15*. [CrossRef]
46. iiMedia Research. *China's Shared Bicycle Industry Monitoring Report for the Second Quarter of 2017*; iiMedia Research: Hongkong, China, 2017.
47. Han, Y.; Dai, H.; Dai, K.; Ma, Y.; Cheng, R. Investigation and Statistical Analysis of Bicycle Sharing on University Campus. *Adv. Appl. Math.* **2017**, *26*, 1018–1026. [CrossRef]

sustainability

Article

Flexible Mobile Hub for E-Bike Sharing and Cruise Tourism: A Case Study

Andrea Bardi [1], Luca Mantecchini [2], Denis Grasso [1], Filippo Paganelli [2,*] and Caterina Malandri [2]

[1] ITL—Fondazione Istituto sui Trasporti e la Logistica, Via dei Mille, 21-40121 Bologna, Italy; andrea.bardi@fondazioneitl.org (A.B.); Denis.Grasso@Regione.Emilia-Romagna.it (D.G.)

[2] DICAM—School of Engineering and Architecture, University of Bologna, Viale del Risorgimento, 2-40136 Bologna, Italy; luca.mantecchini@unibo.it (L.M.); caterina.malandri2@unibo.it (C.M.)

* Correspondence: filippo.paganelli2@unibo.it; Tel.: +39-0512-093-336

Received: 30 June 2019; Accepted: 27 September 2019; Published: 2 October 2019

Abstract: Bike sharing is no longer a novelty in transportation and has now become a mobility solution in its own right. This study investigated the potential scope of application of e-bike sharing solutions for a niche sector such as cruise tourism, the importance of which is growing, with the aim of improving sustainability and reducing pollution levels in cruise ports. A revealed preference survey was administered to cruise tourists, who chose a pilot e-bike service once they had disembarked from the ship to visit the nearby city center, to investigate the main variables affecting satisfaction with the service under investigation. An ordered probit model was specified and calibrated to identify the relationship among the variables influencing e-bike sharing usage by cruise tourists and their satisfaction. Subsequently, the marginal effect of each significant factor was evaluated to quantify its actual impact on the related e-bike sharing satisfaction level. The results obtained are consistent with the literature, but interesting interpretations are provided in terms of the relative importance of significant variables.

Keywords: e-bike sharing; transport sustainability; mobile depot; cruise tourism; ordered probit model

1. Introduction

In the last few years, bike sharing has become one of the most interesting and popular urban mobility options worldwide and, according to some authors [1–3], it is one of the fastest growing transportation solutions in the history of modern transportation. In 2014, the number of cities with bike sharing programs amounted to 855, with a total of 946,000 bikes operating worldwide [4]; at the end of 2018, according to Business Insider, these numbers had increased to 1600 and as much as 18 million, respectively. It is interesting to note that the number of bike sharing programs has doubled since 2014, whereas the number of bikes has increased by about 20 times in the same time span. In particular, a real boom has occurred in China over the last few years, making it the most important market in the world for this transportation mode at present. An example is the Hangzhou public bicycle program, launched in 2008 with 78,000 bicycles distributed among 2960 docking stations [5]. In particular, among the top 15 programs still active, the top 11 are located in China, followed by Paris, London, and Taipei; the Netherlands, as a whole, has the same fleet size as the Paris program (source: https://www.statista.com/chart/14542/bike-sharing-programs-worldwide/). Power-assisted electric bikes—better known as e-bikes—flourished in the early 2000s alongside traditional bikes and quickly gained great popularity. Electric bike sharing pilot projects have been proposed in many countries and by different actors (e.g., universities, port authorities, and municipalities). The positive aspects of e-bikes include the average higher speed allowed by the technology (up to 25 kph) and the reduced fatigue, favoring users such as the elderly/casual bikers and usage on uneven terrain in general. At the same time, capable batteries allow the extension of the potential average range of a trip [6]. However, e-bike components are costly and increase the vehicle weight, almost doubling it in comparison to traditional bikes. As far as the financial sustainability of the service is concerned, bike sharing is also a feasible business model for e-bikes, as the costs are spread over many users, in particular when integrated mobility is considered. The main themes of e-bike sharing concern them being hybrid vehicles, the relationship between range and demand, user interface and business model, and power supply management (recharging station, green power source, and duration). In this respect, it is interesting to note that pilot projects are increasingly proposing additional plug-in or removable batteries to allow the bikes to remain in service while the old battery is plugged in at the charging station.

The importance of Sustainable Urban Mobility Plans (SUMPs) and the widespread use of mobile devices to help users plan their trips, thus enhancing multimodality within local integrated transportation schemes, has resulted in bike sharing being considered a transportation alternative on its own.

Cycling has strategic importance for the sustainable development of cities and has become one of the fundamental parts of urban mobility strategies. The use of bikes as a transportation solution in urban and/or tourist contexts is universally recognized as positive due to the lack of polluting emissions, the reduction of traffic congestion, and the improvement of users' health. Bike sharing also provides a low-carbon solution to the "last mile" problem, playing an important role in bridging the gap among existing transportation networks, and is useful for recreation and tourism-related activities [7]. The presence of an effective bike sharing service can indeed make a city more attractive and easier to visit, strongly motivating tourists to choose it as a holiday destination. Bike trips can also become an integral part of the tourist experience, even when implemented to connect specific points of interest to the city center [8,9], with positive effects on its business model too. Despite that, every administration interested in introducing bike sharing programs for leisure/tourism purposes should take into account the many barriers that could hinder the growth and the development of this solution. The first element that users usually perceive as limiting is the lack of biking infrastructure, followed by the absence of knowledge about how to use bike sharing [10]. In general, all the system's elements need to be well designed in order to provide an efficient service capable of issuing basic but fundamental principles, such as autonomy, ease of utilization, user safety, and equipment security. Financial autonomy is, finally, the most critical goal. Many past projects in this field have failed or needed improvement mainly from the financial perspective [11]. The conflict between the necessity of offering low fares (in order to attract new users and guarantee an accessible public service) and the high overall costs of the service (due to vehicle and station maintenance, general operating costs, bike redistribution operations, etc.) may be solved, on the one hand, by a well-planned ad campaign or other forms of funding [12], and on the other hand, by developing strategies, infrastructures, and technologies capable of making the service more widely available (in particular, for tourism and leisure purposes), providing timely information for easy, satisfactory, and accessible use, even for short periods of time.

This paper sets out to examine two key research questions. First, what are the major driving forces for the development of e-bike sharing programs for cruise tourists? Second, what are the major motivating factors for cruise tourists to participate in e-bike sharing services?

In this study, these questions were investigated by means of a test case consisting of a prototypal container-based e-bike sharing mobile depot, located at the cruise terminal of Ravenna port—a small maritime town in Northern Italy—and by designing and implementing a methodological framework that included (i) a data collection and processing phase performed by means of a questionnaire survey administered to a group of e-bike users, (ii) the development of an ordered response model to identify the significant variables, and (iii) the evaluation of the marginal effects of the relevant outcomes.

The paper is structured as follows: In Section 2, an overview of the bike sharing system (BSS) evolution is provided and the suitability of the BSS applied to cruise tourism is investigated. The methodological framework is then introduced together with the statistical model. Section 3 outlines and discusses the principal results and findings. Concluding remarks on the test case are presented in Section 4.

2. Literature Review

2.1. The Evolution Framework of Bike Sharing

The variety of solutions related to bike sharing can be categorized according to target users (commuters, tourists, etc.), business models, vehicles, and project scale (municipal, regional, pilot, etc.). A broad classification of existing service categories splits the topic into (i) station-based systems, namely, the traditional bike sharing layout for public projects, with a relevant number of stations serving a high number of users for a vast area; (ii) free-floating systems, that is, the layout mostly adopted by private companies over the last few years, where bikes can be collected and dropped off anywhere within the service area by using technology devices for operation and payment; and (iii) bike rental, which is usually operated in tourist and leisure areas, with a single station and mainly during the peak season.

Bike sharing can be viewed as an evolution of bike rental in terms of target users (i.e., leisure mobility); on the other hand, bike sharing is based on the shared usage of a fleet of bicycles, available to users on a self-service, short-term, as-needed basis [12] and, in general terms, intended to be one-way from public spots, without bearing the costs and responsibilities of ownership. Bike stations are usually unattended; thus, users can manage all the different phases of the process (reservation, pick-up, and drop-off) on their own. The stations' network allows users to make point-to-point trips and return the vehicles to different stations. While bike rental can cover long periods, bike sharing is designed for short-term utilization; even the tariff scheme is designed for that purpose, with a certain amount of time granted for free (up to 30 min) and then a fixed fare issued for time unit [13]. Similar to car sharing, bike sharing programs normally cover purchase and maintenance costs, as well as storage and parking responsibility [14]. An important element that further underlines the difference between bike sharing and bike rental is the business model: BSSs can be classified according to their governing financing and managing model (e.g., public, private, or public–private partnership); ownership, operator, and operational model; scale; and range [15]. On the contrary, bike rental is almost always managed by a private owner and the benefits generated by the service (economic, environmental, and social) usually have a negligible effect on the urban community.

Despite the recent globally spreading trend of this transportation alternative, the first trials started in the late 1960s. In particular, five different generations in BSSs can be identified:

1. The "White bikes" project was introduced in Amsterdam, Netherlands in 1965. Bikes were made available free of charge at different locations across the city. Users could ride one bike to their destination and drop it off for the next following user [13]. The program came to an end in a few days, as bikes were either vandalized or appropriated for private use [16].
2. The second generation spread during the 1990s in Denmark and was replicated in many cities all over Europe and North America. In particular, the experiment called "Byciclen" in Copenhagen (1995) consisted of over 1100 vehicles, characterized by solid rubber tires and lenticular wheels located in docking stations throughout the city center and made available in exchange for a coin deposit which was refunded upon return. Despite this, the lack of a time limit for bike usage encouraged the practice of not returning the bikes to their stations.
3. During the third generation, "Smart Bikes" (or "IT-Based System") spread out, following the employment of technological improvements—such as electronically locking racks or bike locks, telecommunication systems, smartcards and fobs, mobile phone access, and on-board computers—which, as a whole, allowed for higher levels of security and the possibility to track the vehicles during the trips [12]. The most important third generation programs were launched in France (Lyon, La Rochelle, Paris, etc.) between 2005 and 2007, with significant results [13]. Apart from security issues, data on trips made it possible to survey users' habits both for research

and logistics purposes [17,18]; in particular, one of the main issues dealt with by contemporary research has been system rebalancing (i.e., relocating bicycles during the night from the destination spots to places where demand is stronger) [19–22].

4. The main innovations concerning the fourth generation have been the introduction of e-bikes and the integration of the service into public transit and car sharing schemes thanks to the use of smartcards, touch screen kiosks, and Global Positioning System (GPS).
5. The further acceleration in technological development and the return of free flow systems in place of docking stations started in China in 2016 and has paved the way for the fifth generation. Users can return the bicycles anywhere, resolving station availability issues, and the remote tracking of the locker and the use of mobile apps allow for precise and efficient collection/drop-off and payment systems.

Nowadays, the majority of active projects worldwide can be classified as either fourth or fifth generation, with docking stations and free flow models developing alongside each other.

2.2. Bike Sharing and Tourism

Since the 20th century, tourism has become one of the major sources of income for many cities. The relationship between tourism and transportation development is inseparable, and a balanced development of both of these aspects affects the local economy as well as nationwide and international competitiveness in many countries [7]. Socioeconomic factors influence transport mode choices for holiday travel. The size of the household or traveling with children, for example, have an impact on private car choice. In addition, a gap between environmental consciousness and travel choice is observable for holiday mobility [23]. Bike tourism can be split between cycle holidays, where cycling is the main purpose (usually in rural or natural regions), and holiday cycling, in which the occasional use of a bicycle is chosen as an alternative mode of transportation for exploring a destination [7]. A simple definition of bike tourism has been proposed, among others, by Han et al. [24]:

- users are away from home;
- the duration of the trip can vary from a single day to several days;
- noncompetitive means of transport;
- recreation/leisure form;
- cycling should be the main purpose of the holiday;
- occurs in an active context.

However, bicycle tourism has remained a marginal niche until the last decade. Indeed, at the end of the 1990s, cycle tourism was estimated to represent around 2–4% of total holidays. Nowadays, this share is rapidly growing, especially in countries characterized by a strong bike-oriented tradition, such as Scandinavian countries. Biking has become the preferred leisure activity of many other groups of people because it makes it possible to enjoy authentic experiences connected with nature and culture that would not be available traveling by car [25].

A bike rental system for tourists traveling by train, coach, underground, and trams would allow for renting and returning bikes close to tourist attractions and main transport routes, which in turn need to be marked and equipped with information facilities to ensure that tourists have a smooth experience in any condition, even without specific skills or proficiency [26]. Increasing numbers of case studies document the success of collaborations between bike clubs/communities, land managers, as well as government and tourism organizations in the design, construction, and maintenance of tracks, larger-scale facilities, opportunities, and data sharing [27].

Contemplation, exploration, stimulus seeking, self-development, physical challenge, and social interaction are the main motivators for tourist cycling. Beyond the wide heterogeneity among bike sharing users in terms of gender, age, and education background, tourists' motivations and needs have to be taken into account from the earliest phases of the decision-making and promotion processes in order to drive the formation of a positively perceived value of bike activities, resulting in satisfaction, desire, and loyalty. Otherwise, factors such as fatigue, misunderstanding of the rules, and knowledge gaps may result in foreign tourists perceiving barriers to hiring a means of transport abroad [7]. Evidence provided by European project reports, press articles, and, to a minor extent, scientific literature demonstrate that e-bikes are used when available, although less usually than traditional bikes. Indeed, e-bike preference is common among those attracted by the technological aspects. Therefore, Cairns et al. [28] conclude that new BSSs should rely preferably on manual bikes that can be easily upgraded to electric in a second step.

From the market side, BSSs have had, as a whole, varying degrees of success. Proper strategies need to integrate transport planning, system design, and choice of business model to develop a system of products and services capable of fulfilling users' needs, and this is even truer for services addressing tourists. A sustainable bike sharing system cannot be split from supportive local policies, infrastructure, assessment of users' motivations, and close cooperation with private stakeholders and developers [29]. Beyond the positive aspects, BSSs, according to de Chardon [30], have failed in (i) fighting social exclusion; (ii) providing a feasible transport alternative due to the reduced size of the fleet; (iii) being transparent as far as purposes, benefits, economic sustainability, and success metrics are concerned; and (iv) building a comprehensive integrated transport plan towards sustainability. In addition, the technological aspects have been unnecessarily stressed to attract investment and advertising rather than complying with mobility topics.

2.3. Bike Sharing and Cruise Tourism

A cruise could generally be defined as a voyage on a ship undertaken wholly for reasons of leisure and recreation [31]. Many private companies (cruise companies) organize trips usually starting from North America, Europe, or China with a variety of different destinations all over the world. Caribbean and Mediterranean cities account for as much as 55% of cruise capacity [32]. Nowadays, this sector is recognized by many authors as one of the most dynamic from the points of view of both maritime transportation activities and tourism. Just to mention some figures, in 2011, around 20 million cruise passengers were counted, with an annual growth rate of around 7% [33]. In parallel, the number and dimensions of ships have also dramatically increased. Currently, the biggest ships are able to carry as many as 6000 passengers [34]. The supply of cruise products has also become more diversified. On one side of the spectrum, there are small-scale adventures or luxury cruises to the most remote and vulnerable marine environments; on the other hand, there are large-scale cruises on vessels equivalent to floating cities that operate to established cruise destinations, such as the Caribbean, the Mediterranean, and Northwest Europe. Other activities (i.e., river cruising and boating) have also gained popularity in several regions. This industry is surely facilitating an improvement in the tourism potential, infrastructure, and social development of a large number of port cities, but it is also producing a negative impact on maritime, urban, socioeconomic, and environmental resources. In fact, in addition to the general global impacts produced by the cruise industry (GHG emissions, water pollution, waste production, etc.), many local effects produced by the arrival of big ships in fragile contexts or small seaside locations have to be considered. Typically, cruises foresee frequent intermediate stops at locations in which tourists can spend only a few hours visiting the place and doing shopping. Many authors [31,32,35] demonstrated that the benefits derived from this kind of tourism, hospitality, and transportation model are usually marginal compared with the social costs generated, which can be up to seven times larger. Thus, it is very important to develop sustainable solutions in order to mitigate the strong impact generated by the arrival of many cruises at seaside locations, with particular reference to the transfer mode from the ship to the city center or other tourist attractions. Traditionally, trolleybuses or other motorized vehicles have been used for this purpose. Offering alternative transportation services—bikes, for example—could be an interesting but challenging solution.

The sustainable behavior of many cruise-goers is challenged by the low familiarity with the place and the reduced comfort of the displacement. In fact, bike sharing seems to be one of the most promising measures to encourage people to choose sustainable means of transportation during their holidays, especially in urban contexts. In particular, tourists seem to appreciate the possibility to use a healthy, enjoyable, and relatively cheap door-to-door transportation mode at the holiday destination. These effects are maximized in cities with a good level of infrastructure, such as segregated bike paths and routes with low traffic volumes dedicated to leisure activities. A study carried out in Copenhagen [7] demonstrated that 67% of the persons interviewed had visited a cycling-friendly city at least once or twice in their life. In the same study, the authors found that cycling was enjoyed both by people coming from cycling-oriented countries (that may use it as a habit) and those from noncycling countries (to whom the possibility to experience a new transport solution was proposed).

Thus, we can say that an efficient bike sharing system is one of the most requested and valued services by travelers nowadays. Considering this, tourists using bike sharing could enhance both the financial and environmental performances of the service, helping to reach a positive revenue/expense ratio [15], while at the same time reducing the negative footprint on urban and natural environments introduced by tourism (i.e., pollution, congestion rates, etc.), which will in turn increase cities' livability and attractiveness in a positive loop.

Even if the literature has so far focused only on behavioral aspects [24] and sustainable mobility features connected with bike sharing, its feasibility for tourist purposes has been explored by means of pilot projects. A pilot project carried out by the Valencia port authority in close cooperation with the municipality relied on both the tourist vocation and the good level of maritime infrastructure of the city. The mobile automated bike depot proposed allowed reaching the city center (6.5 km from the port) by means of 10 electric bikes with an easy and sustainable mobility solution. The project was jointly financed by Valencia port and the European Community and operated for a six-month test period. The service was tailored to younger people and dynamic adults; bikes were made available by means of a dedicated app at the port and a temporary drop-off point in the city center. The e-bikes were equipped with a smart GPS locker system and batteries, the recharge point of which was powered by solar panels. The weak points of this project were the high running costs needed outside the European Community's funds and the limited scope for action, where the advertising campaign and local stakeholder commitment were thwarted by packages offered onboard by cruise companies. The main lesson learned from this project is the paramount importance of commitment from both local stakeholders and cruise companies to make the service attractive and economically consistent.

"Quikbyke" is a 2015 spin-off of a renowned producer of e-bikes with the purpose to promoting their use in tourist locations by means of a solar-powered rental point (called "shop-in-a-box"), which could be quickly removed and transported to another location in order to satisfy the seasonal travelers' flows. The first trials were installed in Omaha, Nebraska, United States during summer 2016 and moved during winter to St. Petersburg, Florida, United States to attract the winter cruise passengers. The fully solar-powered rental point can be run while recharging up to six e-bikes at the same time, with no external energy inputs. The standard size and self-contained design of the box makes the handling easy by ship or truck (the latter solution was also adopted in our test case). Vehicles are equipped with lithium-ion batteries that allow a 40–50 km range. Even if the fares were quite high ($5 for 30 min) in comparison with the Valencia and Ravenna test cases, the project proved itself so successful that the owner is planning to expand through franchising in the Caribbean area, where many cruise companies operate and better environmental conditions exist.

3. Materials and Methods

3.1. Case Study: An E-Bike Sharing Service with a Mobile Hub

The motivation that led to the development of this test case is to increase the attractiveness of the port of Ravenna and its hinterland by fostering an efficient, on-demand transport system, guaranteeing, at the same time, sustainable connections between the cruise passenger terminal and the historical city center. The bike depot analyzed in this paper is a tentative solution which mixes together the positive aspects of each of the three categories mentioned above in order to be accessible and attractive for tourists and tourist locations (where seasonal and discontinuous tourist flows exist), providing user friendliness and the possibility of being easily transported according to the spatial and temporal demand pattern Figure 1.

(**a**) (**b**)

Figure 1. (**a**) Ravenna Corsini port area and city aerial view. (**b**) Port basin and quays of Ravenna Corsini port.

The Ravenna cruise terminal is a 20 min drive from the Ravenna city center (12 km). Nowadays, tourists reach the city center mainly by shuttle services organized by cruise companies. Independent travelers can reach the city center by public transport (bus service), but the stop location is not favorable. The classic bike sharing schemes are not appropriate for developing a service suitable for cruise passengers. In fact, the concentration and variability of demand (both in terms of volume and time), the need to have pick-up and delivery stations close to the ships, and the need for coordination between the service manager and the cruise company have led to the development of a prototypal bike sharing service with a "mobile hub", that is, a service capable of being transported wherever necessary and only when necessary. The Ravenna case study, described in this work, was developed and financed in the framework of the "Moses" Standard + project (Maritime and multimOdal transport Services based on Ea Sea-Way project) financed by the Italy–Croatia CBC Program 2014–2020 and based on the main results of the IPA Adriatic project EA SEA-WAY. The Moses project's objective is to improve the accessibility and mobility of passengers across the Adriatic area and its hinterland through the development of new cross-border, sustainable, and integrated transport services and the improvement of physical infrastructures related to those new services.

The mobile hub depot in Ravenna port was tested in summer 2018. The 20 ft equivalent mobile depot was advertised on-site and could shelter as many as 20 e-bikes (including 1 tricycle tailored to address disabled users' needs), the charging points, and the complementary items of each vehicle (helmets, GPS tracker tools compatible with Google maps and a data storage unit, lithium-ion batteries, and security locks). The bikes had a declared range of up to 50 km, an electric engine on the front wheel, and an integrated shift on the back wheel. While the bikes weighed as much as 20 kg, the tricycle weighed as much as 30 kg. Moreover, the declared recharge time of the batteries was 6 h Figure 2.

Figure 2. Mobile depot for e-bikes at Ravenna port.

3.2. Conceptual Framework and Data Collection

A conceptual framework (data collection, data processing and modeling, and outcome) was proposed in order to explore and analyze the factors influencing the degree of satisfaction connected to the use of e-bike sharing services, as shown in Figure 3. The relationship between service and customer satisfaction has been explored widely in the literature. For example, Pruyn & Smidts [36] studied the negative effect of waiting time, Gallarza & Gil Saurab [37] explored the existence of a "quality–value–satisfaction–loyalty" chain, and De Vos [38] concluded that users traveling with their preferred mode are more satisfied—other things being equal—than those who, in addition, overperceive travel time. The inputs of the model were the potential explanatory variables and the degree of satisfaction. A Revealed Preference (RP) questionnaire was designed and issued to users when returning the bike in order to collect users' characteristics, bike sharing usage, and satisfaction data. The explanatory variables were grouped into individual characteristics (age, gender, and role on board—i.e., passenger or crew member), trip-related factors (use of other public transport modes in the same trip, duration, and travel purpose—i.e., destination being the historical city center or beaches), and personal perception (ease of use, familiarity, reliability, and willingness to pay (WTP)).

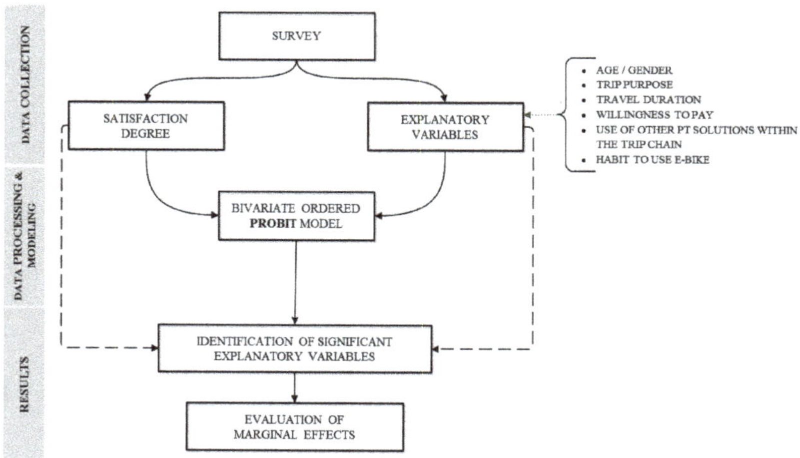

Figure 3. Methodological approach of the survey and data analysis.

As previously described, the main aim of this study was to identify which explanatory variables (individual, travel-related, or perceptual) are significant and their relationship with the satisfaction degree and attractiveness of the examined service. An ordered probit model was developed to identify the variables influencing e-bike sharing usage by cruise tourists and their satisfaction.

The total sample surveyed consists of 120 cruise passengers and crew members of cruise ships moored in the Ravenna cruise terminal in October 2018. The cruise traffic at Ravenna port along the pilot project duration had been harmed by meteorological conditions, so that three out of the eight ships foreseen did not manage to enter the port. This greatly limited the sample size. Two other constraining factors were the reduced extension of the available e-bike fleet (20 vehicles) and the all-inclusive feature of the travel packages bought by cruise-goers, which discouraged the use of services on the spot. The members of the sample were individual travelers from all over the world, of different ages, travel behaviors, and social backgrounds, who stopped only for a few hours to visit the historic center of Ravenna, the beaches, or other tourist attractions.

Special attention was devoted to designing the survey in order to provide comprehensible, easy-to-understand, and easy-to-answer questions, while at the same time ensuring sufficient information would be collected and a reduced nonresponse ratio. It is important to underline that variables widely used in literature on bike sharing surveys were considered here (e.g., trip duration, user friendliness, ease of pick-up/return, etc.). In addition, an estimate of the willingness to pay has been provided, which is, to the best of the authors' knowledge, a key aspect not frequently dealt with in the existing literature.

Table 1 summarizes the variables considered. The bike sharing satisfaction degree was defined as a typical ordinal variable scaled according to a 3-point Likert scale: 0—Bad, 1—Average, and 2—Good. Cronbach's alpha was used as a criterion to measure the reliability of the questionnaire; the result ($\alpha = 0.719$) showed a sufficient level of consistency of the data.

Table 1. Summary statistics for ordered observed variables.

Variable	Description	Value	Frequency (%)
Age group	<30	0	25%
	30–50	1	43%
	>50	2	32%
Gender	Male	1	59%
	Female	0	41%
Pax/Crew	Passenger	1	70%
	Crew staff	0	30%
Trip duration	<1 h	0	28%
	1–3 h	1	36%
	>3 h	2	36%
Destination	Historic center	1	58%
	Beaches	0	42%
Willingness to pay (WTP)	<€10/day	0	31%
	€10–20/day	1	57%
	>€20/day	2	12%
User friendliness and ease of pickup/return	Yes	1	84%
	No	0	16%
Other Public Transport utilization	Yes	1	59%
	No	0	41%
First time e-bike	Yes	1	61%
	No	0	39%
Satisfaction	Bad	0	19%
	Normal	1	35%
	Good	2	46%

3.3. Statistical Model

Discrete outcome modeling techniques are frequently utilized, as the dependent variables consist of categorical or ordered variables. When treating such variables, general statistical models, such as least-squares (LS) regression, suffer from many shortcomings, such as heteroscedasticity, violation of assumption of independence, identically distributed errors, and predicted probabilities outside the unit interval. Even basic statistical analysis, such as ANOVA or the *t*-test, are unreliable because the key assumptions of these models are that the response or dependent variables must be continuous, with normally or roughly normally distributed residuals. A wide body of literature recognizes that linear regression is inappropriate when the dependent variable is categorical or ordered, as in our specific case. An appropriate theoretical model in such a situation is the ordered probit model [39].

Following these premises, an ordered probit model was developed and calibrated to identify factors that potentially affected the satisfaction degree of e-bike sharing with the mobile hub for the cruise tourists. In ordered models, an ordered dependent variable is explained by a number of scaled independent variables and the parameters of the latent model do not have a simple interpretation per se. Rather, the main interest lies in the shift of the predicted discrete ordered outcome distribution as one or more of the regressors change (i.e., the marginal probability effects).

In our particular case, by adopting this approach, we could account for the commonly shared unobserved factors that affect both e-bike sharing usage by cruise tourists and the related satisfaction degree. Subsequently, the marginal effects of each determined factor (significant variables) were evaluated to quantify their actual impacts on the related e-bike sharing satisfaction levels.

Originally developed in biostatistics [40], this model was brought into the social sciences in the 1970s, and recently, it has been widely applied in transportation studies. In greater detail, several applications can be found in vehicle ownership analysis [41], road safety and injury models [42,43], determinants of bicycle choice [44], cyclists' travel behavior [45], and car sharing usage [46].

The model was defined and specified starting from the ordinal data observed for each observation, under the hypothesis that a latent continuous metric driving the ordered responses given by the users existed. The model is expressed with the following well-known form:

$$y_i^* = \beta_i X_i + \varepsilon_i, \quad y_i = k \; if \;\; \mu_{k-1} < y_i^* < \mu_k, \quad k = 0, \ldots, K \tag{1}$$

where y^*_i is the latent, unobserved dependent variable (satisfaction degree); $\mathbf{X}_\mathrm{i}$ is the vector containing the independent (measured) variables; and $\boldsymbol{\beta}_\mathrm{i}$ is the vector of unknown regression coefficients associated with the explanatory variables. While it is not possible to observe the true values of the y^*_i variables, it is instead possible to observe the actual ordinal answers y_i provided by users on the 1,2,3 scale within the thresholds μ_k.

As usual, ε represents the random error term, normally distributed with mean = 0 and variance = 1, $\varepsilon_\mathrm{i} \sim \mathrm{N}(0,1)$. Under these hypotheses, the probability of observing each ordinal response is given by

$$\begin{aligned} P[y_i = 1] &= P\left[\mu_0 < y_i^* < \mu_1\right] = P[\mu_0 < \mathbf{X}_i\boldsymbol{\beta}_i + \epsilon_i < \mu_1] \\ &= P[\mu_0 - \mathbf{X}_i\boldsymbol{\beta}_i < \epsilon_i < \mu_1 - \mathbf{X}_i\boldsymbol{\beta}_i] = \Phi(\mu_1 - \mathbf{X}_i\boldsymbol{\beta}_i) - \Phi(\mu_0 - \mathbf{X}_i\boldsymbol{\beta}_i) \end{aligned} \tag{2}$$

where $\Phi(\cdot)$ is the standard bivariate normal cumulative distribution function. It is straightforward to see that, in general terms,

$$P[y_i = k] = \Phi(\mu_k - \mathbf{X}_i\boldsymbol{\beta}_i) - \Phi(\mu_{k-1} - \mathbf{X}_i\boldsymbol{\beta}_i). \tag{3}$$

The unknown parameters in the model—the coefficients vector $\boldsymbol{\beta}_i$—were estimated by means of the Maximum Likelihood Estimation (MLE) technique, maximizing the log-likelihood function:

$$lnL = \sum_{i=1}^{N} \sum_{k=1}^{K} \delta_{ik} ln\left[\Phi_{i,k} - \Phi_{i,k-1}\right] \tag{4}$$

where $i = 1, \ldots, N$ is the sample size, $\Phi_{i,k} = \Phi[\mu_k - \boldsymbol{X}_i\boldsymbol{\beta}_i]$, and $\Phi_{i,k-1} = \Phi[\mu_{k-1} - \boldsymbol{X}_i\boldsymbol{\beta}_i]$. The estimated coefficients do not allow for immediately quantifying the magnitude of the related explanatory variable effects on the satisfaction but only for evaluating the direction (positive or negative) of the effects on the outcomes. To properly quantify the impacts, the marginal effects of each explanatory variable of interest were calculated as follows:

$$\frac{\partial P(y_i = k)}{\partial \boldsymbol{X}_i} = [\phi(\mu_{k-1} - \boldsymbol{X}_i\boldsymbol{\beta}_i) - \phi(\mu_k - \boldsymbol{X}_i\boldsymbol{\beta}_i)]\boldsymbol{\beta}_i \tag{5}$$

where Φ is the probability density function of the standard normal distribution.

4. Results and Discussion

Following the methodological framework reported in Figure 3, the calibration of the binary ordered probit model allowed us to identify the significant variables and to estimate the marginal effects, that is, the variations of satisfaction level as the regressors change. The results of the calibration are reported in Table 2. Only the variables that were significant (p-value < 0.05) were included in the model.

Prior to discussing the results, it is useful to repeat that this was an initial exploratory analysis carried out during a short pilot project and, thus, the results must be treated with caution, given the dimension of the sample available. The scientific literature on surveys suggests a relationship between the dimension of the sample, population size, desired degree of accuracy, and confidence interval [47]. It is worth noting that, at the time of analysis, the service analyzed was extremely niche although promising; thus, in the authors' opinion, the identification of the attributes that appear to significantly influence travel behavior is still useful from a policy point of view. In addition, the results obtained are consistent with the literature on bike mobility and contribute to shedding light on a rather unexplored research topic. A relevant ratio of users accepting to answer the survey is, other considerations notwithstanding, a significant result for the research scope and justified a reduction in the sample size [47].

Eight significant variables were identified (p-value < 0.05); the magnitude and the sign of the calibrated parameters provide a general evaluation of the impact on the satisfaction level. This analysis was helpful to answer the first research question: What are the major factors acting as drivers for the development of e-bike sharing programs for cruise tourists?

Table 2. Results of the model.

Variable	β	Std. Error	p-Value	Marginal Effect
Age (>50)	0.255	0.097	0.018	0.0671
Passenger	0.106	0.112	0.007	0.0344
User-friendliness and ease of pickup/return	0.548	0.156	0.000	0.1415
Trip duration (>3 h)	0.379	0.174	0.000	0.0825
Destination center	0.504	0.064	0.002	0.1106
WTP (€10–20/day)	0.211	0.104	0.023	0.0508
Other public transport utilization	-0.132	0.080	0.006	-0.0398
First time e-bike	0.444	0.156	0.000	0.0979
Log-likelihood at convergence	−1987.35			
Likelihood ratio index (ρ^2)	0.194			

All the significant variables—except the one related to utilization of other public transport services in the same trip chain—had a positive effect on the satisfaction level.

The estimation of the marginal effects, on the other hand, answered the second research question: What are the major motivators for tourists to join a bike sharing service? In fact, the magnitude of users' satisfaction variation with each parameter allowed us to determine the key aspects to focus on to promote a more efficient and popular service. In detail, e-bike users over 50 years of age showed a positive marginal effect on satisfaction level of around 7%, and this is in line with the evidences available in the literature [48,49]. Concerning trip duration, average usage of >3 h had a higher impact on satisfaction (+8%). In the literature, we found much lower average values for the duration of e-bike trips (as much as 30 min) [6,50,51]; therefore, we assume that satisfaction was linked here to longer trip duration mainly due to the specific case (trip purpose and users' characteristics).

Concerning the WTP, according to the results, a service fare in the range of €10–20 per day had a significant positive effect on satisfaction and an estimated marginal utility of around 5%. This is roughly consistent with average fare values found in the literature, although the average journey time found here may imply higher costs [52–55]

Even the "first-time users" involved in the analysis perceived the service positively (almost +10% marginal effect). This is probably linked to the positive feedback regarding the user friendliness of the e-bike and the innovativeness of the pick-up and delivery solution based on a mobile hub station. In addition to that, the marginal effect on the degree of satisfaction generated by the ease of use was the highest among the significant variables inspected (+14%).

Users also showed a greater propensity to use the e-bike sharing service to reach and visit the historic city center (see Table 1), with an intrinsic marginal effect on satisfaction of +11%.

Interestingly, the use of other public transport modes in the same trip chain showed a marginal drop in satisfaction (−4%), probably due to the lack of integration between the different services in terms of connections, frequencies, and fares; also, this finding is in line with the existing literature [56–58]. On the other hand, the use of integrated multimodal fare schemes is reported in the literature as being a driver of paramount importance for the use of public transport systems by tourists, along with the degree of integration and flexibility of transport systems [59].

5. Conclusions

Over time, bike sharing has come to represent a successful story of sustainable transportation opportunities, the diffusion of which is also relevant and indisputable in those countries which have been so far less sensitive towards environmental concerns [60]. The aim of this paper was to propose and assess the opportunities for an additional niche of operation for e-bike services consisting of a mobile depot which can be easily moved and transported within and across both port and urban areas to respond to demand evolution. Two research issues were proposed: the identification of (i) the major factors driving the development of e-bike sharing programs for cruise tourists, and (ii) the main motivators for cruise tourists to join a service of this kind.

As activities during cruises are usually thoroughly scheduled by the ship owner, the scope of the solution described in this paper is limited only to cases where a partnership/sponsorship exists between local stakeholders, cruise promoters, and technology suppliers or where a limited and floating demand exists, such as in marinas. Under these premises and from a managerial point of view, the main aspects foreseen to make the service effective are the infrastructure (paths and, foremost, a user-friendly collection/drop-off system that is easy to use even for first-time users), transport integration and tailored fare schemes in order to fight segregation of certain users' categories, and a focus on safety and reliability (i.e., information provision, ad campaigns, comfort, and design of docking stations).

The survey conducted during the pilot project in Ravenna involved a diversified sample of users of different ages and nationalities/countries of origin. The use ratio of the service was quite low compared to the population of a typical cruise vessel, but the response ratio was high, since almost all users agreed to answer the survey when returning the vehicle. The statistical analysis showed good satisfaction among users, even when they were not familiar with e-vehicles, which is traditionally perceived as a barrier in the mainstream literature on the topic [61]. The results obtained by the calibration of the binary ordered probit model are consistent with the literature. The promising aspects of this work lie in the recognition of a new niche of application of shared mobility that goes beyond the traditional boundaries of users' average age and WTP, as well as the distance and time traveled threshold usually connected with bike sharing, which can be ascribed to the peculiar vocation of the service. Users appreciated electric-powered vehicles for traversing the surroundings of the port, even when they were not familiar with such vehicles.

This conclusion has a potentially significant scope of application at the policy level, steering the municipality and the stakeholders to foster sustainable transportation schemes linking the port and the city center, notwithstanding the distance. A tailored and effective business model is needed to run this service, and transport integration can be a solution (together with schedule coordination to reduce transfer time loss, which greatly affects users' appreciation). Finally, additional progress on the safety and reliability of the service can be achieved with close cooperation between the local administration and the bike vendor.

Further research may set out to investigate fleet management and technology, which can improve appreciation and feasibility. For example, GPS tracking costs increase with precision requirements; nevertheless, the analysis of GPS tracking data is useful for public administration and transport agencies to map users' travel behavior and paths and thus to design supply opportunities accordingly.

Author Contributions: Conceptualization, L.M. and A.B.; methodology, L.M., F.P., and C.M.; validation, L.M., C.M., and D.G.; investigation, D.G. and A.B.; resources, D.G. and A.B.; data curation, L.M., F.P., and C.M.; writing—original draft preparation, L.M. and F.P.; writing—review and editing, L.M., F.P., C.M., A.B., and D.G.; project administration, A.B. and D.G.; funding acquisition, A.B. and D.G.

Funding: This research received no external funding.

Conflicts of Interest: The authors declare no conflict of interest.

References

1. Campbell, A.A.; Cherry, C.R.; Ryerson, M.S.; Yang, X. Factors influencing the choice of shared bicycles and shared electric bikes in Beijing. *Transp. Res. C* **2016**, *67*, 399–414. [CrossRef]
2. Kou, Z.; Cai, H. Understanding bike sharing travel patterns: An analysis of trip data from eight cities. *Phys. A* **2019**, *515*, 785–797. [CrossRef]
3. Midgley, P. The Role of Smart Bike-sharing Systems in Urban Mobility. *Journeys* **2009**, *2*, 23–31.
4. Fishman, E. Bikeshare: A review of recent literature. *Transp. Rev.* **2016**, *36*, 92–113. [CrossRef]
5. Zhao, J.; Deng, W.; Song, Y. Ridership and effectiveness of bikesharing: The effects of urban features and system characteristics on daily use and turnover rate of public bikes in China. *Transp. Policy* **2014**, *35*, 253–264. [CrossRef]
6. Cherry, C.; Cervero, R. Use characteristics and mode choice behavior of electric bike users in China. *Transp. Policy* **2007**, *14*, 247–257. [CrossRef]
7. Kaplan, S.; Manca, F.; Sick Nielsen, T.A.; Prato, C.G. Intentions to use bike-sharing for holiday cycling: An application of the Theory of Planned Behavior. *Tour. Manag.* **2015**, *47*, 34–46. [CrossRef]
8. Postorino, M.N.; Mantecchini, L.; Paganelli, F. Green airport investments to mitigate externalities: Procedural and technological strategies. In *Sustainable Entrepreneurship and Investments in the Green Economy*, 1st ed.; Vasile, A.J., Nicolò, D., Eds.; IGI Global: Hershey, PA, USA, 2017; pp. 231–256.
9. Malandri, C.; Mantecchini, L.; Postorino, M.N. Airport ground access reliability and resilience of transit networks: A case study. *Transp. Res. Procedia* **2017**, *27*, 1129–1136. [CrossRef]
10. Hentz Leister, E.; Vairo, N.; Sims, D.; Bopp, M. Understanding bike share reach, use, access and function: An exploratory study. *Sustain. Cities Soc.* **2018**, *43*, 191–196. [CrossRef]
11. Sun, F.; Chen, P.; Jiao, J. Promoting public bike-sharing: A lesson from the unsuccessful Pronto System. *Transp. Res. D* **2018**, *63*, 533–547. [CrossRef]
12. Shaneen, S.A.; Guzman, S.; Zhang, H. Bikesharing in Europe, the Americas, and Asia- Past, Present, and Future. *Transp. Res. Rec.* **2010**, *2143*, 159–167. [CrossRef]

13. Zhu, A. Chinese Dock-Less Bike-Sharing Model—The Market Situation and Underlying Implications. Bachelor's Thesis, Ca' Foscari University of Venice, Venice, Italy, 2018.
14. Bakogiannis, E.; Vassi, A.; Siti, M.; Christodoulopoulou, G. Developing a Sustainable Mobility Plan in Piraeus with Special Emphasis on Cycling. *J. Traffic Transp. Eng.* **2016**, *4*, 61–74.
15. Bakogiannis, E.; Avgi, V.; Siti, M.; Christodolopoulou, G. Bike Sharing Systems as a Tool to Increase Sustainable Coastal and Maritime Tourism. The Case of Piraeus. *Reg. Sci. Inq.* **2018**, *10*, 57–70.
16. De Maio, P. Bike-sharing: History, Impacts, Models of Provision, and Future. *J. Public Transp.* **2009**, *12*, 41–56. [CrossRef]
17. Fishman, E.; Washington, S.; Haworth, N. Bike Share: A Synthesis of the Literature. *Transp. Rev.* **2013**, *33*, 148–165. [CrossRef]
18. Si, H.; Shi, J.G.; Wu, G.; Chen, J.; Zhao, X. Mapping the bike sharing research published from 2010 to 2018: A scientometric review. *J. Clean. Prod.* **2019**, *213*, 415–427. [CrossRef]
19. García-Palomares, J.C.; Gutiérrez, J.; Latorre, M. Optimizing the location of stations in bike-sharing programs: A GIS approach. *Appl. Geogr.* **2012**, *35*, 235–246. [CrossRef]
20. Schuijbroek, J.; Hampshire, R.C.; van Hoeve, W.J. Inventory rebalancing and vehicle routing in bike sharing systems. *Eur. J. Oper. Res.* **2017**, *257*, 992–1004. [CrossRef]
21. Caggiani, L.; Camporeale, R.; Marinelli, M.; Ottomanelli, M. User satisfaction-based model for resource allocation in bike-sharing systems. *Transp. Policy* **2019**, *80*, 117–126. [CrossRef]
22. Caggiani, L.; Camporeale, R.; Ottomanelli, M.; Szeto, W.Y. A modeling framework for the dynamic management of free-floating bike-sharing systems. *Transp. Res. Part C Emerg. Technol.* **2018**, *87*, 159–182. [CrossRef]
23. Böhler, S.; Grischkat, S.; Haustein, S.; Hunecke, M. Encouraging environmentally sustainable holiday travel. *Transp. Res. Part A Policy Pract.* **2006**, *40*, 652–670. [CrossRef]
24. Han, H.; Meng, B.; Kim, W. Bike-traveling as a growing phenomenon: Role of attributes, value, satisfaction, desire, and gender in developing loyalty. *Tour. Manag.* **2017**, *59*, 91–103. [CrossRef]
25. Steinbach, J. River Related Tourism in Europe—An Overview. *GeoJournal* **1995**, *35*, 443–458. [CrossRef]
26. Roman, M.; Roman, M. Bicycle Transport as an Opportunity to Develop Urban Tourism—Warsaw Example. *Procedia Soc. Behav. Sci.* **2014**, *151*, 295–301. [CrossRef]
27. Pickering, C.; Leung, Y. Editorial for the Special issue of the Journal of Outdoor Recreation and Tourism on mountain biking. *J. Outdoor Recreat. Tour.* **2016**, *15*, iii. [CrossRef]
28. Cairns, S.; Behrendt, F.; Raffo, D.; Beaumont, C.; Kiefer, C. Electrically assisted bikes: Potential impacts on travel behavior. *Transp. Res. Part A Policy Pract.* **2017**, *103*, 327–342. [CrossRef]
29. Zhang, L.; Zhang, J.; Duan, Z.; Bryde, D. Sustainable bike-sharing systems: Characteristics and commonalities across cases in urban China. *J. Clean. Prod.* **2015**, *97*, 124–133. [CrossRef]
30. Médard de Chardon, C. The contradictions of bike-share benefits, purposes and outcomes. *Transp. Res. Part A Policy Pract.* **2019**, *121*, 401–419. [CrossRef]
31. Polat, N. Technical Innovations in Cruise Tourism and Results of Sustainability. *Procedia Soc. Behav. Sci.* **2015**, *195*, 438–445. [CrossRef]
32. Rosa-Jiménez, C.; Perea-Medina, B.; Andrade, M.J.; Nebot, N. An examination of the territorial imbalance of the cruising activity in the main Mediterranean port destinations: Effects on sustainable transport. *J. Transp. Geogr.* **2018**, *68*, 94–101. [CrossRef]
33. Lamers, M.; Eijgelaar, E.; Amelung, B. The Environmental Challenges of Cruise Tourism: Impacts and Governance. In *The Handbook of Tourism and Sustainability*; Routledge: London, UK, 2015.
34. Perez Hoogkamer, L. Assessing and Managing Cruise Ship Tourism in Historic Port Cities: Case Study Charleston, South Carolina. Ph.D. Thesis, Columbia University, Washington, DC, USA, 2013.
35. MacNeill, T.; Wozniak, D. The economic, social, and environmental impacts of cruise tourism. *Tour. Manag.* **2018**, *66*, 387–404. [CrossRef]
36. Pruyn, A.; Smidts, A. Effects of waiting on the satisfaction with the service: Beyond objective time measures. *Int. J. Res. Mark.* **1998**, *15*, 321–334. [CrossRef]
37. Gallarza, M.G.; Gil Saurab, I. Value dimensions, perceived value, satisfaction and loyalty: An investigation of university students' travel behavior. *Tour. Manag.* **2006**, *27*, 437–452. [CrossRef]
38. De Vos, J. Do people travel with their preferred travel mode? Analyzing the extent of travel mode dissonance and its effect on travel satisfaction. *Transp. Res. A* **2018**, *117*, 261–274.

39. Greene, W.H. *Econometric Analysis*, 5th ed.; Prentice Hall: Upper Saddle River, NJ, USA, 2003.
40. Aitchison, J.; Silvey, S. The Generalization of Probit Analysis to the Case of Multiple Responses. *Biometrika* **1957**, *44*, 131–140. [CrossRef]
41. Chu, Y.L. Automobile ownership analysis using ordered probit models. *Transp. Res. Rec.* **2002**, *1805*, 60–67. [CrossRef]
42. Yamamoto, T.; Hashiji, J.; Shankar, V.N. Underreporting in traffic accident data, bias in parameters and the structure of injury severity models. *Accid. Anal. Prev.* **2008**, *40*, 1320–1329. [CrossRef]
43. Heydari, S.; Fu, L.; Miranda-Moreno, L.F.; Jopseph, L. Using a flexible multivariate latent class approach to model correlated outcomes: A joint analysis of pedestrian and cyclist injuries. *Anal. Met. Accid. Res.* **2017**, *13*, 16–27. [CrossRef]
44. Wang, C.H.; Akar, G.; Guldmann, J.M. Do your neighbors affect your bicycling choice? A spatial probit model for bicycling to The Ohio State University. *J. Transp. Geogr.* **2015**, *42*, 122–130. [CrossRef]
45. Fu, L.; Farber, S. Bicycling frequency: A study of preferences and travel behavior in Salt Lake City, Utah. *Transp. Res. A* **2017**, *101*, 30–50. [CrossRef]
46. Becker, H.; Loder, A.; Schmid, B.; Axhausen, K.W. Modeling car-sharing membership as a mobility tool: A multivariate Probit approach with latent variables. *Travel Behav. Soc.* **2017**, *8*, 26–36. [CrossRef]
47. Henry, G.T. *Practical Sampling, Applied Social Research Methods*; SAGE Publications Ltd.: Thousand Oaks, CA, USA, 1990; Volume 21. [CrossRef]
48. Dill, J.; Rose, G. E-bikes and transportation policy: Insights from early adopters. *Transp. Res. Rec.* **2012**, *2314*, 1–6. [CrossRef]
49. Hiselius, L.W.; Svensson, Å. E-bike use in Sweden—CO2 effects due to modal change and municipal promotion strategies. *J. Clean. Prod.* **2017**, *141*, 818–824. [CrossRef]
50. Fyhri, A.; Fearnley, N. Effects of e-bikes on bicycle use and mode share. *Transp. Res. D* **2015**, *36*, 45–52. [CrossRef]
51. Lopez, A.J.; Astegiano, P.; Gautama, S.; Ochoa, D.; Tampère, C.; Beckx, C. Unveiling e-bike potential for commuting trips from GPS traces. *ISPRS Int. J. Geo-Inf.* **2017**, *6*, 190. [CrossRef]
52. Audikana, A.; Ravalet, E.; Baranger, V.; Kaufmann, V. Implementing bikesharing systems in small cities: Evidence from the Swiss experience. *Transp. Policy* **2017**, *55*, 18–28. [CrossRef]
53. Schröder, J.O.; Weiß, C.; Kagerbauer, M.; Reiß, N.; Reuter, C.; Schürmann, R.; Pfisterer, S. Developing and evaluating intermodal e-sharing services—A multi-method approach. *Transp. Res. Proc.* **2014**, *4*, 199–212. [CrossRef]
54. Manzi, G.; Saibene, G. Are they telling the truth? Revealing hidden traits of satisfaction with a public bike-sharing service. *Int. J. Sustain. Transp.* **2018**, *12*, 253–270. [CrossRef]
55. Li, W.; Kamargianni, M. Providing quantified evidence to policy makers for promoting bike-sharing in heavily air-polluted cities: A mode choice model and policy simulation for Taiyuan-China. *Transp. Res. A* **2018**, *111*, 277–291. [CrossRef]
56. Whalen, K.E.; Páez, A.; Carrasco, J.A. Mode choice of university students commuting to school and the role of active travel. *J. Transp. Geogr.* **2013**, *31*, 132–142. [CrossRef]
57. González, R.M.; Martínez-Budría, E.; Díaz-Hernández, J.J.; Esquivel, A. Explanatory factors of distorted perceptions of travel time in tram. *Transp. Res. F* **2015**, *30*, 107–114. [CrossRef]
58. Schweizer, J.; Rupi, F. Performance evaluation of extreme bicycle scenarios. *Procedia-Soc. Behav. Sci.* **2014**, *111*, 508–517. [CrossRef]
59. Le-Klaehn, D.T.; Hall, C.M. Tourist use of public transport at destinations—A review. *Curr. Issues Tour.* **2015**, *18*, 785–803. [CrossRef]
60. Paganelli, F. Urban Mobility and Transportation. In *Sustainable Cities and Communities, Encyclopedia of the UN Sustainable Development Goals*; Filho, W.L., Azul, A.M., Brandli, L., Özuyar, P.G., Wall, T., Eds.; Springer International Publishing: Berlin/Heidelberg, Germany, 2019; in press.
61. Swiers, R.; Pritchard, C.; Gee, I. A cross sectional survey of attitudes, behaviours, barriers and motivators to cycling in University students. *J. Transp. Health* **2017**, *6*, 379–385. [CrossRef]

sustainability

MDPI

Article

Analysis of Network Structure of Urban Bike-Sharing System: A Case Study Based on Real-Time Data of a Public Bicycle System

Yi Yao [1], Yifang Zhang [1,*], Lixin Tian [1], Nianxing Zhou [2], Zhilin Li [3] and Minggang Wang [1,4]

1 School of Mathematical Sciences, Nanjing Normal University, Nanjing 210023, China; 05302@njnu.edu.cn (Y.Y.); tianlixin@njnu.edu.cn (L.T.); magic821204@sina.com (M.W.)
2 School of Geography, Nanjing Normal University, Nanjing 210023, China; zhounianxing@263.net
3 Department of Mathematics, North Carolina State University, NC 27695, USA; zhilin@ncsu.edu
4 Department of Mathematics, Nanjing Normal University Taizhou College, Taizhou 225300, China
* Correspondence: 180938024@stu.njnu.edu.cn

Received: 12 July 2019; Accepted: 27 September 2019; Published: 30 September 2019

Abstract: To better understand the characteristics of a bike-sharing system, we applied complex network methods to analyze the relationship between stations within the bike-sharing system. Firstly, using Gephi software, we constructed the public bicycle networks of different urban areas based on the real-time data of the Nanjing public bicycle system. Secondly, we analyzed and compared degree, strength, radiation distance, and community structure of the networks to understand the internal relations of the public bicycle system. The results showed that there were many stations with low usage of public bicycles. Furthermore, there was a geographical division between high-demand and low-demand areas for public bicycles. The usage of public bicycles at a station was not only related to land use but also related to the usage of bicycles at stations nearby. Moreover, the average service coverage of the public bicycle system was consistent with the original intention of "the first and last mile", and public bicycles could meet different travel needs.

Keywords: sustainable mode of transportation; bike-sharing system; public bicycle; complex network; network structure

1. Introduction

Bike-sharing systems (BSSs) not only facilitate people with the trouble of "the first and last mile" but also provide them with a sustainable and carbon-free mode of transportation. In 1965, an NGO called Provo established a public bicycle system (PBS) to reduce air pollution and relieve traffic congestion in Amsterdam, which is regarded as the prototype of BSSs [1].

With economic booming and social modernization, bicycles are not the top choice in most countries. Nevertheless, due to increasingly serious air pollution and traffic congestion, public bicycle, a station-based bike-sharing, has gradually become a significant transportation alternative once again. However, untimely rebalancing operations and unreasonable station distributions not only reduce users' satisfaction but also increase the scheduling and maintenance costs, which results in waste of public resources. Therefore, studies on PBSs are beneficial to solve the issue of "unavailable bicycles and unavailable docks" and can improve the social and economic benefits of PBSs.

Scholars have carried out a lot of research on BSSs and achieved fruitful results. Characteristics of users and stations, influencing factors, and built environment are all main research contents. Users' gender [2–7], age [3,6–8], income and education [2,5,6,9,10], weather conditions [11–13], infrastructure construction, population density, built environment [7,12–19], and relationship with other means of transportation [20–33] have been confirmed to affect the usage of bike-sharing. On the other hand, in order to solve the problem of rebalancing, optimization of BSSs has also become another hot issue [34–43]. As far as research methods are concerned, statistical methods such as correlation analysis, clustering analysis, and regression analysis are adopted to analyze characteristics and influencing factors [3,7,9,13,14,19,20,22,23,44], and optimization algorithms are used to solve rebalancing operations for BSSs [35,38–40,42,43].

The existing researches mainly discuss BSSs from the aspects of the relationship between BSSs and external environments, while studies from the perspective of the relationship between internal stations of BSSs are insufficient. To fill this gap, we built public bicycle networks (PBNs) using Gephi software according to complex network methods, aiming to analyze internal correlation characteristics of BSSs.

The remainder of the paper is organized as follows. Section 2 reviews the relevant literature. Section 3 provides the data and methodology, and Section 4 analyzes the results. Lastly, Section 5 presents conclusions.

2. Literature Review

2.1. Relationship between BSSs and External Environments

Relationship between BSSs and external environments is a hot issue.

Weather conditions are an important influencing factor for the use of public bicycles. More bicycles were used in good weather, while rain, snow, low temperature, and gales resulted in less use [11–13].

Infrastructure construction and population also profoundly influence the use of bike-sharing. In Montreal, increasing the number of stations played a more active role in cycling than increasing the number of parking piles [12]. More public bicycle lanes promoted the use of public bicycles [14]. In general, station capacity, zonal population, employment density, commercial land, and daily life places nearby were positively correlated with cycling [7,13,15–19].

The relationship between BSSs and other transportation systems is getting more and more attention. Scholars analyzed the impact of bike-sharing on walking, cars, public transportation, and subways. Many results showed that bike-sharing replaced walking and public transportation to some extent, and bike-sharing was expected to play a more important role in reducing car use [2,20,21,25–29]. Subway stations were important starting and ending stations of BSSs, and the increase of cycling flow would lead to an increase of average daily subway flow [21,23,32]. Some researchers found that sudden change in public transport services, such as a public transit disruption or a strike, would affect the use of bicycles significantly [30,31]. As for free-floating BSSs and station-based BSSs, Li et al. pointed out that the former had a greater effect on the latter especially on weekdays and on young people [24]. Luo et al. found that greenhouse gas emissions factor of free-floating BSSs was 82% higher than station-based BSSs based on life cycle assessment [33].

2.2. Internal Relationship of BSSs

Some studies of BSSs carried out clustering analysis on the stations, which essentially explored the relationship between the stations and also the internal relationship of the systems. The usage pattern of stations was identified by the clustering method, and the clustering results were influenced by station locations and socioeconomic situations [7,23,45–47]. More bike-sharing movements were observed within a cluster than between clusters in Lyon [48]. However, clustering analysis mainly distinguishes the similarity of stations and does not highlight the relationship between them.

Recently, some scholars have begun to focus on BSSs using network analysis methods. Rixey pointed out that station network effects were extremely important to ridership levels, with a robust, statistically significant relationship within systems. Station network variables were put forward to test the effects of public bicycle station network density, distribution, and size on cycling. The results revealed that there was a positive correlation between the number of stations within 4800 m of a given station and the number of people cycling monthly [9]. Saberi et al. combined geospatial methods with complex network methods to better understand the interdependence between BSSs and public transportation systems in London [31]. Austwick et al. explored BSSs in different cities by network analysis, and they found that the strength rank curve for the top 50 stations in each system displayed a similar scaling law. Furthermore, community detection in the network could identify local use [49]. Lin et al. proposed a novel graph convolutional neural network with data-driven graph filter model to predict station-level hourly demand with fairly good prediction accuracy. Graph network analysis showed that this model could not only gather more information from more stations but also utilize more underlying correlations between stations [50].

Although the existing research has provided us a lot of insights, there is potential to deepen the knowledge of BSSs. (1) The relationship between BSSs and external environment has been more in-depth, while the internal correlation of BSSs needs to be expanded. Complex network analysis is applied to study the internal correlation between nodes according to their connection relationship, which is relatively under-utilized to understand BSSs [49]. (2) How to construct a bike-sharing network and select appropriate statistical analysis indicators is an important issue that needs to be discussed. As research on the internal relationship of BSSs by complex network analysis is still inadequate, this problem is the premise of understanding the characteristics of BSSs. (3) How to select representative areas for comparative analysis to gain an in-depth understanding of BSSs is also another problem that we should pay much attention to. Because of different population density, employment density, built environment, and land use in different areas, characteristics of BSSs also vary. That is, conclusions and suggestions put forward according to the information in certain regions are not necessarily applicable to other regions. As a result, regional information should be gathered for system operators to possess a more comprehensive knowledge of BSSs.

According to the spatiotemporal real-time data of the PBS in Nanjing, we first constructed the public bicycle networks (PBNs) of different urban areas with Gephi software. Then, we not only analyzed statistical indicators such as degree and strength but also defined "radiation distance". Finally, we analyzed internal correlation characteristics between stations and compared characteristics in different regions.

3. Data and Methodology

We chose Nanjing PBS in China as our case study. Nanjing is the center of politics, economy, education, and culture in Eastern China and also the provincial capital of Jiangsu Province, with an area of 6587.02 km^2. At the end of 2016, the total population was 8.27 million with the urban population density 1484 people/km^2. There were 10,402 public transport vehicles in operation, with a network of 10,476.3 km in length and a total of 1776 million passengers, ranked 6th in China in 2016. Behind the economic prosperity and the development of modern transportation, air pollution cannot be ignored. In 2016, the annual average concentration of pm10 was 0.085 mg/m^3, with exhaust emissions of 773.482 billion m^3 and soot emissions of 48,600 tons [51]. Therefore, the study of Nanjing PBS is of positive significance to improve the efficiency of PBSs in modern metropolis with high population density, heavy traffic, and air pollution.

Nanjing PBS has more than 600 stations almost covering the main city since Nanjing Public Bicycle Company was founded in 2015. People pay a deposit for a public bicycle card. The charge is zero for the first two hours, one Chinese yuan for the third hour, and three Chinese yuan per hour for the fourth hour and above. A user picks up a bicycle at a station and returns it in the parking pile at the destination station to complete this trip.

3.1. Data

Nanjing Public Bicycle Company provided us with 662,007 anonymous daily data from 20 March 2016 to 26 March 2016, including user ID, renting (returning) stations, renting (returning) time, longitude and latitude of returning stations (see Table 1). Some situations such as bicycle and timer malfunctions would lead to inaccurate cycling time records. Thus, the data that indicated less than 2 min or more than 2 h usage were excluded, and the number of actual valid data was 593,582. In this week, 131,695 users used public bicycles, with an average of 4.5 times, the average duration of 15.77 min, and the average cycling distance of 1.43 km, which was consistent with "the first and last mile".

Table 1. Sample data of Nanjing public bicycle system (PBS).

User ID	Renting Time	Returning Time	Renting Station	Returning Station	Longitude of Returning Station	Latitude of Returning Station
10***	2016/3/20 0:09	2016/3/20 0:20	1118228	1216847	118.7629	32.03357
22***	2016/3/21 16:18	2016/3/21 16:24	1118406	1119129	118.7621	32.04811
10***	2016/3/24 19:51	2016/3/24 19:59	1119305	1211847	118.7544	32.041

Gulou District is the central area and also the political, economic, educational, and cultural center of Nanjing. The resident population of Gulou District is 1,249,400, ranked 1st in the six main urban districts of Nanjing, and the area is 53 km^2, with a population density of 23,574 people/km^2. In 2016, the total production value of Gulou District was 112.303 billion Chinese yuan, ranked 1st, and the per capita annual disposable income was 54,791 Chinese yuan. High population density and employment density lead to high demand for public bicycles. Qixia District is located in a remote area with a large area and low population density. The resident population of Qixia District is 693,300, ranked 3rd, and the area is 395.44 km^2, with a population density of 1753 people/ km^2. In 2016, the total production value of Qixia District was 92.723 billion Chinese yuan, ranked 2nd, and the per capita annual disposable income was 48,379 Chinese yuan. Low population density may be the main reason for low demand for public bicycles. Gulou District and Qixia District, one as the center of economics and politics and the other as the emerging district of culture and education, are different in geographical location, regional functions, distribution of stations, and demand for public bicycles. Therefore, it is reasonable to select these two districts to analyze and compare the internal correlation characteristics of PBNs.

3.2. Complex Network Theory

Complex network theory explores topological structures and properties of a network utilizing graph theory and statistical physics, which can show the nature of the network without depending on the specific location of nodes and specific shape of edges. In a network, nodes represent research objects, and edges connecting two nodes represent their relationship. Now, complex network theory has been successfully applied in financial markets [52–54], transportation systems [9,31,49,50,55–57], and energy fields [58–61].

We treat a station as a node. If a user rents a public bicycle at station A and returns it to station B, there is a directed edge pointing from A to B. The weight of this edge is equal to the number of cycling records from A to B, which can show how close the relationship between A and B is. Hence, a PBN is a directed weighted network.

3.2.1. Degree

For a PBN, public bicycles rented from station i are returned to various destination stations, and the number of these destination stations is defined as the out-degree k_i^{out} of node i. Public bicycles rented from various origin stations are returned to station i, and the number of these origin stations is defined as the in-degree k_i^{in} of node i. The calculation formulas of k_i^{out} and k_i^{in} are as follows:

$$k_i^{out} = \sum_{j=1}^{N} a_{ij}, \quad k_i^{in} = \sum_{j=1}^{N} a_{ji} \tag{1}$$

where a_{ij} denotes an element of adjacency matrix, that is, if there is a directed edge from station i to station j, $a_{ij} = 1$; otherwise, $a_{ij} = 0$. N represents the number of nodes.

3.2.2. Strength

For a PBN, the out-strength s_i^{out} represents the number of public bicycles rented from station i, and the in-strength s_i^{in} represents the number of public bicycles returned to station i. The calculation formulas of s_i^{out} and s_i^{in} are as follows:

$$s_i^{out} = \sum_{j=1}^{N} w_{ij}, \quad s_i^{in} = \sum_{j=1}^{N} w_{ji} \tag{2}$$

where w_{ij} represents the weight of the edge from station i to station j.

3.2.3. Radiation Distance

Compared with general complex networks, PBNs have their own way of calculating path length. For example, if one user rents a public bicycle from station A and returns it to station B and another user rents one from station B and returns it to station C, there is a path from station A to station C according to the concept of "path" in complex network theory. However, we cannot explain this path as a user starting from station A and going to station C via station B. Furthermore, there may be no records renting one from station A and returning it to station C. Therefore, applying "path length" of complex network theory to PBNs may be of no significance. As a result, we define "radiation distance" to measure the distance between stations (nodes).

In a PBN, we define the radiation distance from station A to station B as the actual distance between them, and the radiation distance of station A is the average radiation distance from station A to the other stations. Hence, the radiation distance of a PBN is the average radiation distance of all the stations.

3.2.4. Community Structure

In a network, nodes in the same community are closely connected. Modularity Q is used to evaluate the quality of community division, and community division with the maximum value of modularity is optimal. For a directed weighted complex network, modularity is calculated as follows:

$$Q = \frac{1}{W} \sum_{i,j} (w_{ij} - \frac{s_i^{out} s_j^{in}}{W}) \; \delta(M_i, M_j) \tag{3}$$

where W is the sum of all the weights, M_i is the community that station i belongs to and M_j is the community that station j belongs to. If station i and station j belong to the same community, $\delta(M_i, M_j) = 1$, otherwise, $\delta(M_i, M_j) = 0$.

4. Results

4.1. The Construction of the PBN

We first treated each public bicycle station in Gulou District as a node, and then added the stations in other districts that had public bicycle renting or returning with the nodes into the network. Hence, there were totally 587 stations (nodes) in the PBN of Gulou District. There were directed edges between the nodes, and the weights were the number of public bicycles rented or returned.

4.2. The Analysis of Out-Degree and In-Degree

The average in-degree of the PBN of Gulou District was 57 with the maximum being 287. Since the in-degree of each station was numerically dispersed, we performed interval segmentation to better understand its distribution. We first divided the values of in-degree by the interval length of 20. Then, we considered the integer closest to the weighted average of in-degree and the number of the stations within the interval as the representative of the interval value. In this way, we got the in-degree distribution, see Figure 1a. In a log–log plot, the least square estimation was applied to estimate the in-degree distribution, and the regression equation was $y = -1.2221x + 2.3808$ with the coefficient of determination $R^2 = 0.897$, which showed a good fitting effect (see Figure 1b).

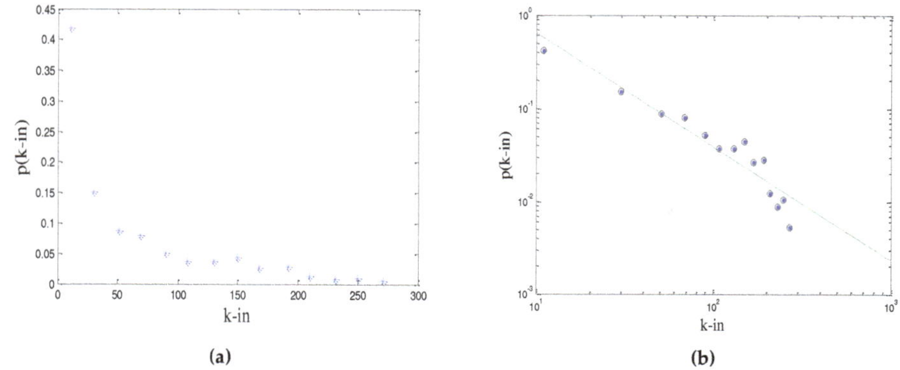

Figure 1. (**a**) The in-degree distribution of the public bicycle network (PBN) of Gulou District; (**b**) the in-degree distribution estimation in a log-log plot.

The average out-degree of the PBN of Gulou District was 57 with the maximum being 289. Figure 2a shows the out-degree distribution. In a log–log plot, the regression equation was $y = -1.1262x + 1.9299$ with the coefficient of determination $R^2 = 0.876$ (see Figure 2b). As a result, the in-degree and out-degree distribution of the PBN of Gulou District followed the power-law distribution with the power exponents of 1.22 and 1.13, respectively, which were smaller than power exponents of scale-free networks between 2–3. According to the data, there were 245 stations with an in-degree of less than 20, accounting for 41.7% of all the stations, and there were 235 stations with an out-degree of less than 20, accounting for 40.0%. These results indicated that the percentage of the stations with small in-degree or out-degree was lower than that of typical scale-free networks.

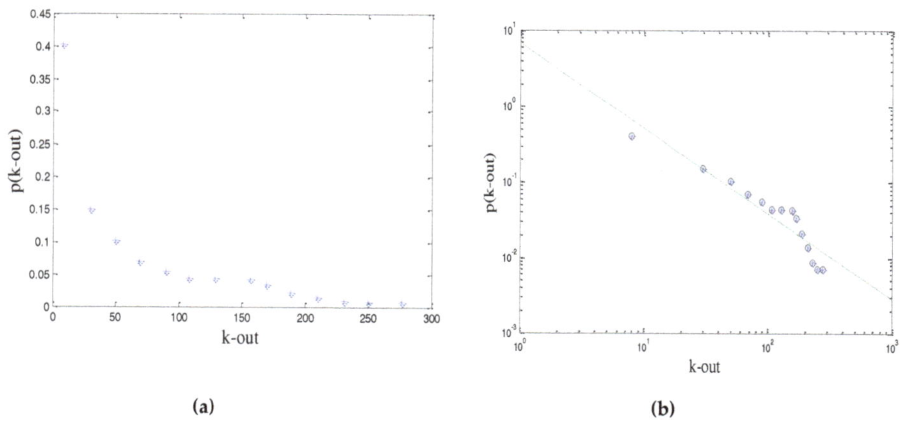

Figure 2. (**a**) The out-degree distribution of the PBN of Gulou District; (**b**) the out-degree distribution estimation in a log-log plot.

There were 225 stations (nodes) in the PBN of Qixia District. The average in-degree and out-degree were both 11, and the maximum in-degree and out-degree were 93 and 87, respectively, which were smaller than those of the PBN of Gulou District. Geographically, Gulou District borders Qixia, Xuanwu,

Qinhuai, Yuhuatai, and Jianye districts, while Qixia District borders Gulou and Xuanwu districts. From the perspective of station size, there were 175 stations in Gulou District and 41 stations in Qixia District. Therefore, there were big gaps between the two PBNs in station connections and station size.

The in-degree distribution and out-degree distribution of the PBN in Qixia District were similar to long tail distribution (see Figure 3) and they followed the power-law distribution with the power exponents of 1.33 and 1.21, respectively. The power exponents were larger than those of the PBN of Gulou District, which probably meant idle situation of stations in Qixia District was more serious than that in Gulou District.

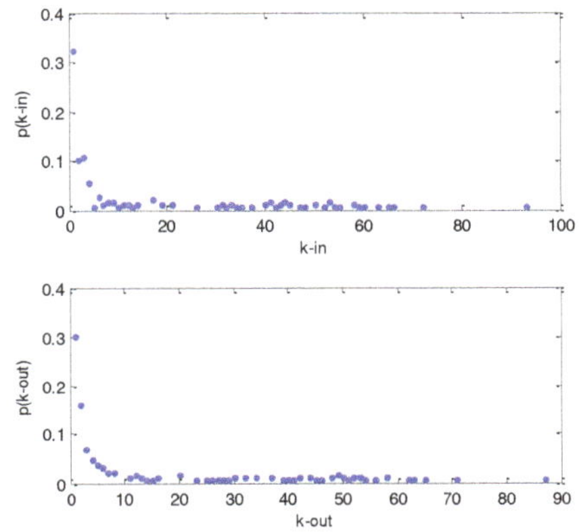

Figure 3. The in-degree and out-degree distribution of the PBN of Qixia District.

4.3. The Analysis of Out-Strength and In-Strength

Table 2 lists the top ten stations of the PBN of Gulou District based on in-strength and based on out-strength, and the number in brackets after the station name is the station number. Nine of the top ten based on in-strength and the top ten based on out-strength were the same except Longjiang Stadium East Station and Mingcheng Century Park East Gate Station. At the same time, Table 3 lists the bottom ten stations, and the number in brackets after the station name is the station number. The bottom ten based on in-strength and the bottom ten based on out-strength were nearly the same except the "724 Institute" Station and Shengshi Garden Station. This result indicated that renting flow and returning flow at a station was almost equivalent. Moreover, the strength of the top ten was much stronger than that of the bottom ten, which showed that the usage of public bicycles at different stations varied greatly.

Figure 4 shows the geographical distribution of these stations. The red and purple marks correspond to the top ten and the bottom ten stations in Tables 2 and 3, respectively. There are 11 different stations in Table 2, so there are 11 top ten stations in Figure 4 without distinguishing in-strength from out-strength, and the same is true for the bottom ten. There were regional differences in the use of public bicycles. The bottom ten stations were mainly distributed in the north of Metro Line 4, while the top ten were in the south. According to the land use, the commercial land, for example, Xinjiekou Huaqiao Road and Fenghuang business area as well as other cultural and recreational facilities, sports facilities, and educational institutions were concentrated in the south of Metro Line 4. The distribution of stations may indicate that areas close to central business districts had a higher demand for public bicycles due to their high population density and employment density, that is, the more social and economic activities, the more use of public bicycles. Hence, adequate and timely supply of public bicycles around these areas is the priority.

Table 2. The top ten stations of the PBN of Gulou District.

No.	Station	In-Strength	Station	Out-Strength
1	Qingliangmen Suguo Supermarket(1)	3111	Qingliangmen Suguo Supermarket(1)	3262
2	Suning Global Trade City(2)	2851	Hetai International Building(5)	2758
3	International Service Outsourcing Building(3)	2817	Yingchunli West Gate(4)	2647
4	Yingchunli West Gate(4)	2723	Nanjing University of Chinese Medicine South Station(9)	2646
5	Hetai International Building(5)	2468	Suning Global Trade City(2)	2600
6	Jiaheyuan(6)	2390	International Service Outsourcing Building(3)	2387
7	Zhonghai Fenghuangxian South Gate(7)	2237	Jiaheyuan(6)	2142
8	National University Science Park(8)	2178	Zhonghai Fenghuangxian South Gate(7)	2090
9	Nanjing University of Chinese Medicine South Station(9)	2136	Mingcheng Century Park East Gate(11)	2043
10	Longjiang Stadium East Station(10)	2104	National University Science Park(8)	1871

Table 3. The bottom ten stations of the PBN of Gulou District.

No.	Station	In-Strength	Station	Out-Strength
1	No. 5 Ningxia Road(1)	104	Qingjiang Xiyuan North Gate(2)	105
2	Qingjiang Xiyuan North Gate(2)	122	No. 5 Ningxia Road(1)	113
3	Tianfeigong Primary School(3)	159	Tianfeigong Primary School(3)	155
4	Biancheng Shijia West Gate(4)	163	Crown Diamond Double Star South(7)	187
5	No.35 Xikang Road(5)	173	Xikang Hotel(6)	191
6	Xikang Hotel(6)	192	Biancheng Shijia West Gate(4)	194
7	Crown Diamond Double Star South(7)	219	Gate 4 of Tianzhenghubin(10)	202
8	724 Institute(8)	219	No.35 Xikang Road(5)	206
9	Gate 1of Tianjinxincun(9)	219	Shengshi Garden(11)	206
10	Gate 4 of Tianzhenghubin(10)	226	Gate 1of Tianjinxincun(9)	249

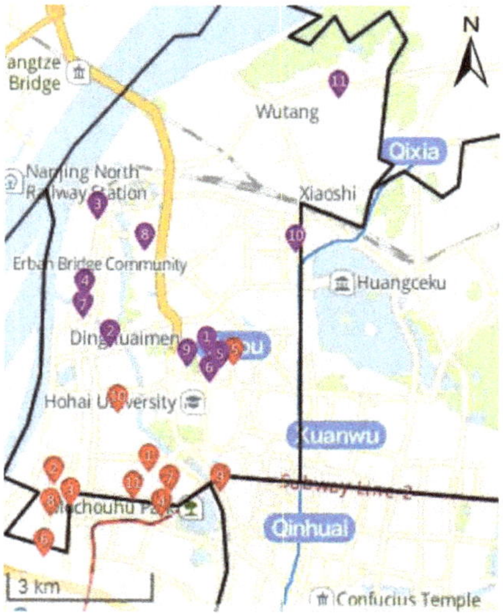

Figure 4. The distribution of the top ten and the bottom ten stations of the PBN of Gulou District.

We should point out that many stations with strong strength were close to subway stations, such as Jiaheyuan Station and Jiqingmen Street Subway Station, Yingchunli West Gate Station and

Hanzhongmen Subway Station, Nanjing University of Chinese Medicine South Station and Shanghai Road Subway Station, which indicated that subway stations were important starting and destination stations for cycling. Therefore, the construction of bicycle stations around subway stations is an important measure to promote the usage of public bicycles.

It was worth noting that there was still low usage of public bicycles even in densely populated areas. There were several residential areas around Qingjiang Xiyuan North Gate Station, and Qingjiang Xiyuan West Gate Station was also nearby. According to the records, almost 60% of the stations associated with these two stations were the same. In fact, Qingjiang Xiyuan West Gate Station was closer to the bus stop and subway station, so the usage of public bicycles was higher than that of Qingjiang Xiyuan North Gate Station. Except for station location, competitive relationship with other stations also affected the usage of public bicycles. Therefore, unreasonable station setting not only reduced the efficiency of PBSs but also led to a waste of public resources.

In directed weighted networks, nodes with strong strength are worth studying. We analyzed Qingliangmen Suguo Supermarket Station, which had the strongest in-strength and out-strength of the PBN of Gulou District.

Figure 5 shows the station distribution around Qingliangmen Suguo Supermarket Station. On the one hand, Mochou Lake Subway Station West Station, Lixue Primary School Station, Huayang Jiayuan North Gate Station, and Yingchunli West Gate Station contributed the most to the in-strength of Qingliangmen Suguo Supermarket Station. On the other hand, the most frequent destinations of this station were Mochou Lake Subway Station West Station, Huayang Jiayuan East Gate Station, Lixue Primary School Station, Jinxin Garden North Gate Station, and Suning Qianqiu Garden East Gate Station. The results showed that if an area where supermarkets, residential areas, subway stations, and schools were concentrated, it was always the main area for the use of public bicycles, which was consistent with the existing research results [7]. Furthermore, in-strength and out-strength of the stations around Qingliangmen Suguo Supermarket Station were above the average level of the network. The observed data showed that the usage of public bicycles at a station was not only related to the built environment but also affected by the usage of bicycles at stations nearby. Ensuring the availability of public bicycles and parking piles of these key stations would have a direct impact on the use of public bicycles throughout the whole area.

As a popular supermarket in Gulou District, Qingliangmen Suguo Supermarket is adjacent to 10 residential areas, including Huayang Jiayuan, Jinxin Garden, Yingchunli, and Suning Qianqiu Garden, and many users cycle to and from the communities and the supermarket every day. Table 4 lists the comparison of the distances and time required from these communities to the supermarket by different travel modes. The comparison indicated that public bicycle was a time-saving means of transportation when the distance was about 1000 m. Convenience, speed, and economy are the advantages of public bicycles for short-distance travel.

Table 4. Comparison of different travel modes.

Travel Mode	Distance from Huayang Jiayuan (Meter)	Travel Time (Minute)	Distance from Jinxin Garden (Meter)	Travel Time (Minute)	Distance from Yingchunli (Meter)	Travel Time (Minute)	Distance from Suning Qianqiu Garden (Meter)	Travel Time (Minute)
Walking	570	8	1000	>10	1200	>10	960	>10
Driving	1800	9	1300	6	1000	6	1500	9
Cycling	600	4	1100	7	1000	6	1000	6

The top ten stations of the PBN of Qixia District based on in-strength were also the top ten stations based on out-strength (see Table 5), and the same was true for the bottom ten (see Table 6). Figure 6 shows the distribution of these stations. The red and purple marks correspond to the top ten and the bottom ten in Tables 5 and 6, respectively. Most of the top ten stations were located around Maigaoqiao Subway Station and along Heyan Road, the major economic area in Qixia District. In particular, the in-strength and out-strength of Gate 1 of Maigaoqiao Subway Station were significantly stronger than the other stations due to the origin station of Metro Line 1. At the same time, six stations within 1200 m around Maigaoqiao Subway Station had strong in-strength and out-strength, forming an area of high demand for public bicycles. Among the six stations, Qiancaomingyuan South Station and Jinghongyuan South Station contributed the most to in-strength and out-strength, which indicated that public bicycle was an effective means of transportation between communities and important transportation hubs. As a result, public bicycles can bring people more convenience if the connection of public bicycle stations with other transportation stations is well projected in areas with sparse distribution of public transportation. In addition, the usage of public bicycles in Qixia District was less than that in Gulou District.

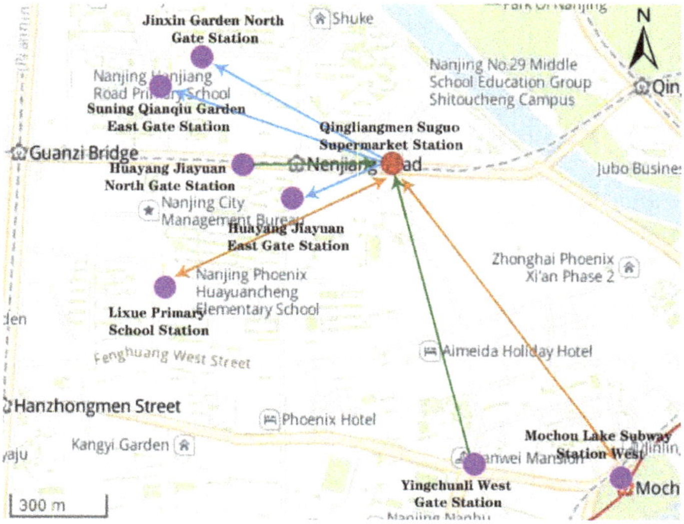

Figure 5. The station distribution around Qingliangmen Suguo Supermarket Station.

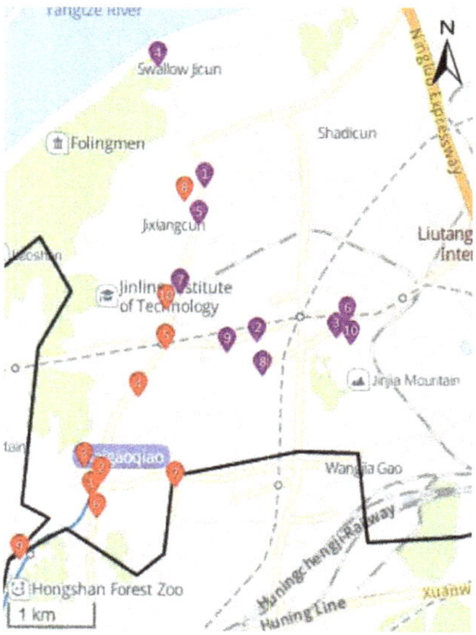

Figure 6. The distribution of the top ten and the bottom ten stations of PBN of Qixia District.

Table 5. The top ten stations of the PBN of Qixia District.

No.	Station	In-Strength	Station	Out-Strength
1	Gate 1of Maigaoqiao Subway(1)	1727	Gate 1of Maigaoqiao Subway(1)	1753
2	Maigaoqiao Bus Station(2)	1063	Maigaoqiao Zijin Rural Commercial Bank(3)	1003
3	Maigaoqiao Zijin Rural Commercial Bank(3)	1037	Maigaoqiao Bus Station(2)	970
4	Heyan Road Community East(4)	862	Heyan Road Community East(4)	912
5	Xiaozhuang International Plaza West(5)	818	Mufu Mountain Villa(10)	808
6	Jiangsu Province Hospital on Integration of Chinese and Western Medicine West(6)	714	Shengli Village East(8)	788
7	Lanting Yayuan Suguo Supermarket(7)	714	Lanting Yayuan Suguo Supermarket(7)	720
8	Shengli Village East(8)	700	Jiangsu Province Hospital on Integration of Chinese and Western Medicine West(6)	673
9	Hongshan Zoo North(9)	681	Xiaozhuang International Plaza West(5)	646
10	Mufu Mountain Villa(10)	608	Hongshan Zoo North(9)	643

Table 6. The bottom ten stations of the PBN of Qixia District.

No.	Station	In-Strength	Station	Out-Strength
1	Jinyuan Department Store East(1)	105	Jinyuan Department Store East(1)	89
2	Social Security Administration of Qixia District(2)	106	Traffic bureau of Qixia South(4)	106
3	Wanshou Garden North(3)	117	Social Security Administration of Qixia District(2)	119
4	Traffic bureau of Qixia South(4)	155	Wanshou Garden North(3)	122
5	Jinqu Village East(5)	161	Jinqu Village East(5)	152
6	Dafa Yanlanwan West(6)	166	Dafa Yanlanwan West(6)	180
7	Yanhua Garden North(7)	226	Yanziji Park(9)	235
8	Gaoli Auto Parts Company East(8)	238	Yanhua Garden North(7)	245
9	Yanziji Park(9)	248	Industrial and Commercial Bureau of Yanziji(10)	266
10	Industrial and Commercial Bureau of Yanziji(10)	262	Gaoli Auto Parts Company East(8)	285

The stations with strong or weak strength in the PBNs had a clear division in the distribution. Areas where supermarkets, residential areas, schools, and subway stations were gathered were always areas with high demand for public bicycles. The usage of public bicycles at a station was not only related to the concentration of social and economic activities but also affected by the usage of bicycles at stations nearby.

4.4. *The Analysis of Radiation Distance*

The actual distances between stations were calculated based on longitude and latitude of the stations. The average radiation distance of the PBN of Gulou District was 1.59 km, which was consistent with the original intention of "the first and last mile". Because of the dense distribution of public bicycle stations in Gulou District, it was easier to obtain public bicycle services in shorter distances. Furthermore, 11.3% of the cycling distances were more than 3 km, 0.13% were more than 10 km, and the maximum distance was 17 km, which indicated that public bicycles were not only used for short-distance travel but also for long-distance travel because of convenience, economy, and exercise purposes.

The average radiation distance of the PBN of Qixia District was 1.87 km, which was longer than that of Gulou District. According to the records, 15% of the cycling distances were more than 3 km and the proportion at Yanziji Park Station was even more than 30%. In fact, the area of Qixia District was seven times that of Gulou District, and there were fewer stations in Qixia District, which led to more long-distance cycling.

4.5. The Analysis of Community Structure

According to Louvain algorithm, the PBN of Gulou District was divided into five communities numbered 1 to 5. As explained in network construction, the PBN of Gulou District also contained stations in other districts that had bicycle renting and returning with stations in Gulou District. In order to understand the relationship between public bicycle stations within Gulou District, we removed stations in other districts, only leaving stations in Gulou District (see Figure 7).

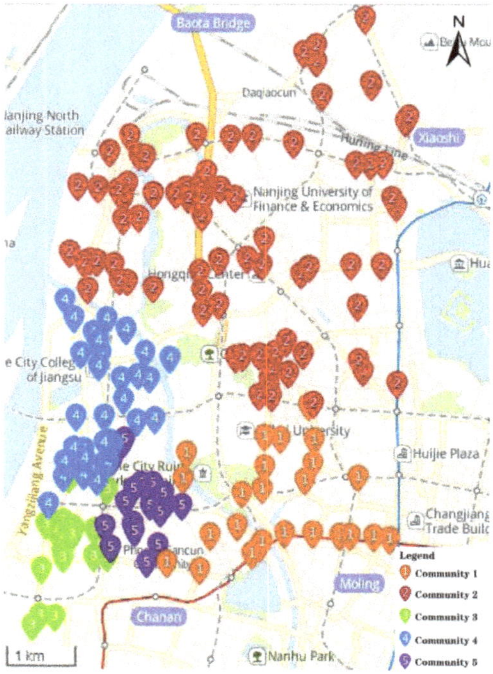

Figure 7. The community distribution of the PBN of Gulou District.

Community 1 marked in orange was located in the southeast of Gulou District. This community was around Xinjiekou Huaqiao Road area, which is an important business district in Gulou District mentioned above and also an area with some strong strength stations. There were a few stations in this community, and the stations were sparsely distributed. There were no stations with weak strength, which indicated that Community 1 was an active one for the use of public bicycles.

Community 2 marked in red was located in the north of Gulou District. It contained the most stations, which were sparsely distributed. More than 90% of these stations had weaker strength than the average, which showed that Community 2 was mainly composed of stations with lower usage of public bicycles. No. 5 Ningxia Road Station, Tianfeigong Primary School Station, and other stations within the bottom ten stations in Table 3 belonged to this community.

Community 3 marked in green was located in the southwest of Gulou District. There were a few stations, and 70% of them had stronger strength than the average. This community contained many key stations, forming an area with high demand for public bicycles.

Community 4 marked in blue was located in the west of Gulou District. There were many stations and 65% of them had weaker strength than the average. The usage of public bicycles of Community 4 was only higher than that of Community 2.

Community 5 marked in purple was adjacent to Community 1, Community 3, and Community 4. It contained the fewest stations, which were densely distributed. Qingliangmen Suguo Supermarket Station, the station with maximum strength of the network, belonged to this community. Furthermore, Lixue Primary School Station, Huayang Jiayuan North Gate Station, and Jinxin Garden North Gate Station, which contributed the most to the strength of Qingliangmen Suguo Supermarket Station, also belonged to this community. Community 5 was an active one for the use of public bicycles.

Consistent with the analysis of Figure 5, the inactive Community 2 and Community 4 were in the north of Metro Line 4, while the active Community 1, Community 3, and Community 5 were in the south. The significant difference in the geographical distribution of communities confirmed that the stations with strong strength tended to cluster with stations with strong strength, and vice versa.

The PBN of Qixia District was divided into four communities (see Figure 8). Community 2 marked in orange had the most use of public bicycles, and Gate 1 of Maigaoqiao Subway Station belonged to this community. Community 4 marked in blue had the least use. Due to small population density, vast geographical area, and sparse stations, the usage of public bicycles was significantly lower than that of Gulou District. How to improve usage is an important issue for system operators.

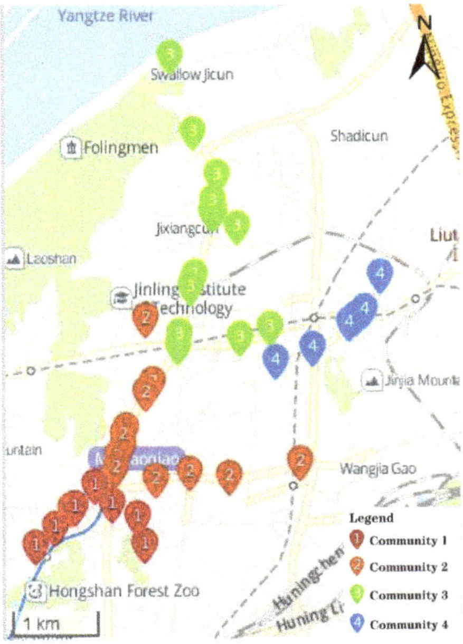

Figure 8. The community distribution of the PBN of Qixia District.

5. Conclusion

Based on complex network theory and geographic visualization method, we constructed the regional PBNs using Gephi software and analyzed degree, strength, radiation distance, and community structure to understand the internal relationship of the PBSs, and we got the following conclusions:

(1) We may understand the usage of Nanjing PBS through the analysis of degree distribution and strength. The degree distribution followed the power-law distribution with the power exponents between 1 to 2, and there were still many stations with low usage of public bicycles.
(2) We understand the usage of stations and their internal relationship through the analysis of strength and community structure. Stations with strong or weak strength had a clear geographical distribution, and the cycling flow at stations in different areas varied greatly. The areas with more social and economic activities were also the areas with more use of public bicycles [7], which was confirmed again by our regional study using complex network analysis. We also found that the usage of public bicycles at some stations was not only related to land use but also related to the usage of bicycles at stations nearby.
(3) We understand the role of public bicycles through the analysis of radiation distance and strength. The average radiation distance of the PBS was consistent with the original design intention of "the first and last mile", and cycling distance was greater in remote areas. The observed data showed that public bicycles were not only served for short-distance travel but also long-distance travel. Cycling between residential areas and subway stations, between residential areas and supermarkets, was very common, and subway stations were important origin-destination stations of public bicycles.

Changes in the structure of PBNs over time will be studied in the future.

Author Contributions: Conceptualization, Y.Y.; Data curation, L.T.; Project administration, Y.Y.; Supervision, Y.Y.; Visualization, N.Z.; Writing – original draft, Y.Z.; Writing – review & editing, Y.Y., Z.L. and M.W.

Funding: The research is supported partially by the following: The National Natural Science Foundation of China (71690242, 71503132, 71811520710), the National Social Science Fund of China (No.18BTJ040), Qing Lan Project of Jiangsu Province, China (2017), Six talent peaks project in Jiangsu Province, China (JY-055), China Postdoctoral Science Foundation (2013M541697).

Conflicts of Interest: The authors declare no conflict of interest.

References

1. DeMaio, P. Bike sharing: History, impacts, models of provision, and future. *J. Public Transp.* **2009**, *12*, 41–56. [CrossRef]
2. Fishman, E.; Washington, S.; Haworth, N.; Mazzei, A. Barriers to bikesharing: An analysis from Melbourne and Brisbane. *J. Transp. Geogr.* **2014**, *41*, 325–337. [CrossRef]
3. Vogel, M.; Hamon, R.; Lozenguez, G.; Merchez, L.; Abry, P.; Barnier, J.; Borgnat, P.; Flandrin, P.; Mallon, I.; Robardet, C. From bicycle sharing system movements to users: A typology of Vélo'v cyclists in Lyon based on large-scale behavioural dataset. *J. Transp. Geogr.* **2014**, *41*, 280–291. [CrossRef]
4. Goodman, A.; Cheshire, J. Inequality in the London bicycle sharing system revisited: Impacts of extending the scheme to poorer areas but then doubling prices. *J. Transp. Geogr.* **2014**, *41*, 272–279. [CrossRef]
5. Shaheen, S.A.; Cohen, A.P.; Martin, E.W. Public bike sharing in north America: Early operator understanding and emerging trends. *Transp. Res. Rec. J. Transp. Res. Board.* **2013**, *2387*, 83–92. [CrossRef]
6. Barbour, N.; Zhang, Y.; Mannering, F. A statistical analysis of bike sharing usage and its potential as an auto-trip substitute. *J. Transp. Health* **2019**, *12*, 253–262. [CrossRef]
7. Yao, Y.; Jiang, X.; Li, Z. Spatiotemporal characteristics of green travel: A classification study on a public bicycle system. *J. Clean. Prod.* **2019**, *238*, 117892. [CrossRef]
8. Buck, D.; Buehler, R.; Happ, P.; Rawls, B.; Chung, P.; Borecki, N. Are bikeshare users different from regular cyclists? *Transp. Res. Rec. J. Transp. Res. Board.* **2013**, *2387*, 112–119. [CrossRef]
9. Rixey, R. Station-level forecasting of bike sharing ridership: Station network effects in three U.S. systems. *Transp. Res. Rec. J. Transp. Res. Board.* **2013**, *2387*, 46–55. [CrossRef]
10. Woodcock, J.; Tainio, M.; Cheshire, J.; O'Brie, O.; Goodman, A. Health effects of the London bicycle sharing system: Health impact modeling study. *Br. Med. J.* **2014**, *348*, g425. [CrossRef] [PubMed]

11. Corcoran, J.; Li, T.B.; Rohde, D.; Charles-Edwards, E.; Mateo-Babiano, D. Spatio-temporal patterns of a Public Bicycle Sharing Program: The effect of weather and calendar events. *J. Transp. Geogr.* **2014**, *41*, 292–305. [CrossRef]
12. Faghih-Imani, A.; Eluru, N.; El-Geneidy, A.M.; Rabbat, M.; Haq, U. How land-use and urban form impact bicycle flows: Evidence from the bicycle-sharing system (BIXI) in Montreal. *J. Transp. Geogr.* **2014**, *41*, 306–314. [CrossRef]
13. El-Assi, W.; Mahmoud, M.S.; Habib, K.N. Effects of built environment and weather on bike sharing demand: A station level analysis of commercial bike sharing in Toronto. *Transportation* **2017**, *44*, 589–613. [CrossRef]
14. Buck, D.; Buehler, R. Bike lanes and other determinants of capital bike share trips. In Proceedings of the Transportation research board 91st annual meeting, Washington, DC, USA, 22–26 January 2012.
15. Wang, X.; Lindsey, G.; Schoner, J.E.; Harrison, A. Modeling bike share station activity: Effects of nearby businesses and jobs on trips to and from stations. *J. Urban Plann. Dev.* **2016**, *142*, 4015001. [CrossRef]
16. Wang, Y.; Chau, C.K.; Ng, W.Y.; Leung, T.M. A review on the effects of physical built environment attributes on enhancing walking and cycling activity levels within residential neighborhoods. *Cities* **2016**, *50*, 1–15. [CrossRef]
17. Faghih-Imani, A.; Hampshir, R.; Lavanya, M.; Eluru, N. An empirical analysis of bike sharing usage and rebalancing: Evidence from Barcelona and Seville. *Transp. Res. A.* **2017**, *97*, 177–191. [CrossRef]
18. Luo, S.; Zhen, F.; Yin, Q. How Built Environment Influence Public Bicycle Usage: Evidence from the Bicycle Sharing System in Qiaobei Area, Nanjing. *Sci. Geogr. Sin.* **2018**, *38*, 332–341. [CrossRef]
19. Liu, H.C.; Lin, J.J. Associations of built environments with spatiotemporal patterns of public bicycle use. *J. Transp. Geogr.* **2019**, *74*, 299–312. [CrossRef]
20. Martin, E.W.; Shaheen, S.A. Evaluating public transit modal shift dynamics in response to bikesharing: A tale of two U.S. cities. *J. Transp. Geogr.* **2014**, *41*, 315–324. [CrossRef]
21. Ma, T.; Liu, C.; Erdoan, S. Bicycle sharing and public transit: Does capital bike share affect metrorail ridership in Washington, D.C. *Transp. Res. Rec. J. Transp. Res. Board* **2015**, *2534*, 1–9. [CrossRef]
22. Klingen, J. Do metro interruptions increase the demand for public rental bicycles? Evidence from Paris. *Transp. Res. Part A* **2019**, *123*, 216–228. [CrossRef]
23. Gu, T.Q.; Kim, I.; Currie, G. Measuring immediate impacts of a new mass transit system on an existing bike-share system in China. *Transp. Res. Part A* **2019**, *124*, 20–39. [CrossRef]
24. Li, W.Y.; Tian, L.X.; Gao, X.Y.; Batool, H. Effects of dockless bike-sharing system on public bike system: Case study in Nanjing, China. *Energy Procedia* **2019**, *158*, 3754–3759. [CrossRef]
25. Fishman, E.; Washington, S.; Haworth, N. Bike share: A synthesis of the literature. *Transp. Rev.* **2013**, *33*, 148–165. [CrossRef]
26. Fishman, E.; Washington, S.; Haworth, N. Bike share's impact on car use Evidence from the United States, Great Britain, and Australia. *Transp. Res. Part D* **2014**, *31*, 13–20. [CrossRef]
27. Shaheen, S.; Martin, E.; Cohen, A.P.; Finson, R. *Public Bikesharing in North America: Early Operator and User Understanding*; Mineta Transportation Institute: San Jose, CA, USA, 2012.
28. Shaheen, S.A.; Martin, E.W.; Chan, N.D.; Cohen, A.P.; Pogodzinski, M. *Public Bikesharing in North America during a Period of Rapid Expansion: Understanding Business Models, Industry Trends and User Impacts*; Mineta Transportation Institute: San Jose, CA, USA, 2014.
29. Chen, J.; Zhang, Y.; Zhang, R.; Cheng, X.; Yan, F.Y. Analyzing users' attitudes and behavior of free-floating bike sharing: An investigating of Nanjing. *Transp. Res. Procedia* **2019**, *39*, 634–645. [CrossRef]
30. Fuller, D.; Luan, H.; Buote, R.; Auchinclossd, A.H. Impact of a public transit strike on public bicycle share use: An interrupted time series natural experiment study. *J. Transp. Health* **2019**, *13*, 137–142. [CrossRef]
31. Saberi, M.; Ghamami, M.; Gu, Y.; Shojaei, M.H.; Fishman, E. Understanding the impacts of a public transit disruption on bicycle sharing mobility patterns: A case of Tube strike in London. *J. Transp. Geogr.* **2018**, *66*, 154–166. [CrossRef]
32. Bachand-Marleau, J.; Lee, B.H.Y.; El-Geneidy, A.M. Better understanding of factors influencing likelihood of using shared bicycle systems and frequency of use. *Transp. Res. Rec. J. Transp. Res. Board.* **2012**, *2314*, 66–71. [CrossRef]
33. Luo, H.; Kou, Z.Y.; Zhao, F.; Cai, H. Comparative life cycle assessment of station-based and dock-less bike sharing systems. *Resour. Conserv. Recycl.* **2019**, *146*, 180–189. [CrossRef]

34. Fricker, C.; Gast, N. Incentives and Redistribution in Homogeneous Bike-Sharing Systems with Stations of Finite Capacity. *Euro J. Transp. Logist.* **2016**, *5*, 261–291. [CrossRef]
35. Chemla, D.; Meunier, F.; Wolfler Calvo, R. Bike sharing systems: Solving the static rebalancing problem. *Discr. Optim.* **2013**, *10*, 120–146. [CrossRef]
36. Alvarez-Valdes, R.; Belenguer, J.M.; Benavent, E.; Bermudez, J.D.; Muñoz, F.; Vercher, E.; Verdejo, F. Optimizing the level of service quality of a bikesharing system. *Omega* **2016**, *62*, 163–175. [CrossRef]
37. Ghosh, S.; Varakantham, P.; Adulyasak, Y.; Jaillet, P. Dynamic repositioning to reduce lost demand in bike sharing systems. *J. Artif. Intell. Res.* **2017**, *58*, 387–430. [CrossRef]
38. Park, C.; Young Sohn, S. An Optimization Approach for the Placement of Bicycle-sharing stations to Reduce Short Car Trips: An Application to the City of Seoul. *Transp. Res. Part A* **2017**, *105*, 154–166. [CrossRef]
39. Ho, S.C.; Szeto, W.Y. A hybrid large neighborhood search for the static multivehicle bike-repositioning problem. *Transp. Res. Part B* **2017**, *95*, 340–363. [CrossRef]
40. Zhang, D.; Yu, C.; Desai, J.; Lau, H.Y.K.; Srivathsan, S. A time-space network flow approach to dynamic repositioning in bicycle sharing systems. *Transp. Res. Part B.* **2017**, *103*, 188–207. [CrossRef]
41. Scarinci, R.; Markov, I.; Bierlaire, M. Network design of a transport system based on accelerating moving walkways. *Transp. Res. Part C.* **2017**, *80*, 310–328. [CrossRef]
42. Cruz, F.; Subramanian, A.; Bruck, B.P.; Iori, M. A heuristic algorithm for a single vehicle static bike sharing rebalancing problem. *Comput. Oper. Res.* **2017**, *79*, 19–33. [CrossRef]
43. Legros, B. Dynamic repositioning strategy in a bike-sharing system; how to prioritize and how to rebalance a bike station. *Eur. J. Oper. Res.* **2019**, *272*, 740–753. [CrossRef]
44. Feng, Y.L.; Affonso, R.C.; Zolghadri, M. Analysis of bike sharing system by clustering: The Velib' case. *IFAC PapersOnLine* **2017**, *50*, 12422–12427. [CrossRef]
45. Froehlich, J.; Neumann, J.; Oliver, N. Sensing and predicting the pulse of the city through shared bicycling. Int. Joint Conf. *Artif. Intell.* **2009**, *9*, 1420–1426.
46. Come, E.; Oukhellou, L. Model-based count series clustering for bike sharing system usage mining: A case study with the Velib' system of Paris. *ACM Trans. Intell. Syst. Technol.* **2014**, *5*, 39–65. [CrossRef]
47. Zhou, X. Understanding spatiotemporal patterns of biking behavior by analyzing massive bike sharing data in Chicago. *PLoS ONE* **2015**, *10*, 1–20. [CrossRef] [PubMed]
48. Borgnat, P.; Abry, P.; Flandrin, P.; Robardet, C.; Rouquier, J.B.; Fleury, E. Shared bicycles in a city: A signal processing and data analysis perspective. *Adv. Complex Syst.* **2011**, *14*, 415–438. [CrossRef]
49. Austwick, M.Z.; O'Brien, O.; Strano, E.; Viana, M. The structure of spatial networks and communities in bicycle sharing systems. *PLoS ONE* **2013**, *8*, e74685. [CrossRef]
50. Lin, L.; He, Z.; Peeta, S. Predicting station-level hourly demand in a large-scale bike-sharing network: A graph convolutional neural network approach. *Transp. Res. Part C: Emerg. Technol.* **2018**, *97*, 258–276. [CrossRef]
51. Nanjing Statistics Bureau, Survey Office of the National Bureau of Statistics in Nanjing. *Statistical Yearbook of Nanjing*; China Statistics Press: Beijing, China, 2017.
52. Banerjee, A.; Chandrasekhar, A.G.; Duflo, E.; Jackson, M.O. The diffusion of microfinance. *Science* **2013**, *341*, 1236498. [CrossRef]
53. Kim, H.J.; Lee, Y.; Kahng, B.; Kim, I.M. Weighted scale free network in financial correlations. *J. Phys. Soc. Jpn.* **2002**, *71*, 2133–2136. [CrossRef]
54. Onnela, J.P.; Kaski, K.; Kertesz, J. Clustering and information in correlation based financial networks. *Eur. Phys. J. B* **2004**, *38*, 353–362. [CrossRef]
55. Yang, Y.; Liu, Y.; Zhou, M.; Li, F.; Sun, C. Robustness assessment of urban rail transit based on complex network theory: A case study of the Beijing Subway. *Saf. Sci.* **2015**, *79*, 149–162. [CrossRef]
56. Zhang, X.; Ren, Y.; Huang, B.; Han, Y. Analysis of time-varying characteristics of bus weighted complex network in Qingdao based on boarding passenger volume. *Phys. A Stat. Mech. Appl.* **2018**, *506*, 376–394. [CrossRef]
57. Hossain, M.M.; Alam, S. A complex network approach towards modeling and analysis of the Australian Airport Network. *J. Air Transp. Manag.* **2017**, *60*, 1–9. [CrossRef]
58. Fichera, A.; Frasca, M.; Volpe, R. Complex networks for the integration of distributed energy systems in urban areas. *Appl. Energy* **2017**, *193*, 336–345. [CrossRef]

59. Tsolas, S.D.; Karim, M.N.; Hasan, M.F. Optimization of water-energy nexus: A network representation-based graphical approach. *Appl. Energy* **2018**, *224*, 230–250. [CrossRef]
60. Wang, M.; Chen, Y.; Tian, L.; Jiang, S.; Tian, Z.; Du, R. Fluctuation behavior analysis of international crude oil and gasoline price based on complex network perspective. *Appl. Energy* **2016**, *175*, 109–127. [CrossRef]
61. Zhang, W.B.; Tian, L.X.; Yao, Y.; Tian, Z.H.; Wang, M.G.; Zhang, G.Y. Dynamic evolution characteristics of European union emissions trade system price from high price period to low price period. *J. Clean. Prod.* **2019**, *224*, 188–197. [CrossRef]

sustainability

MDPI

Article

Cluster Analysis of Public Bike Sharing Systems for Categorization

Tamás Mátrai * and János Tóth

Department of Transport Technology and Economics, Budapest University of Technology and Economics, Stoczek utca 2, H-1111 Budapest, Hungary; toth.janos@mail.bme.hu

* Correspondence: tamas.matrai@mail.bme.hu; Tel.: +36-20-260-6171

Received: 14 June 2020; Accepted: 3 July 2020; Published: 8 July 2020

Abstract: The world population will reach 9.8 billion by 2050, with increased urbanization. Cycling is one of the fastest developing sustainable transport solutions. With the spread of public bike sharing (PBS) systems, it is very important to understand the differences between systems. This article focuses on the clustering of different bike sharing systems around the world. The lack of a comprehensive database about PBS systems in the world does not allow comparing or evaluating them. Therefore, the first step was to gather data about existing systems. The existing systems could be categorized by grouping criterions, and then typical models can be defined. Our assumption was that 90% of the systems could be classified into four clusters. We used clustering techniques and statistical analysis to create these clusters. However, our estimation proved to be too optimistic, therefore, we only used four distinct clusters (public, private, mixed, other) and the results were acceptable. The analysis of the different clusters and the identification of their common features is the next step of this line of research; however, some general characteristics of the proposed clusters are described. The result is a general method that could identify the type of a PBS system.

Keywords: public bike sharing; cluster analysis; categorization; data collection

---

1. Introduction

According to the UN forecast, the world population in mid-2017 was about 7.6 billion people, and by 2050 it is predicted to reach 9.8 billion. Along with this, urbanization is expected to increase [1,2]. Cycling is one of the fastest developing sustainable transport solutions [3–6]. Modernized and urban lifestyles have faded away physical activity of everyday life and this has resulted in a threat to population health caused by sedentary lifestyles [7]. It is estimated that physical inactivity causes 21–25% of breast and colon cancer and even greater proportions are estimated for diabetes (27%) and ischemic heart disease (30%) [8].

Public bike sharing (PBS) systems, also known as "Public-Use Bicycles", "Bicycle Transit," "Bikesharing", or "Smart Bikes," can be defined as a short-term urban bicycle rental schemes that allow bicycles to be picked up at any self-service bicycle station and returned to any other bicycle station, consisting in point-to-point trips [9]. Basically, people use bicycles on an "as-needed" basis, without the responsibility of the bicycle ownership [10]. Nowadays, different type of PBS systems start to spread all around the world, which can be operated without the docking stations, hence called dockless systems [11,12]. With the spread of public bike sharing systems, it is very important to understand the differences between systems [10,13–16]. Without understanding the differences neither the impact of these systems can be calculated, nor is high-quality decision support possible.

We developed a complete framework during a doctoral research for analyzing, comparing, and categorizing public bike sharing systems, as such a comprehensive system is still missing from the literature [17]. The first level of our framework is to collect data about existing systems and perform a cluster analysis. Then, a SWOT (strengths, weaknesses, opportunities, and threats) analysis for each cluster is compiled based on the examined systems. The third step is to create a benchmark tool, which supports the evaluation of systems. At the fourth level, impact analysis and impact assessment are carried out [18–21].

The present article deals with the clustering of different bike sharing systems around the world (i.e., it concerns the first level of our framework). The lack of a comprehensive database about PBS systems in the world does not allow for a simple comparison or evaluation of the systems [22]. Furthermore, the original goal of the creation of a PBS system is quite often unclear [23]. Without knowing the initial goal, the success of the system cannot be evaluated. A systematic literature review and scientometric analysis was conducted by Si et al. [17] from most of the bike-sharing-related articles between 2010 and 2018 from which it is clear that the researchers main focus was not on business models.

Several articles analyze the value creation of a bike sharing system [10,24–26], although all of them start from the assumption that there are several distinct business models for bike sharing. DeMaio [16] introduced several examples of model provision in his article, but there was no clear definition of the different models. Other articles [24,25,27–29] are using the business model canvas [30] approach or at least some of its elements, but these are not provide an easy to use categorization.

Our initial idea was to apply an unsupervised machine learning algorithm to a dataset, which should lead us to findings related to business models. This approach was applied in other industries like the Spanish scientific journals [31] or electric mobility [32] successfully. The cluster analysis methodology was not up to now applied in the field of PBS business models, but we collected a large dataset, which can be used to this purpose.

The goal of the clustering process is to create groups (clusters of objects) of the dataset, in a way that: (i) the objects in a given cluster are similar as much as possible; and (ii) the objects belonging to different clusters are highly different [33].The cluster analysis usually applied in the domain of spatial studies related to public bike sharing (e.g., [34–37]). In this field, the studies mostly focus on the distribution of bikes or stations.

Our main assumption is that a large proportion (i.e., 90%) of the public bike sharing systems around the world could be classified into one of the four clusters. These clusters are formulated based on the type of the owner and the type of the operator. A SWOT analysis based on this categorization could help PBS project promoters and owners to develop higher-quality systems. The clustering methodology proposed by the authors contributes, among others, to reducing a large number of primary data to several basic categories that can be treated as subjects for further analysis in the public bike sharing domain.

2. Methodology

Our research followed the steps described in Figure 1.

The first step was data collection, which was followed by the initial dataset analysis. Then, the first cluster analysis based on expert opinion was conducted. The statistical tests and regression analysis were applied in order to select the parameters for the second cluster analysis. In the end, both internal and external cluster validation techniques were applied. During the analysis of the results, we compared three scenarios to each other, where different parameters were considered:

- The entire dataset (all collected 64 parameters),
- Selected parameters based on multinomial regression,
- Only operator and owner parameters.

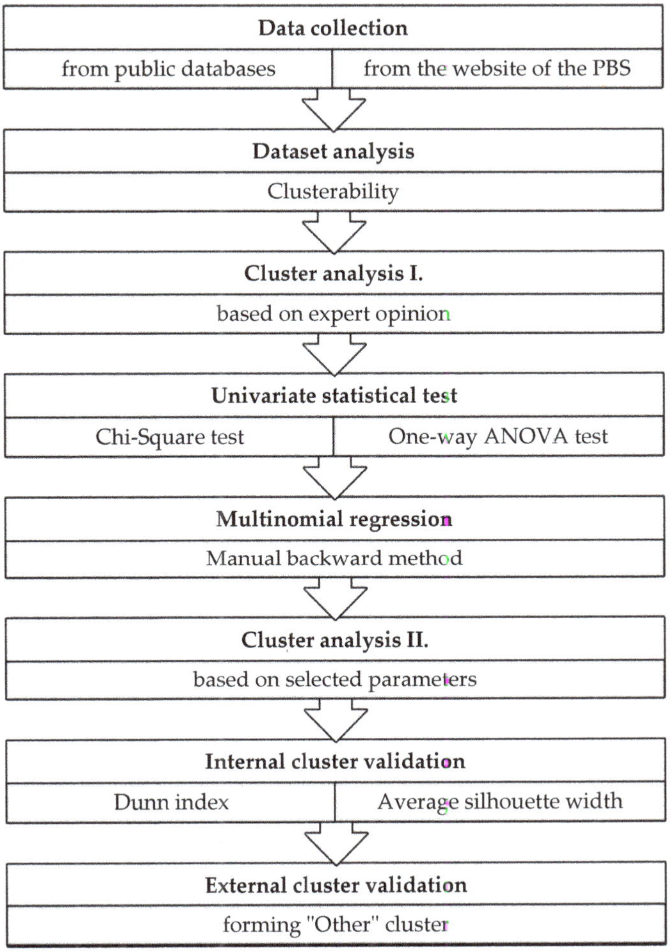

Figure 1. Flow-chart for the cluster analysis.

2.1. Data Collection, Database

The original idea was to collect 80 parameters on 125 systems around the world. The collection of this data was based on open web databases and the webpages of the different bike sharing systems. We assumed that the data of the bike sharing system website are up-to-date and accurate. Our starting point was the collection of systems by Meddin [38]; this database contained 2124 active systems at the beginning of 2019. We selected the 125 systems based on the following criteria:

- 50 systems from Europe, 50 systems from Asia, 5 systems from Australia and New Zealand, and 20 systems from the Americas,
- 3/4 should be docked systems, while the remaining 1/4 is dockless.

After a 6-month-long collection period, we had to reduce the dataset to 64 systems and 64 parameters. We made the decision that to exclude the dockless systems (n = 31) from the analysis and only focus on docked bike sharing. There were several systems (n = 30) where—despite all efforts—we did not reach the minimum viable information. These systems were excluded from the analysis so as to not distort the results. Out of the originally desired 80 parameters, we had to exclude some due to the lack of available data. For example, we intended to gather information about the goals of the different systems, although it was not possible since very few system declare their initial goal, as Ricci pointed out earlier [23].

The final database was grouped around the following main topics:

- Location of the systems (Continent, Country, City, etc.),
- Contextual data (climate, start year of operation, size of the city, size of the service area, population, income, topology of the city, etc.),
- Data about the system (owner, operator, number of bikes, number of stations, etc.),
- Fare system (access fee, usage fee, deposit etc.),
- Data related to the system operations (when it is closed, how a bike can be hired, etc.),
- Derived data (bike density, station density, etc.).

2.2. Dataset Analysis

The first step was to visualize the dataset in a two-dimensional space. As the dataset itself contains several parameters, a principal component analysis (PCA) algorithm was used to reduce the number of dimensions. The algorithm presents the results in a scattered plot diagram, which gives us an easily understandable visual representation of the dataset [33].

There are several methods to calculate the distance between each pair of observations. Gower distance [39] is one of the few measures that are capable of handling both categorical and continuous variables, therefore this method was used for our calculation. The dissimilarity between two variables is the weighted mean of the contributions of each variable. This automatically implies that a particular standardization process is applied to each variable.

The daisy function from the cluster package [40] is suitable for calculating Gower distances in R. The result of computation of these distances is known as a dissimilarity matrix. The Gower distance can be described with the following Equation (1).

$$d_{ij} = \frac{\sum_{k=1}^{p} \omega_k * \delta_{ij}^{(k)} * d_{ij}^{(k)}}{\sum_{k=1}^{p} \omega_k * \delta_{ij}^{(k)}} \quad (1)$$

where d_{ij} is a weighted mean, ω_k is the weight, $\delta_{ij}^{(k)}$ is the 0–1 weight, which becomes zero when the variable $x[,k]$ is missing in either or both rows (i and j) or when the variable is asymmetric binary and both values are zero and in all other situations it is 1, and $d_{ij}^{(k)}$ – k^{th} variable contribution to the total distance

We analyzed the entire dataset from the cluster tendency point of view. During the visual assessment of clustering tendency (VAT approach), we used the following steps:

1. Compute the dissimilarity matrix for the data set using Gower distance.
2. Reorder the dissimilarity matrix so that similar objects get close to one another, which results in an ordered dissimilarity matrix.
3. The ordered dissimilarity matrix is converted into an image for visual inspection.

The color level is proportional to the value of the dissimilarity between observations. The observations in the same cluster are displayed in a consecutive order [41].

After the visual inspection, we also used the statistical method called Hopkins statistic to evaluate clusterability. This method measures the probability if a dataset was generated by a uniform distribution, so it tests the spatial randomness of the data. The calculations are the following:

- Get a random sample from the original real dataset.
- Compute a distance from each point to each nearest neighbor of the original real dataset.
- Generate a random dataset based on uniform distribution with the same variation as the original real dataset.

- Compute a distance from each point to each nearest neighbor of the random dataset.
- Calculate the Hopkins statistics (H) as the mean nearest neighbor distance in the random dataset divided by the sum of the mean nearest neighbor distance in the original real and the random dataset.

The formula of Hopkins statistics can be defined as below (2):

$$H = \frac{\sum_{i=1}^{n} y_i}{\sum_{i=1}^{n} x_i + \sum_{i=1}^{n} y_i} \quad (2)$$

where H is the Hopkins statistics, y_i is the nearest neighbor distance in the random dataset, x_i is the nearest neighbor distance in the real dataset, and n is number of sample points in the dataset.

The null hypothesis is that the original real dataset is uniformly distributed (i.e., there are no meaningful clusters). The alternative hypothesis is that the dataset is not uniformly distributed. (i.e., there can be find meaningful clusters). If the Hopkins statistics is close to 1, we can reject the null hypothesis and conclude that there is significant clusterability. A higher than 75% value indicates a clusterability at the 90% confidence level.

2.3. Clustering Based on Expert Opinion

Our main hypothesis was that most of the PBS systems can be clustered based on the owner type and the operator type. Therefore, during data collection, we used two owner categories: Public and Private, while in the type of operator we created 4 categories: Advertising company, Private Company, Service provider, and Public. Based on these types, we created 4 clusters, which can be seen in Table 1. This categorization was based on the expert opinion of the two authors.

Table 1. Clustering logic based on the operator and the owner.

Cluster Number	Owner	Operator	Comment
1	Public	Advertising company	An advertising company provides services to a public institution, the income for the advertising company might not be realized from the direct user fees, but from some advertising spaces around the city
3	Public	Service provider	A service provider operates a system on behalf of the public institution, the income for these service providers can be an availability payment.
4	Public	Public	A public institution operates the system or sets up a public company for operation, the income is directly from the user fees.
2	Private	Private company	The owner of the system is the same private company as the operator, the incomes are from the user fees and from the advertisements.

2.4. Univariate Statistical Tests

In order to determine which of the 64 parameters should be included in a multivariate regression model, some preselection is required [42]. As the dependent variables are both categorical and continuous, while the independent variable is categorical, we had to use two types of statistical tests. We used the SPSS statistical software for these tests.

We used the Pearson's chi-square test to discover whether there is a relationship between two categorical variables. As all the variables were measured at an ordinal or nominal level (i.e., were categorical data) and both variables consist of at least two independent groups, the test was applicable. The null hypothesis was that Variable 1 (Cluster) is dependent from Variable 2 (all other categorical variables) [43].

We used the one-way ANOVA test to determine if there is a statistical difference between the means of independent groups and the population. The independent variable (cluster in our case) divides the dataset into mutually exclusive groups. We used this test where the dependent variables were continuous. The null hypothesis was that all group means are equal, while the alternative hypothesis was that at least one of the group means is not equal to the others. As the one-way ANOVA is an omnibus test, we do not know which of the groups are different [44].

We selected a higher significance level for both tests not to eliminate the possible candidates from the multivariate regression analysis as it was suggested by Bursac et al. (2008). If the p-value was less than our chosen significance level ($\alpha = 0.25$), we rejected the null hypothesis, and concluded that there is an association between our two variables, therefore we selected the dependent variable for further tests [42].

2.5. Multinomial Regression

We used multinomial logistic regression to predict the nominal dependent variable (cluster of the PBS system) based on the preselected independent variables (both categorical and continuous ones). This also allows to have interaction between the independent variables to predict the dependent one. We used the SPSS statistical software for this.

The applicability of this method is based on the following assumptions:

- The dependent variable is a nominal one and should be mutually exclusive.
- There are two or more independent nominal or continuous variables.
- There should be no multicollinearity.
- There need to be a relationship between any continuous independent variable and the logit transformation of the dependent one.
- There should be no outliers.

We checked the entire dataset for the first 3 assumptions. The multicollinearity assumption was continuously tested for each different model and the rest was automatically tested in SPSS. As the software is not capable of running any automated model selection processes due to categorical variable, we decided to use the backward method and computed each step manually. First, we eliminated those independent parameters where we believed that the relationship to the dependent one would only be statistical, but there is no real reason to be related (e.g., start of operation, country etc.). Then, we added all remaining parameters to the model. We selected the variables with multicollinearity and eliminated one of them based on the significance. We reduced the model until we got a statistically significant one.

2.6. Cluster Analysis for Selected Parameters

We used a clustering method for creating associated groups from the dataset. We used the same method with different parameter sets. We decided to use a k-medoids algorithm, which belongs to the k-means clustering approaches. The most commonly used method is the partitioning around medoids (PAM) algorithm [45]. The PAM algorithm is based on the search of k representative medoids in the dataset and then it clusters the remaining dataset around them. As it does not use the means of the cluster, this method is less sensitive to outliers. The method consists of two phases: The build phase and the swap phase. In the build phase, the first step is the selection of k medoids. The second step is the calculation of the dissimilarity matrix, while the third step is the assignment of each observation into the closest medoids (therefore cluster), based on the calculated distance. In the swap phase, the fourth step is to check if swapping the current medoid of the cluster to any other object in the given cluster is reducing the average dissimilarity. If this happens, the cluster medoid should be changed to the new object and we must go back to the third step and start over again. If none of the medoids change in the fourth step the procedure stops.

We used the R software [46] and the factoextra package [47] to compute the clustering. We used Gower distance to calculate the dissimilarity of the variables.

2.7. Internal Cluster Validation

In order to determine how good the clustering is, we applied internal cluster validation statistics, which uses the internal information of each cluster without external data. All the different statistics measure the compactness, the separation, and the connectedness of the different clusters [40,48].

- The *average distance between clusters* measures the separation of clusters; as the average distance increases, so does the separation.
- The *average distance within cluster objects* measures the compactness of the clusters; as it decreases, the compactness increases.
- The *average silhouette width* also measures the separation between clusters. Each silhouette width coefficient is close to 1 if the object is in the right cluster, 0 means that the object is between clusters and −1 means that the object is entirely in the wrong cluster. So, we want the average to be as close to 1 as possible.
- The *Pearson Gamma* or *normalized gamma* coefficient shows the correlation between distances and a 01-vector where 0 means same cluster and 1 means different clusters.
- The *Dunn index* may be calculated in two ways, but in both cases, the Dunn index should be maximized: In the first version, the minimum separation divided by the maximum diameter; in the second way, the minimum average dissimilarity between two cluster divided by the maximum average within cluster dissimilarity.

In addition to the statistical indexes, we can also use visual methods to explore the results of clustering. The first possibility is to visualize the clusters with PCA in a two-dimensional space. The other option is the silhouette plot, where the diagram shows the silhouette coefficient for each object in an ordered way separated for each cluster.

2.8. External Cluster Validation

During the external cluster validation, we can compare two cluster validation techniques to each other. As in this research we created an expert based categorization as well as the wider parameter-based cluster using PAM method, we can compare the two categorizations to each other. The external cluster validation parameters measure how the external cluster number is matched to the clustered one.

The Rand index [49] measures the similarity between two clusters; its range is from −1 (no common value) to 1 (completely the same). The Variation Index described by Meila [50] is also a valuable tool to measure the similarity of the two clusters.

3. Results and Discussion

The initial phase of our research was to collect the necessary data for our clustering analysis. We shared all the data which collected for this purpose online [51].

We presented the results in three different scenarios below. In the first case, we always made the assessment on the entire dataset. The second presented scenario is the one with the selected parameters based on multinomial regression. The third case is when we only use the operator and the owner parameters. We used Gower distance here as the measure of dissimilarity of the different objects.

We visualized the raw dataset in a two-dimensional space using PCA methodology (Figure 2). The two axes have no specific meaning, they only provide an artificial scale for visualization purposes. Although the scaling and the axes of the figures are not the same, it is viable for comparing the resulting patterns to each other. The dataset with all parameters is less clusterable than the one with selected parameters. In the last one, only four datapoints are visible, since the entire dataset is clustered into these four points.

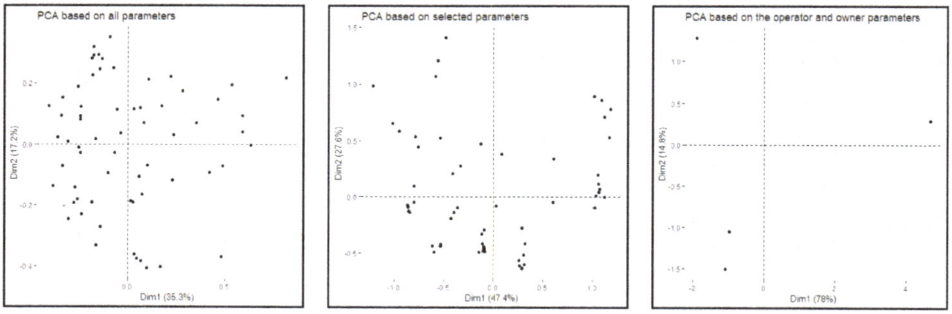

(**a**) (**b**) (**c**)

Figure 2. Dataset visualized with principal component analysis (PCA) (**a**) based on all parameters; (**b**) based on the selected parameters; (**c**) based on the operator and owner parameters.

The same conclusion can be drawn from the heatmap resulting from the VAT approach (see Figure 3) as well as from the Hopkins statistics. The factoextra package [47] implements $H_{alt} = 1 - H$ as the definition of H provided in the methodology section. H_{alt} = 0.2661683 proved to be for all parameters (scenario 1), while H_{alt} = 0.1565344 for the selected parameters (scenario 2). We used the seed number 123 for the calculation of Hopkins statistics.

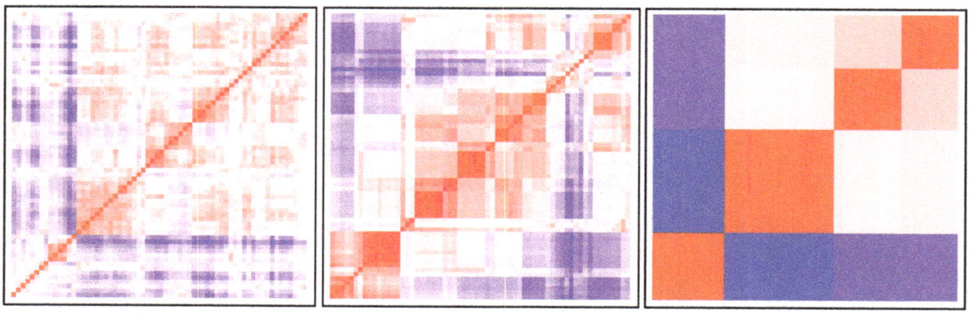

(**a**) (**b**) (**c**)

Figure 3. Visualization of the dissimilarity matrix (**a**) based on all parameters; (**b**) based on the selected parameters; (**c**) based on the operator and owner parameters.

We used the SPSS for the preselection of the parameters for the multivariate regression. Table 2 contains those variables whose p-value is lower than the chosen significance level ($\alpha = 0.25$) in the Chi-square test.

Table 3 contains those variables whose p-value is lower than the chosen significance level ($\alpha = 0.25$) in the one-way ANOVA test. Parameter names can be found next to the dataset description in [51].

After seven iterations with the multinomial regression, the model with the following parameters were selected:

- Factor: Int_PT_fare, Int_user_card, Wout_registration, Deposit_short, Mobile_station.
- Continuous: First_30, Agglo_coverage, Station_density, Long_aff, E_bike_density, Bikes.

We ran the cluster analysis in the R software using the PAM method for all three scenarios with $k = 4$. As shown in Figure 4, the clustering is better with the selected parameters.

Table 2. Results of the Chi-square tests.

Parameter	Value	Asymptotic Significance (2-Sided)
Owner	63.574	0.000
Operator	183.798	0.000
User_card_dock	15.654	0.001
Wout_registration	12.026	0.007
User_card_terminal	11.718	0.008
Int_user_card	11.156	0.011
Deposit_short	11.148	0.011
Continent	19.562	0.021
Credit_card_terminal	9.109	0.028
Mobile_station	8.045	0.045
App_bike	7.999	0.046
Country	126.487	0.051
Hour_closed	6.995	0.072
Diff_station	19.487	0.077
Code_dock	6.133	0.105
Code_terminal	6.106	0.107
Diff_renting_option	25.603	0.109
Deposit_long	6.050	0.109
Int_PT_Fare	5.415	0.144
Code_bike	4.287	0.232

Table 3. Results of the one-way ANOVA tests.

Parameter	F	Significance
Operation	13.773	0.000
Deposit_short_EUR	4.367	0.007
Service_area	3.877	0.013
Deposit_long_EUR	3.793	0.014
E_bikes_dens	3.686	0.016
Station	3.433	0.022
Bikes	3.365	0.024
Docks	3.141	0.031
City_size	2.337	0.082
Population	2.096	0.109
Long_att	1.884	0.142
Long_aff	1.818	0.154
First_30	1.776	0.160
Agglo_coverage	1.555	0.209
Station_density	1.449	0.237

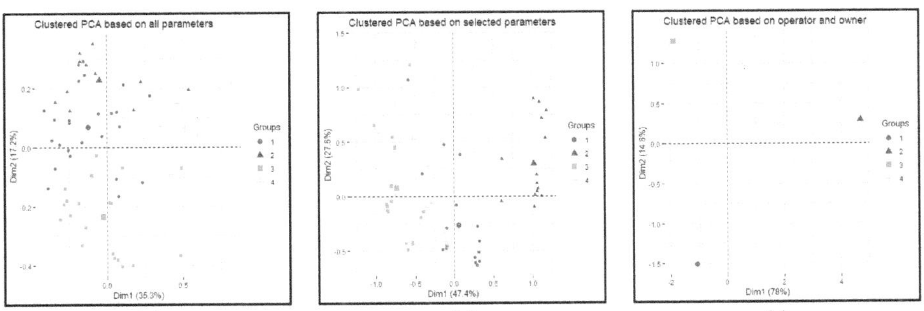

(a) (b) (c)

Figure 4. Clustered dataset visualized with PCA (**a**) based on all parameters; (**b**) based on the selected parameters; (**c**) based on the operator and owner parameters.

The results of the clustering can be described with the average silhouette width of each cluster (Table 4) and the silhouette plots Figure 5). The average silhouette width increases to 1 in the absolutely clustered scenario. The cluster based on the selected parameters has no negative data in cluster 1 and 2, which means a good clustering result.

Table 4. Cluster size and average silhouette width in three different scenarios.

Cluster Id.	Based on All Parameters		Based on the Selected Parameters		Based on the Operator and Owner Parameters	
	Cluster Size	Average Silhouette Width	Cluster Size	Average Silhouette Width	Cluster Size	Average Silhouette Width
1	25	0.04	19	0.33	14	1
2	11	0.10	15	0.60	15	1
3	20	0.12	17	0.22	23	1
4	8	0.17	13	0.24	12	1

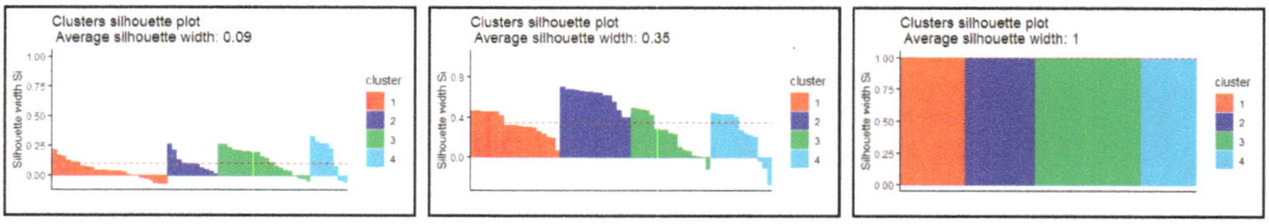

(**a**) (**b**) (**c**)

Figure 5. Silhouette plot for clustering (**a**) based on all parameters; (**b**) based on the selected parameters; (**c**) based on the operator and owner parameters.

The internal cluster validation statistics shows similar results (Table 5). Where it is appropriate, the owner and operator scenario has the theoretical minimum or maximum values. There are some exceptions e.g., the first Dunn index shows worse results for the selected parameter scenario than the all-parameters one.

Table 5. Internal cluster validation statistics for the three different alternatives.

	Based on All Parameters	Based on the Selected Parameters	Based on the Operator and Owner Parameters
Number of observations	64	64	64
Average distance between clusters	0.2543182	0.330922	0.7448368
Average distance within cluster	0.2315138	0.1898745	0
Average silhouette width	0.01317239	0.356468	1
Pearson Gamma coefficient	0.1585032	0.4660481	0.8330978
Dunn index, first version	0.2005322	0.1838311	∞
Dunn index, second version	0.8049529	1.323103	∞

External cluster validation was based on the clustering related to the expert opinion. Therefore, the last column of Table 6 is just for reference purposes; obviously, it has no meaning besides that the statistical calculation is working.

Table 6. External cluster validation statistics for the three different alternatives.

	Based on All Parameters	Based on the Selected Parameters	Based on the Operator and Owner Parameters
Corrected Rand index	0.07757761	0.4368271	1
Variation Index	2.22212	1.217804	4.440892×10^{-16}

The results of the clustering algorithm for the selected parameter can be seen in Figure 6. There is a distinct cluster which is clearly different from the others, while the remaining three are somewhat overlapping.

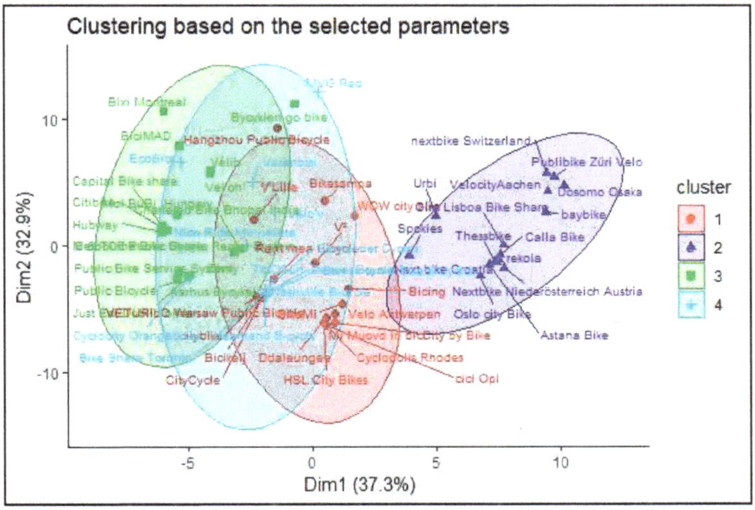

Figure 6. Clustering visualized with PCA based on the selected parameters.

We compared expert-based clustering with parameter-based clustering. Out of 64 systems, 42 clustered into the same cluster as the expert based method suggests and the remaining 22 (35%) were misplaced by the clustering algorithm. Based on our results, we decided that these 22 systems should be categorized as a fifth cluster, called Other. The clustering was correct for the systems, which were owned by a private company: No system with such parameters was missing, and none was misplaced to this cluster. The public systems showed similarly good results, as there were only 3 out of 12 that were misplaced and even these 3 showed some similarity with the selected clusters (e.g., Bixi Montreal is a public system, but has similarity with a commercial one, the two others were Chinese systems, where the difference between a public or a private company is sometimes hard to spot).

Eleven systems were misplaced into the cluster where the operator is supposed to be an advertising company, although all these systems are operated by a service provider. It was also true for the other way around: Four systems out of five were misplaced into the cluster with a service provider as operator, although they have an advertising company operator. This might indicate that the service provider and the advertising company business model and those system characteristics are not as distinct from each other as the purely public and purely private models.

If we only use three big clusters (purely public, purely private, and mixed), we end up only with the miscategorization of 6 systems out of 64. The analysis of the different clusters and the identification of their common features is the next step of this line of research, however some general characteristics of the proposed four clusters can be described here as a starting point. The basis of these descriptions is both the categorization described in Section 2.3 and the results of the clustering exercise:

- Cluster 1 (Public systems): Both the operator and owner of the system are public institutions. The owner is usually a city or one of its companies. The operator can be the same organization or a new one created for this specific purpose. The income is coming directly from the user fees, but usually it requires subsidization. The goal of such a scheme usually is to provide an alternative transport mode or educate the citizens rather than profit making. Typical example: MVG Rad (Munich).

- Cluster 2 (Private systems): Both the operator and owner of the system are private companies, usually the same one. The income is coming from the user fees as well as advertisements. The goal here is clearly profit making, therefore a more cost-efficient operation is envisaged. Furthermore, some limitation related to the network or the users can be applied. Typical example: NextBike Croatia.
- Cluster 3 (Mixed systems): The owner of the system is a public entity (usually a city or a transport operator), while the operator is a private company. The goal of the owner is usually to provide wider transport choices to citizens, while a private company is providing a service. There are two distinct business models for the private company based on the main source of income. In both cases, the user fees are collected for the owner, hence the financial risk from the usage is on the owner side. In the first type, the service provider gets a service fee (availability payment) based on a service level agreement. In the second type, the operator can use different advertising spaces around the city to cover the expenses of the system operation. However, from the system point of view, there are not enough distinct features of these two subtypes to cluster them separately. Typical example: MOL-Bubi (Budapest-subtype 1); Velib (Paris-subtype 2).
- Cluster 4 (other systems): There are several systems that can be categorized by an unsupervised algorithm to one of the above clusters based on the expert knowledge of the authors cannot fit well with them. The reasons are usually hard to spot, but for instance, it can be that a public company design a system with clear profit-making goals, or a private company acts similarly towards a public entity. There are especially some outliers in the Chinese systems, due to the specificities of the country political structure.

Additionally, there are some limitations with the current methodology. As it was stated above, collecting all parameters from the different systems is a time-consuming process. Furthermore, the current data become outdated very quickly as not only new systems emerge, but also the technology changes. This research does not consider dockless schemes, although they have become more and more popular in recent years [52]. At the same time, these systems are almost all profit-oriented, privately funded systems, which can be easily put under the same, distinct category. The other problem with this type of applied, data-driven approach is that an error in data collection can cause problems in interpreting the results. This was one of the reasons that we chose a clustering method that is less vulnerable to outliers.

4. Conclusions

In this study, we developed a method for categorizing public bike sharing systems, which consisted of 8 steps:

1. Data collection,
2. Dataset analysis,
3. Cluster analysis I.,
4. Univariate statistical tests,
5. Multinomial regression,
6. Cluster analysis II.,
7. Internal cluster validation,
8. External cluster validation.

During data collection, we faced several problems, therefore only 64 parameters were collected from 64 systems around the world [51]. The dataset analysis showed that the dataset is clusterable. We selected the operator and owner type as initial parameters for expert based clustering, from which we created four clusters.

We preselected 19 factor type and 15 continuous type parameters for multinomial regression, based on the univariate statistical tests, out of which five factor parameters and six continuous parameters were selected for the final model.

We reran the PAM-approach-based clustering again with the selected parameters, which resulted in better fit than in the case of all parameters. Forty-two systems were assigned to the correct clusters, the remaining 22 were misplaced by the clustering algorithm. Our initial assumption was too optimistic, as only 65% of the systems could be clustered with this method. Thirty-five percent fall into the "Other" category. At the same time, if we only use three main clusters (public, private, and mixed), the error is reduced to 6 systems out of 64.

This can be arguably a correct solution as the service provider and the advertising company business model might not be separated. So, there are four proposed clusters: Public systems, private systems, mixed systems, and other systems.

This article describes the basic characteristics of these clusters, however analyzing the characteristics in details of the different clusters is the next step of this ongoing research. Additionally, future research work will be devoted to overcoming some of the limitations of the presented methodology. One of the main limitations is the data availability; if new, reliable data become available (e.g., usage data, travel pattern, financial data), the current methodology can be expanded to cover this. Another development path can be the inclusion of the dockless schemes to the current analysis, which was neglected in this article due to the lack of reliable data.

This article can help for those who would like to apply the clustering methodology in a different domain. At the same time, it can provide a basis for further research in the public bike sharing domain, as the proposed methodology can be applied for a different set of PBS systems. A newly designed system can be categorized based on the owner and operator, which can help to find similar systems and identify problems and best practices in the earl stage. In other words, this paper can provide significant added value for researchers and academics as well as policy makers and practitioners.

Author Contributions: Conceptualization T.M. and J.T.; methodology T.M. and J.T.; writing—original draft preparation T.M. and J.T.; writing—review and editing, T.M. and J.T.; visualization, T.M.; supervision, J.T. All authors have read and agreed to the published version of the manuscript.

Funding: This research received no external funding.

Conflicts of Interest: The authors declare no conflict of interest.

References

1. United Nations, Department of Economic and Social Affairs Population Division. *World Population Prospects The 2012 Revision Volume I: Comprehensive Tables*; (ST/ESA/SER.A/336); United Nations, Department of Economic and Social Affairs: New York, NY, USA, 2013.
2. United Nations, Department of Economic and Social Affairs; Population Division. *World Urbanization Prospects: The 2018 Revision*; (ST/ESA/SER.A/420); United Nations, Department of Economic and Social Affairs: New York, NY, USA, 2019.
3. Fenton, B.; Nash, A.; Wedderburn, M. *Walking, Cycling and Congestion—Implementer's Guide to Using the FLOW Tools for Multimodal Assessments*; FLOW Consortium: Brussels, Belgium, 2018.
4. Rudolph, F.; Mátrai, T. Congestion from a Multimodal Perspective. *Period. Polytech. Transp. Eng.* **2018**, *46*, 215–221. [CrossRef]
5. Szabó, Z.; Török, Á. Spatial Econometrics—Usage in Transportation Sciences: A Review Article. *Period. Polytech. Transp. Eng.* **2019**. [CrossRef]
6. Pupavac, D.; Maršanić, R.; Krpan, L. Elasticity of Demand in Urban Traffic Case Study: City of Rijeka. *Period. Polytech. Transp. Eng.* **2019**. [CrossRef]
7. Garrard, J. *Active Transport: Children and Young People, An Overview of Recent Evidence*; Vic Health: Melbourne, VI, Australia, 2009.
8. World Health Organization. *Global Health Risks: Mortality and Burden of Disease Attributable to Selected Major Risks*; World Health Organization: Geneva, Switzerland, 2009.
9. Department for City Planning New York. *Bike-Share—Opportunities in New York City*; Department for City Planning New York: New York, NY, USA, 2009.

10. Shaheen, S.; Guzman, S.; Zhang, H. Bikesharing in Europe, the Americas, and Asia: Past, Present, and Future. *Transp. Res. Rec. J. Transp. Res. Board* **2010**, *2143*, 159–167. [CrossRef]
11. Luo, H.; Kou, Z.; Zhao, F.; Cai, H. Comparative life cycle assessment of station-based and dock-less bike sharing systems. *Resour. Conserv. Recycl.* **2019**. [CrossRef]
12. Fishman, E. Bikeshare: A Review of Recent Literature. *Transp. Rev.* **2016**, *36*, 92–113. [CrossRef]
13. Shaheen, S.A.; Martin, E.W.; Chan, N.D. Public Bikesharing in North America: Early Operator and User Understanding, MTI Report 11-19. *Mineta Transp. Inst. Publ.* **2012**. [CrossRef]
14. Fishman, E.; Washington, S.; Haworth, N. Barriers and facilitators to public bicycle scheme use: A qualitative approach. *Transp. Res. Part F Traffic Psychol. Behav.* **2012**, *15*, 686–698. [CrossRef]
15. Fishman, E.; Washington, S.; Haworth, N. Bike Share: A Synthesis of the Literature. *Transp. Rev.* **2013**, *33*, 148–165. [CrossRef]
16. DeMaio, P. Bike-Sharing: History, Impacts, Models of Provision, and Future. *J. Publ. Transp.* **2009**, *12*, 41–56. [CrossRef]
17. Wu, G.; Si, H.; Chen, J.; Shi, J.; Zhao, X. Mapping the bike sharing research published from 2010 to 2018: A scientometric review. *J. Clean. Prod.* **2018**, *213*, 415–427. [CrossRef]
18. Mátrai, T.; Tóth, J. Comparative Assessment of Public Bike Sharing Systems. *Transp. Res. Procedia* **2016**, *14*, 2344–2351. [CrossRef]
19. Tóth, J.; Mátrai, T. Újszerű, nem motorizált közlekedési megoldások. In Proceedings of the Közlekedéstudományi Konferencia, Győr, Hungary, 26–27 March 2015.
20. Mátrai, T.; Mándoki, P.; Tóth, J. Benchmarking tool for bike sharing systems. In *Proceedings of the Third International Conference On Traffic And Transport Engineering (ICTTE), Belgrade, Serbia, 24–25 November 2016*; Cokorilo, O., Ed.; Scientific Research Center Ltd.: Belgrade, Serbia; pp. 550–558.
21. Mátrai, T.; Tóth, J. Categorization of bike sharing systems: A case study from Budapest. In Proceedings of the European Conference on Mobility Management, Athens, Greece, 1–3 June 2016.
22. Médard de Chardon, C. The contradictions of bike-share benefits, purposes and outcomes. *Transp. Res. Part A Policy Pract.* **2019**, *121*, 401–419. [CrossRef]
23. Ricci, M. Bike sharing: A review of evidence on impacts and processes of implementation and operation. *Res. Transp. Bus. Manag.* **2015**, *15*, 28–38. [CrossRef]
24. Winslow, J.; Mont, O. Bicycle Sharing: Sustainable Value Creation and Institutionalisation Strategies in Barcelona. *Sustainability* **2019**, *11*, 728. [CrossRef]
25. Cohen, B.; Kietzmann, J. Ride On! Mobility Business Models for the Sharing Economy. *Organiz. Environ.* **2014**, *27*, 279–296. [CrossRef]
26. Hamidi, Z.; Camporeale, R.; Caggiani, L. Inequalities in access to bike-and-ride opportunities: Findings for the city of Malmö. *Transp. Res. Part A Policy Pract.* **2019**, *130*, 673–688. [CrossRef]
27. Zhang, L.; Zhang, J.; Duan, Z.; Bryde, D. Sustainable bike-sharing systems: Characteristics and commonalities across cases in urban China. *J. Clean. Prod.* **2015**, *97*, 124–133. [CrossRef]
28. Van Waes, A.; Farla, J.; Frenken, K.; de Jong, J.P.J.; Raven, R. Business model innovation and socio-technical transitions. A new prospective framework with an application to bike sharing. *J. Clean. Prod.* **2018**, *195*, 1300–1312. [CrossRef]
29. Gao, P.; Li, J. Understanding sustainable business model: A framework and a case study of the bike-sharing industry. *J. Clean. Prod.* **2020**, *267*, 122229. [CrossRef]
30. Osterwalder, A.; Pigneur, Y.; Tucci, C.L. Clarifying Business Models: Origins, Present, and Future of the Concept. *Commun. Assoc. Inform. Syst.* **2005**, *16*, 1. [CrossRef]
31. Claudio-González, M.G.; Martín-Baranera, M.; Villarroya, A. A cluster analysis of the business models of Spanish journals. *Learn. Publ.* **2016**, *29*, 239–248. [CrossRef]
32. Engel, C.; Haude, J.; Kühl, N. Business Model Clustering: A network-based approach in the field of e-mobility services. In Proceedings of the 2nd Karlsruhe Service Summit, Karlsruhe, Germany, 25–26 February 2016; pp. 11–25.
33. Kassambara, A. *Practical Guide to Cluster Analysis in R: Unsupervised Machine Learning*, 1st ed.; STHDA: Marseille, France, 2017; ISBN 1542462703.
34. Boufidis, N.; Nikiforiadis, A.; Chrysostomou, K.; Aifadopoulou, G. Development of a station-level demand prediction and visualization tool to support bike-sharing systems' operators. *Transp. Res. Procedia* **2020**, *47*, 51–58. [CrossRef]

35. Wei, X.; Luo, S.; Nie, Y.M. Diffusion behavior in a docked bike-sharing system. *Transp. Res. Part C Emerg. Technol.* **2019**, *107*, 510–524. [CrossRef]
36. Jia, W.; Tan, Y.; Liu, L.; Li, J.; Zhang, H.; Zhao, K. Hierarchical prediction based on two-level Gaussian mixture model clustering for bike-sharing system. *Knowl. Based Syst.* **2019**, *178*, 84–97. [CrossRef]
37. Kou, Z.; Cai, H. Understanding bike sharing travel patterns: An analysis of trip data from eight cities. *Phys. A Stat. Mech. Appl.* **2019**. [CrossRef]
38. Meddin, R. Bike Sharing Map. Available online: www.bikesharingmap.com (accessed on 28 February 2020).
39. Gower, J.C. A General Coefficient of Similarity and Some of Its Properties. *Biometrics* **1971**, *27*, 857. [CrossRef]
40. Maechler, M.; Rousseeuw, P.; Struyf, A.; Hubert, M.; Hornik, K. *Cluster: Cluster Analysis Basics and Extensions*, R Package version 2.0.8; R Core Team: Vienna, Austria, 2019.
41. Bezdek, J.C.; Hathaway, R.J. VAT: A tool for visual assessment of (cluster) tendency. In Proceedings of the International Joint Conference on Neural Networks, Honolulu, HI, USA, 12–17 May 2002; pp. 2225–2230. [CrossRef]
42. Bursac, Z.; Gauss, C.H.; Williams, D.K.; Hosmer, D.W. Purposeful selection of variables in logistic regression. *Sour. Code Biol. Med.* **2008**, *3*. [CrossRef]
43. Yeager, K. SPSS Tutorials: Chi-Square Test of Independence. Available online: https://libguides.library.kent.edu/SPSS/ChiSquare (accessed on 31 August 2019).
44. Yeager, K. SPSS Tutorials: One-Way ANOVA. Available online: https://libguides.library.kent.edu/spss/onewayanova (accessed on 31 August 2019).
45. Kaufman, L.; Rousseeuw, P.J. *Finding Groups in Data: An Introduction to Cluster Analysis*; Wiley Series in Probability and Statistics; John Wiley & Sons, Inc.: Hoboken, NJ, USA, 1990; ISBN 9780470316801.
46. R Core Team. *R: A Language and Environment for Statistical Computing*; R Core Team: Vienna, Austria, 2017.
47. Kassambara, A.; Fabian, M. *Factoextra: Extract and Visualize the Results of Multivariate Data Analyses*; STHDA: Marseille, France, 2017.
48. Halkidi, M.; Batistakis, Y.; Vazirgiannis, M. On clustering validation techniques. *J. Intel. Inform. Syst.* **2001**, *17*, 107–145. [CrossRef]
49. Gordon, A.D. *Classification*; Chapman and Hall: London, UK; CRC: Boca Raton, FL, USA, 1999; ISBN 9780367805302.
50. Meilă, M. Comparing clusterings-an information based distance. *J. Multivar. Anal.* **2007**, *98*, 873–895. [CrossRef]
51. Mátrai, T. Public bike sharing systems database. *Mendeley Data* **2019**. [CrossRef]
52. Caggiani, L.; Camporeale, R.; Ottomanelli, M.; Szeto, W.Y. A modeling framework for the dynamic management of free-floating bike-sharing systems. *Transp. Res. Part C Emerg. Technol.* **2018**, *87*, 159–182. [CrossRef]

sustainability

MDPI

Article

Evaluating the Efficiency of Bike-Sharing Stations with Data Envelopment Analysis

Leonardo Caggiani [1], Rosalia Camporeale [2,3,*], Zahra Hamidi [3,4] and Chunli Zhao [2,3]

[1] Department of Civil, Environmental, Land, Building Engineering and Chemistry, Polytechnic University of Bari, Via Edoardo Orabona 4, 70125 Bari, Italy; leonardo.caggiani@poliba.it
[2] Division of Transport and Roads, Department of Technology and Society, Lund University, P.O. Box 118, 22100 Lund, Sweden; chunli.zhao@tft.lth.se
[3] K2—The Swedish Knowledge Centre for Public Transport, Bruksgatan 8, 22236 Lund, Sweden
[4] Department of Urban Studies, Malmö University, Niagara, Nordenskiöldsgatan 1, 21119 Malmö, Sweden; zahra.hamidi@mau.se
* Correspondence: rosalia.camporeale@tft.lth.se

Abstract: This paper focuses on the efficiency evaluation of bike-sharing systems (BSSs) and develops an approach based on data envelopment analysis (DEA) to support the decisions regarding the performance evaluation of BSS stations. The proposed methodology is applied and tested for the Malmöbybike BSS in Malmö, Sweden. This was done by employing spatial analyses and data about the BSS usage trends as well as taking into account transport, land use, and socioeconomic context of the case study. The results of the application demonstrate consistency with the literature and highlight meaningful associations between the station relative efficiency and the urban context. More specifically, the paper provides in-depth knowledge about the preprocessing data, selection of input and output variables, and the underlying analytical approach to be potentially applied to other cases and urban contexts. Overall, the DEA-based methodology presented in this study could assist decision-makers and planners with developing operational strategies for planning and management of BSS stations and networks.

Keywords: BSS station efficiency; data envelopment analysis; spatial analysis in transport; bike-sharing system; bike-sharing station

Citation: Caggiani, L.; Camporeale, R.; Hamidi, Z.; Zhao, C. Evaluating the Efficiency of Bike-Sharing Stations with Data Envelopment Analysis. *Sustainability* **2021**, *13*, 881. https://doi.org/10.3390/su13020881

Received: 31 December 2020
Accepted: 13 January 2021
Published: 17 January 2021

1. Introduction

A bike-sharing system (BSS) is considered an alternative to cars. It is a measure designed to inspire modal shift from short car trips to cycling and intermodal. BSS primary function, typically regarded as a last-mile solution for metropolitan areas, has motivated the investments to provide such services in cities around the world [1,2]. Two main types of BSS exist in cities today, the conventional BSS and the free-floating BSS. The conventional BSS requires the passengers to borrow and return the bicycle from/to fixed stations. Compared to the conventional BSS, free-floating BSS has been recently introduced and it does not have fixed stations for picking up and dropping off bicycles; users are allowed to park the bikes potentially "everywhere" (or within areas with geo-fenced boundaries) as close as possible to their destinations [3]. Both BSS types enable the possibility for the passengers to cycle in a city without owning a bike. This study focuses on conventional BSS.

The first bicycle-sharing scheme was introduced in Amsterdam, the Netherlands in 1965 and it was followed by a station-based BSS implemented in Denmark in 1991 [4]. The first Swedish BSS was a pilot project introduced in Gothenburg in 2005 which operated exclusively in the northern part of the city. The project led to the development of the current BSS in Gothenburg, Styr and Ställ, which was launched in 2010 with 300 bicycles distributed in 20 stations (operating between April and October) [5] and expanded in 2020 to provide 1750 bicycles in 135 stations available throughout the year [6]. Similar systems have also been implemented in other major Swedish cities including Stockholm and Malmö. In 2019, Linköping, the fourth largest city in Sweden, launched the LinBike program as the first Swedish BSS with e-bikes (100 bikes and 17 charging stations). The system, instead of fixed stations, employs recent BSS technologies such as geofencing to define GPS-based virtual zones where users can access or leave the rental bikes [7]. In recent years, in addition to the rapid emergence of these systems in the larger Swedish cities, there has been a growing interest in the development of regional BSSs that could provide viable bike-sharing services across several smaller cities [8].

As a measure to reduce car use and emissions, BSSs offer a set of advantages which explains their widespread adoption by many cities around the world [9,10]. They provide improved accessibility to cycling thus could increase cycling mode share in general [10]; BSSs are a possible last/first-mile mode for connecting to public transport services or can be used as single-mode for shorter journeys. In terms of costs, using rental bikes is often cheaper than renting a car and it is not necessarily more expensive than buying a ticket for public transport for the equivalent travel distance in an urban area. Overall, BSS is considered an affordable, convenient, sustainable, and healthy transport alternative, hence gaining the attention of the cities committed to social, economic, and environmental sustainability [11]. It is worth noting that BSS popularity has not declined due to the recent COVID-19 pandemic but rather has grown considering the reported increase in trip duration and distance compared to the nonpandemic time [12,13], which strengths its potential for future mobility.

Despite the wide-ranging possible benefits and the global popularity of BSSs, there have been cases of financial or operational failures that were mostly caused by mismanagement or under-designed implementations of these systems [14]. Due to inflexible standardized business models or lack of strategies tailored to the local context, such systems typically face issues such as underuse, misplaced bicycles, vandalism and theft, unusable or dysfunctional devices, impractical or unreliable service, sluggish expansion, and lack of adequate cycling infrastructure [15,16]. Previous studies in a Swedish context suggest that the pressure to deliver a commercially viable and profitable service presents challenges to the success of BSSs as it may result in creating sociotechnical configurations that fall short in delivering long-term sustainability benefits [14]. Other studies highlight public acceptance as another relevant factor for the success of BSSs in both Swedish [5] and global contexts [16]. Nikitas [16] maintains that while BSSs are often widely appreciated by users and nonusers, without long-term support and investment from the local authorities such public acceptance may not translate into an actual usage hence failing in achieving sustainability goals.

In general, it is challenging to provide effective BSS in cities since a range of behavioral aspects, as well as technical and organizational factors, can impact the usage of a bike-sharing system. From the BSS planning perspective, station location, the membership, the accessibility of the stations, the number of bikes and racks in each station, the redistribution of bikes during the rush hours, the technology used for building and operation of the system, as well as the attractiveness of the service are considered significant for an effective BSS [4,15,17–19]. In terms of land use, similarly to other travel modes, activity patterns and urban form influence BSS users' travel behaviors. Previous research suggests that population density, job density, as well as cycling infrastructure are all crucial for passengers' choice of traveling by shared bicycles [15,20].

Even though the knowledge about the factors associated with the usage of the BSSs has been fairly studied within the research on shared-bike systems, the topic of station efficiency and its determinant factors has been understudied and under-analyzed [21]. Similarly, in transport practice, typical bike-sharing strategies do not involve scientifically backed and evidence-based measures of station efficiency. In the absence of a station efficiency analysis, it is difficult to identify and eliminate the bottlenecks in a BSS effectively. This was while BSSs have been increasingly planned and implemented to meet mobility needs in an environmentally sustainable way, hence a growing relevance and importance of dealing with the efficiency analysis of shared-bicycle stations.

The objective of this study is to propose and test a method to evaluate the relative efficiency of each shared-bicycle station within a given system and identify its determinants to establish an operational strategy for public BSSs. The proposed method will not only evaluate the efficiency of shared-bicycle stations but also consider the influence of the external variables, thereby contributing to the literature as a methodology for analyzing the efficient operation of BSS stations and the management of the shared-bicycle systems.

The method is proposed and tested through carrying out an analysis of the comparative efficiency of bike-sharing stations, putting forward a general methodology to apply potentially to any context and proposing a numerical application for the city of Malmö, Sweden. The efficiency measures are calculated by a nonparametric approach known as data envelopment analysis (DEA), showing its particular applicability to BSSs. The evaluation result is expected to help in reallocating the existing resources and assist policymakers when deciding where to allocate new stations (planning stages). In this way, it is possible to discover those stations that work better, that are more *efficient* according to the considered parameters, and optimize the system with low costs, i.e., reallocating racks where they are more needed (moving them from less used to more used stations, for instance).

The paper is structured as follows. Section 2 provides the introduction of the proposed DEA methodology from a general perspective, specifying the variables that, according to literature and planning guides, mostly characterize BSSs. Section 3 details the study material and method for the application of DEA to the BSS in Malmö, Sweden, including the detailed description of the explanatory analysis on the dataset to identify a subselection of significant variables. Section 4 presents and discusses the obtained results in Malmö. Section 5 concludes the paper with final remarks and reflections on the proposed approach and its implications.

2. Proposed Methodology

The methodology presented in this section allows at first to define the input and output variables that mostly characterize BSS stations. More specifically, inputs refer to BSS station, built environment, and population-related variables; outputs refer to station usage trends and are based on the trips done by using the system. Data related to BSS usage has to be cleaned and prepared before applying DEA (i.e., removing anomalies that can indicate temporary malfunctioning of the system of broken bicycles/stations) and be able to calculate the efficiency of each station.

Furthermore, to obtain a sufficient differentiation between the efficiency scores and remove from the analysis any potential outliers among the pool of BSS stations (DEA is sensitive to outliers), we propose to use Robust CoPlot (more details in Section 3.4). Robust CoPlot allows choosing inputs, outputs, and stations more significant for the studied context, considering the available data.

After this preliminary data preparation, DEA can be applied to determine the different degrees of efficiency associated with each BSS station. In the following subsections, we provide a more detailed description of the DEA methodology and the inputs/outputs that we suggest to include in the analysis. The data cleaning, elaboration and variable selection are more extensively described when presenting the case study (Section 3).

2.1. Data Envelopment Analysis (DEA)

Mathematically, DEA is a linear programming-based model for evaluating the relative efficiency of a set of decision making units (DMUs) which are homogeneous in the sense that they use the same types of resources (inputs) to produce the same kinds of goods or services (outputs) [22]. DEA evaluates the efficiency of each DMU relative to an estimated production possibility frontier determined by all DMUs. It has been used in several contexts (including education systems, health care units, agricultural production, military logistics, etc.); however, when analyzing the areas approached thus far, energy and transportation have the highest number of applied studies [23].

The application of the method in the transport sector is widespread, especially in the evaluation of airports, ports, railways, and urban transport companies [24,25]. In this paper, we suggest applying DEA to evaluate the relative efficiency of bike-sharing stations: hence, each DMU, in this case, corresponds with a bike-sharing station of a selected system.

To our knowledge, only two recent studies present an application of DEA in the bike-sharing research. The first one, from Hong et al. [21], is applied to a station-based BSS, but it does not include any external variable in the first stage of the model. The second one, from Chang and Wei [26], uses DEA to evaluate and determine the optimal bike-sharing parking points for free-floating bicycles. We believe that the application of DEA to shared systems, although unconventional, is an interesting line of upcoming research that is worthy of further investigations.

DEA does not require any functional relationship between inputs and outputs, although it is important to provide their accurate measurements to apply it successfully. This means that only those variables that could appropriately capture the nuances in the efficiency of the DMUs have to be selected as inputs and outputs.

Since the DEA model employed in this paper relies on the standard input-oriented CCR model [22], the DMUs that, at the result of the application, obtain efficiency values equal to 1 are considered efficient. On the other end, efficiency scores less than 1 denote some inefficiencies of the considered DMU.

Note that to obtain sufficient differentiation between the efficiency scores, the number DMUs should not be too small when compared to the total number of inputs and outputs. In the literature, there is no theoretical treatment that gives a unique suggestion on this issue, but there are different rules of thumb. In this paper, we follow the recommendation by Dyson et al. [27], keeping the number of DMUs greater than or equal to twice the product between the number of inputs and that of outputs.

2.2. *List of Inputs to Include in the Model*

Input variables for DEA represent the aspects that impact the usage of the BSS and travel behavior in general and may explain the differences in the performance of the stations. To include such aspects in the DEA model, they need to be quantified and recorded as a set of variables. Nevertheless, other relevant qualitative parameters, such as weather and seasonal conditions, that may influence the use of BSS network as a whole, could play an important role in the step of interpreting the result.

In this study, a set of input variables were identified based on the review of literature on the usage of the BSS and travel behavior. In particular, the research by Ewing and Cervero [28] and the review study carried out by Eren and Uz [18] were used as key literature for establishing the list of the input variables which are described in Table 1 below.

Table 1. Suggested input variables for measuring the efficiency of the bike-sharing system (BSS) stations using DEA.

Input Variables	Rationality and Description of the Variables
BSS Station Related Variables	
Station age	The variable is relevant for more complex/old systems, particularly if the system has been developed during several stages and groups of stations have been added at different points in time. It can be measured according to the age context of the stations.
Visibility of stations	Visibility of the stations should consider if they are placed next to public transport, or in green areas (i.e., partially hidden by trees/bushes), or in a well-lit environment [29–31]. It can be measured taking into account the involved elements, i.e., by assessing the distance to the bus stops/metro stations, and/or the area of the bushes around the stations, etc.
Density of BSS station	The proximity of BSS stations to each other contribute to the increasing demand for BSS services [32,33]. Different buffers have been suggested for effective BSSs in various contexts [18].

Table 1. *Cont.*

Input Variables	Rationality and Description of the Variables
	BSS Station Related Variables
	Built environment variables
Bicycle infrastructure	Increasing the usage of bicycles requires good bicycle infrastructure [34]. The proximity of BSS stations to the cycling infrastructure impacts the motivation of cycling [35]. This variable can be measured computing the total length of bike lanes within the catchment area of each station, possibly weighted by the type of the bike lanes (e.g., separated paths versus paths shared with traffic).
Street connectivity	Street connectivity reflects the level of infrastructure and traffic safety in the network surrounding a BSS station [36,37]. The variable can be applied or not according to the context, it can be measured calculating the number of intersections and the density of the road network in the area.
Public transport (PT) impact factors	BSS likely promotes the mode share of public transport by serving as a feeder mode for PT [38,39], and vice versa, the provision of the PT service can impact the usage of BSS. Three dimensions related to the public transport can be measured: (1) distance to the PT stations (i.e., bus stops, railways stations); (2) level of provision, which can be measured by the number of stops and stations, number of bus lines, ride frequency; (3) price scheme and approach for accessing to PT and BSS services, e.g., using a smart card for accessing to both services with a fair price is likely to increase the usage of BSS service [28,36,40,41].
Land Use	Land use impacts the demand for trips and affects the choice of travel modes. Residential areas, public and commercial areas, green areas in the city and outskirt, and mixed level of land use are the main parameters for measuring the impact of land use [28,41].
Slope (morphology of the territory)	Slope is one of the main barriers for motivating cyclists to cycle and it strongly affects bicycle usage [18]. It can be measured by assessing the level of slope in specific streets, and the portion of the streets with a certain slope within the city and catchment area [42]. It should be included/considered as a parameter in the general model formulation especially for those cities that have hilly topographies.
	Population related variables
Population size	Population size in the catchment area is an important factor that influences the usage of the BSS service [43]. It can be measured by calculating the number of individuals residing in the catchment area.
Sociodemographic	Age, gender, education, income, employment, ownership of transit mode are the individual factors that most impact the travel behavior [18,41,43]; therefore, these are the parameters within the catchment area suggested to be measured.

2.3. List of Outputs to Include in the Model

The outputs are needed in the model to analyze the performance of BSS stations and calculate generation/attraction factors connected to (the usage of) each station. We propose the following three classes of indicators (five outputs in total), all able to appropriately capture the nuances in the efficiency of bike-sharing stations.

The usage trend of each BSS station shows a cyclical trend, i.e., a pattern that repeats itself after a certain time interval Δt. Here, we suggest calculating the output indicators as daily averages ($\Delta t = 24$ h). Note that the output values have to be normalized according to the number of racks of the largest BSS station in the analyzed system, meaning that each station score is adjusted for the number of racks available at that station (this is the reason why we did not include them among the inputs of the model).

- *Station daily amplitude*: The station daily amplitude is a way to express the daily variation of the number of bicycles in each station. A higher value (higher amplitude) corresponds to a station that is more regularly used throughout the day. We suggest calculating the amplitude of each station using the fast Fourier transform [44]. Fast Fourier transforms are mathematical calculations that convert a domain waveform (amplitude versus time) into a series of discrete waves in the frequency domain. The daily amplitude for each station can be calculated starting from the bicycle variations (usage trends) in ΔT, obtaining their frequency domain using the fast Fourier transform, and assessing the (daily) amplitude value for frequency (cycles/day) = 1.

- *Station prevalence:* This indicator is a proxy for the share of bicycle trips that start (departure prevalence) or end (arrival prevalence) in each station. Given *n* BSS stations in the system, we count the number of trips starting in each stations s_i (picked-up bicycles) during Δt. Then, the stations are ordered from the one that originates more trips (assigning it a score equal to *n*) to the one that originates less trips (score = 1). The scores are assigned progressively, i.e., the second one in the list has *n-1*, the third one *n-2*, and so forth. This process is repeated for every day Δt in the timeframe ΔT of the analysis (since every station may show a different behavior according to Δt), and the daily scores assigned to each station s_i are summed. From these final scores, an average daily value is calculated, dividing the total score assigned to each station for the days Δt included in ΔT. This is the station prevalence calculated for the departures from each station (departure prevalence); the same reasoning can be applied looking at the arrivals (i.e., repeating the calculations for the number of bicycles dropped off in each station during Δt and then obtaining the average arrival prevalence in ΔT).
- *Station attractiveness*: attractiveness is understood as a way to assess how appealing the station is for BSS users compared to the other stations in the network. More specifically, we propose to distinguish an active attractiveness from a passive one, considering the trips that connect each BSS station with the other stations in the network. The unit of these indicators is km/day, associated with each station. To calculate the active station attractiveness, we compute how many trips start in the origin station s_i in ΔT, and we multiply each trip for the kilometers (real network distance, shortest path) necessary to reach the destination station. Then, this value is divided according to how many days are included in ΔT, to obtain an average daily value (km/day). The same (opposite) reasoning is applied to calculate the passive station attractiveness, i.e., how many trips have their destination in s_i in ΔT, computing again an average daily value (km/day). Note that round trips (that is, those trips having both origin and destination in s_i) should not be included in the calculations.

3. Case Study: Malmöbybike

3.1. Context Description and Related Variables

Malmö, with more than 344,000 inhabitants [45], is the third-largest urban area in Sweden. The central-northern part (city center) has the highest population concentration, while smaller urban agglomerations exist in the southwest and eastern parts (Figure 1). As illustrated in Figure 2, the public transportation network follows a similar configuration and is concentrated in areas with higher population density. The cycling infrastructure (Figure 3) includes a bike path network with 520 km of completely separated (from motor vehicle traffic) bike paths and prioritized bike paths shared with other road users [46]. In 2016, Malmöbybike (i.e., the Malmö BSS) started operating with 50 stations in the central areas of the city; during 2019, the network expanded to a total of 100 stations. The recent travel survey conducted in 2018 indicates that the modal share of cycling and public transport in Malmö are, respectively, 25.5% and 25.4% [47].

The spatial data about the population statistics and the built environment characteristics in Malmö were extracted from multiple sources including Statistics Sweden (SCB) [45], Lantmäteriet (Swedish mapping, cadastral and land registration authority) [48] and Trafikverket (Swedish Transport Administration) [49]. The population size data were in a grid format of 100 × 100 m; while other socioeconomic data (such as employment status, education level, income level, etc.) were available with two different cell sizes (250 × 250 m for urban areas and 1000 × 1000 m for suburban areas). Land use data available by Lantmäteriet were employed to map three types of land use namely residential, public and commercial, green areas. Moreover, the transport-related geodata captures the existing cycling infrastructure as well as the public transport network including bus stops and train stations [50].

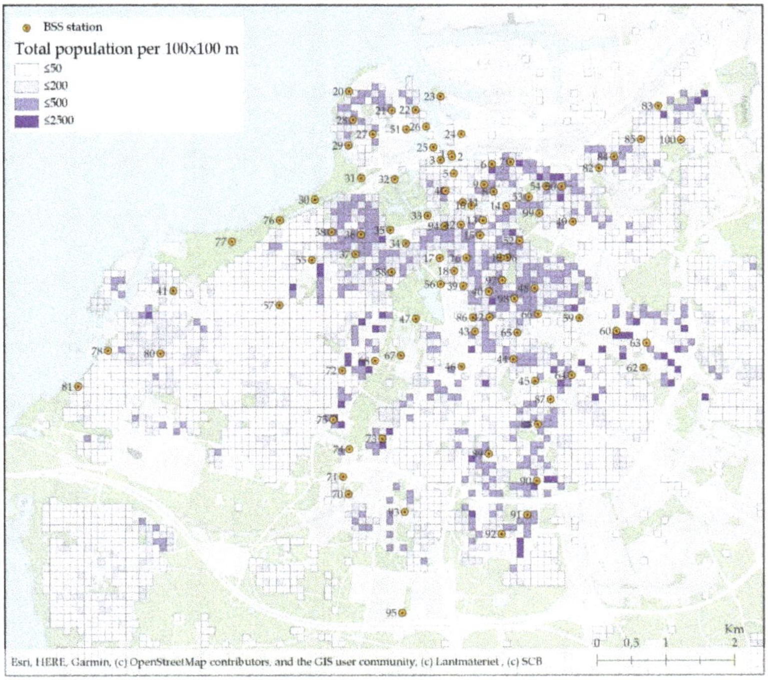

Figure 1. Population distribution in Malmö (number of inhabitants per 100 × 100 m).

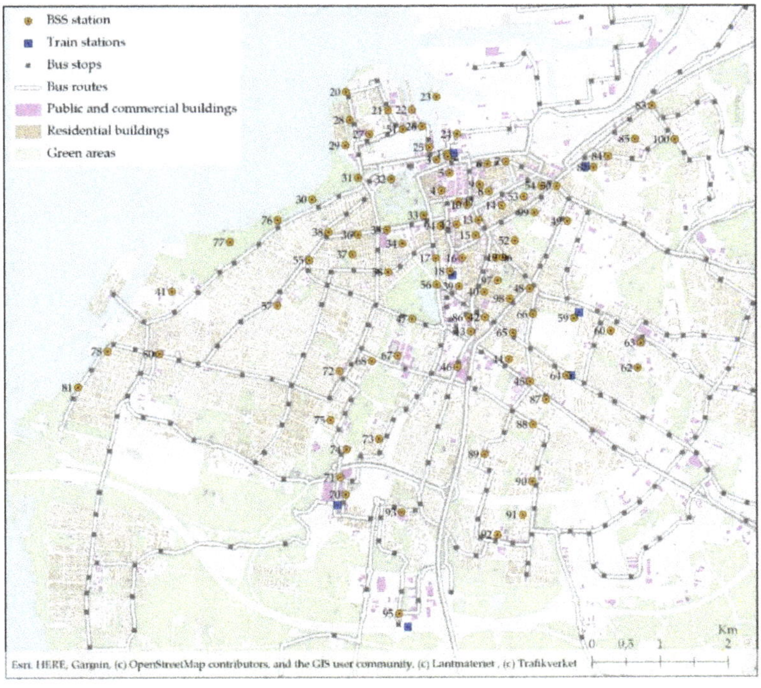

Figure 2. Map of land use and public transport network in Malmö.

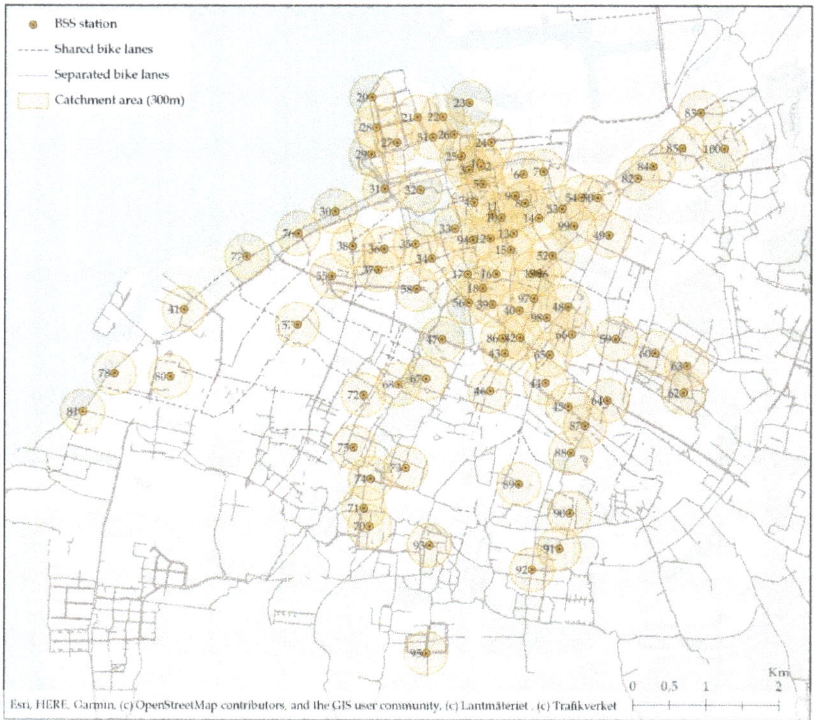

Figure 3. Map of the cycling infrastructure in Malmö.

3.2. BSS Data Description and Preparation

The available dataset on Malmöbybike (January 2018–July 2020) was provided by Clear Channel [51]. It covers all the OD trips in the system during this timeframe, and it makes it possible to have detailed information about the usage of the system, allowing different analyses and data aggregations.

For this application, we selected one-month data, ΔT = June 2020, i.e., the month that has registered the largest number of movements (64,763 trips) in the available dataset. At that date, 100 BSS stations were built and operating in the network. According to Weather spark [52], the average daylight time in June is 17.5 h, with an average temperature of 28 °C; the summer vacation in Sweden usually starts from the last week of June. This background offers an attractive condition for having outdoor activities. Regarding the restriction related to the COVID-19 pandemic, in June 2020 Sweden has restricted the social gathering in restaurants and public spaces (that should not exceed 50 people) and advised everyone to keep social distance in outdoor activities.

Out of the 100 stations, five of them (namely, stations no. 21, 61, 62, 69, and 79) have not been used at all during June; hence, they were removed from the dataset. As far as concerns those stations that have been partially used during the month (i.e., due to malfunctioning in some days), they were excluded only if they had not been used for more than 50% of the observation time (station no. 41 was removed in this stage). The reason is that we were performing a monthly (ΔT) efficiency analysis, determining which stations have been more efficient in the considered period; minor malfunctioning of the stations should be part of the calculations.

An additional data cleaning was performed concerning those bikes that have been used longer than 1 h (i.e., picked up, and not dropped off by 60 min). According to the Malmöbybike terms of use [53], a bike should be used for a maximum of 60 min at a time, and in the case that a bike is not returned within an hour the user would be charged a fine.

Therefore, it is assumed that the trips longer than 60 min are due to bikes that are broken or not functioning correctly. The result of data cleaning was a dataset with 94 stations and 63,338 OD-trips.

Considering the previous research [54] as well as the contextual conditions in Malmö (e.g., the urban area size, the MalmöbyBike coverage area), a radius R = 300 m was considered acceptable to define the *catchment area* (buffer) around each BSS station.

The selected input and output variables are explained and listed in the following Section 3.3.

3.3. Specification of Inputs and Outputs

Based on the input variables suggested in Section 2.2, we used publicly available statistical data to calculate the following list of input variables to apply DEA to the Malmöbybike BSS (Table 2). Note that all the numbers in the input final table are non-negative; the zero values were eliminated by adding a small positive constant, to meet the "positivity" requirement of DEA [55].

Table 2. Description of input categories and variables notations for the DEA applied to Malmöbybike bike-sharing system.

Input Variables	Description of the Variables	DEA Notation
	BSS Station Related Variables	
Density of BSS stations	Number of the BSS stations within 1 km radius from each BSS station	I1 Density of BSS stations (within 1 km)
	Built environmental variables (within the *catchment area*)	
Land use	The area of each land use category. Three types of land use are calculated: residential; public and commercial; green areas [48].	I2 Green areas (km^2) I3 Residential (km^2) I4 Public + Commercial (km^2)
Bicycle infrastructure	The total length of bike lanes. We computed separated bike lanes and shared bike lanes. Separated bike lanes refer to the designated road space clearly defined by signs and regulations that space should be only used for cycling; shared bike lanes are the road spaces shared with pedestrian or cars but recommended for cycling in the interest of creating a more continuous cycling network across the city [49].	I5 Separated bike lanes (m) I6 Shared bike lanes (m)
Public transport impact factors	The number of tracks/bus lines passing by each station/bus stop, to have a proxy of the actual connectivity granted by the public transport system [56].	I7 Number of tracks I8 Number of bus lines
	Population related variables (within the *catchment area*)	
Population size	The average number of residents. Since each catchment area is delimited by a circle, and the population is available in a grid format, we calculated the portion of the area of each element of the grid (square) falling within the circle, and the corresponding share of population assuming a uniform population density in each element of the grid. Provided in grid format (2018), 100 × 100 m [57].	I9 Population size
Age	Population aged 16–64, in grid format (2018), 250 × 250 m urban area, 1000 × 1000 m in suburban areas [57].	I10 Population aged 16–64

Table 2. *Cont.*

Input Variables	Description of the Variables	DEA Notation
	BSS Station Related Variables	
Employment	For the population aged 20–64 (only 2017), two categories of employment are measured: Employed, Unemployed [58].	I11 Unemployed I12 Employed
Household disposable income (2018)	Household disposable income (2018) is measured in four levels: low, medium-low, medium-high, high. The indicator is adjusted considering the consumption units in a household meaning it accounts for different household compositions [59].	I13 Income: low I14 Income: medium-low I15 Income medium-high I16 Income: high
Education	For the population aged 25–64 (2018), four levels of education are measured: Compulsory education, Upper secondary, Post-secondary, less than 3 years and Post-secondary, 3 years or longer, including graduate and postgraduate education [58].	I17 Education: level 1 I18 Education: level 2 I19 Education: level 3 I20 Education: level 4

Although the station age was listed among the suggested input variables (Section 2.2), we did not include this variable for the case study of Malmöbybike. The decision was made since the system is fairly recent, and it has been mainly built in two steps (50 stations in 2016 and 50 more stations in 2019). As previously explained, since DEA provides a relative efficiency of each station, it is important to provide indicators able to capture in a nuanced way the differences among stations from a certain perspective. The (50 + 50) BSS stations have not been opened simultaneously, but gradually over the year(s). Since the information about the exact days/weeks/months of operation of each station is not available and the *Station age* input would have had only two values (the two known years: 2016 and 2019), it was not added to the model.

In the following Table 3, some descriptive statistics (mean, median, minimum, maximum, standard deviation) of the input variables used in this analysis are provided.

Table 3. Descriptive statistics of input variables for the DEA applied to Malmöbybike bike-sharing system (94 DMUs, 20 inputs).

	I1	I2	I3	I4	I5	I6	I7	I8	I9	I10
Mean	12.07	0.04	34,054.29	29,626.77	1880.77	371.21	0.65	5.88	2610.06	1701.69
Median	11.00	0.03	31,174.89	21,718.77	1827.97	261.96	0.00	2.00	2480.00	1634.00
Std. dev.	7.31	0.04	23,024.89	26,015.10	861.93	394.64	2.12	8.19	1853.44	1226.74
Minimum	1.00	0.0001	1.00	42.47	1.00	1.00	0.0001	0.0001	1.00	30.00
Maximum	26.00	0.24	87,317.85	110,052.80	4468.33	1911.09	10.00	50.00	7292.00	5230.00
	I11	I12	I13	I14	I15	I16	I17	I18	I19	I20
Mean	566.44	1155.98	424.91	286.49	266.84	243.72	170.39	454.59	279.72	593.02
Median	533.50	1000.00	406.50	262.00	197.50	163.50	113.00	452.50	240.00	419.50
Std. dev.	429.96	862.35	329.60	215.67	216.59	207.98	164.89	307.23	209.39	505.34
Minimum	1.00	1.00	1.00	1.00	1.00	1.00	1.00	1.00	1.00	1.00
Maximum	1650.00	3451.00	1232.00	943.00	902.00	936.00	851.00	1108.00	845.00	1891.00

Regarding the output calculation, notation and descriptive statistics are summarized in the following Table 4.

Table 4. Variable notations and descriptive statistics of output variables for the DEA applied to Malmöbybike bike-sharing system. (94 DMUs, five outputs).

Output Variables	DEA Notation	Mean	Median	Std. Dev.	Minimum	Maximum
Station daily amplitude	O1	2.01	1.54	1.57	0.12	6.70
Station arrival prevalence	O2	54.07	55.08	24.99	11.77	86.23
Station departure prevalence	O3	54.06	55.85	25.07	10.03	88.07
Station passive attractiveness	O4	45.82	36.19	30.05	5.85	122.65
Station active attractiveness	O5	45.82	38.06	29.34	6.40	122.24

If the calculation of station prevalence (O2 and O3) and attractiveness (O4 and O5) is straightforward following the description of Section 2.3, we provided a more detailed explanation for the assessment of the station daily amplitude O1 using the fast Fourier transform.

Using the Clear Channel database [51] for the Malmöbybike BSS, it was possible to obtain the usage trend of each station in ΔT (June 2020). We did not have any information about bicycle relocations among stations performed by the operator; hence, we made an assumption looking at the available data, which indicates origin and destination of each bike-sharing trip in the network. If the bicycle b_k is in the station s_i at a certain time h_1, but the previously registered trip (ended at h_2) in the system does not have s_i as the destination station, we assumed that relocation happened in the time interval h_1-h_2, more specifically at the midpoint h_3 (so that the time interval h_2-h_3 has the same length of h_3-h_1).

After obtaining the final usage trends (i.e., the bicycle variations) in ΔT taking into account relocations as just described, the fast Fourier transform was applied to convert the time domain waveforms to the frequency domain. The value of each station daily amplitude is the one corresponding to frequency (cycles/day) = 1 (Figure 6).

The following Figures 4–6 show a practical example for two bike-sharing stations in the system.

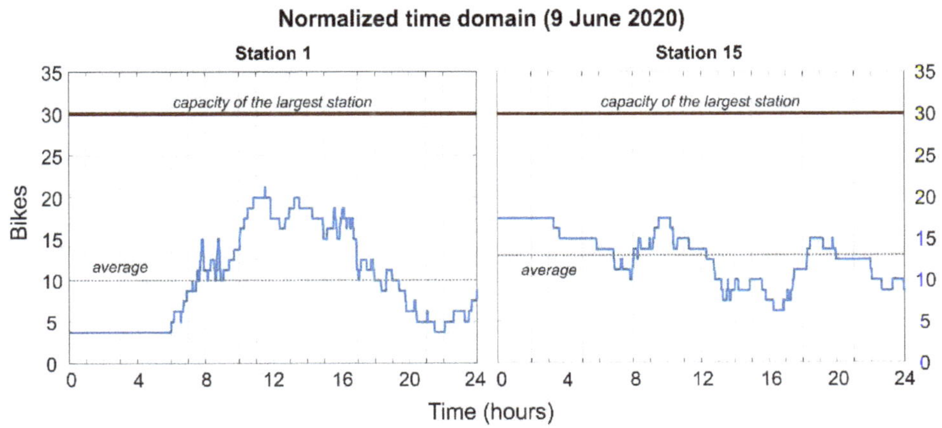

Figure 4. Normalized time-domain (according to the number of racks of the largest BSS station) waveforms for the bike-sharing stations 1 and 15 of the Malmöbybike system on the 9th of June 2020. The daily usage trends (with bicycle variations) can be visualized in blue.

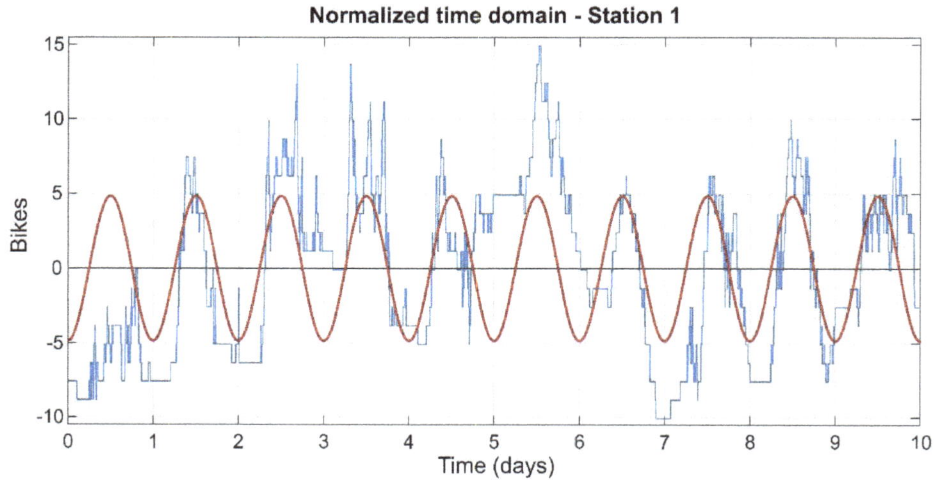

Figure 5. Time-domain waveforms (in blue) for the bike-sharing station 1 of the Malmöbybike system over 10 days included in the analyzed time interval ΔT. A cyclical (daily) periodical behavior can be detected (in red).

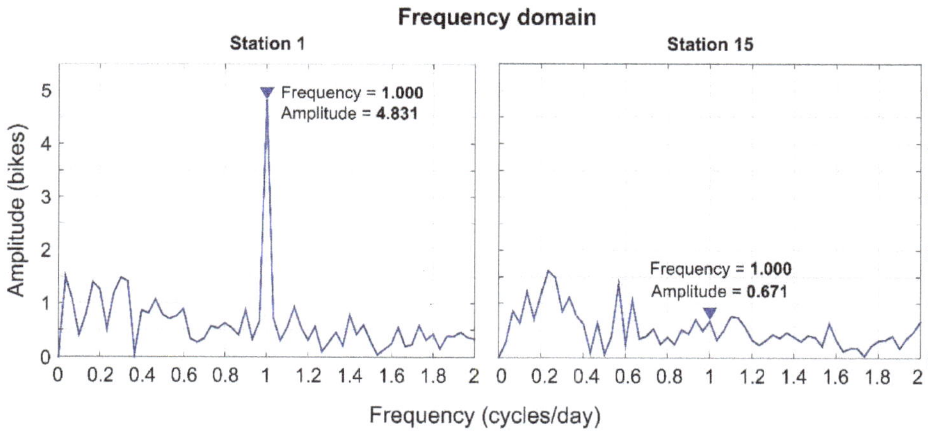

Figure 6. Frequency domain for the bike-sharing stations 1 and 15 of the Malmöbybike system. The station daily amplitude is the one corresponding to frequency (cycles/day) = 1.

Transforming the temporal domain (trend over time of the number of bikes in each station) into the frequency domain allows finding signal periodicity that otherwise would not be easy to identify. Figure 6 highlights a series of peaks representing the different amplitudes of the periodicities identified using the fast Fourier transform. Larger amplitudes show the prevailing periodicities.

We chose to visualize the stations 1 and 15 (in Figures 4 and 6) since they are representatives of the different behaviors that the stations in the Malmöbybike system had during ΔT. Some of them (39.4% of the BSS stations) show a peak corresponding to frequency (cycles/day) = 1 (such as the one shown in Figure 6, Station 1): this means that a typical (daily) periodic behavior (Δt = 24 h) was detected for these stations (look at the corresponding time domain, Figure 4, station 1; Figure 5, over 10 days of observations).

The other stations (look at the representative trend of Station 15, Figure 6) show a smaller amplitude corresponding to frequency (cycles/day) = 1, and peak(s) at lower frequencies (i.e., with cycles longer than 24 h).

The highest frequency peak that was found in the entire database for all the BSS stations is the one corresponding to frequency (cycles/day) = 1, that is, the smallest cyclical temporal unit that can be detected in the system corresponds to Δt.

3.4. Inputs, Outputs, and Station Selection

Since DEA is sensitive to outliers [60] and CoPlot has been often used as a supplemental tool to cluster analysis, DEA and outlier detection methods in the literature [61–63], we decided to suggest its application to the proposed analysis [64–66]. Additionally, this analysis allows reducing the number of variables/DMUs to obtain a sufficient differentiation between the efficiency scores, while following the rule of Dyson et al. [27].

We propose to use Robust CoPlot, an adaptation of multidimensional scaling (MDS) that facilitates rich interpretation of multivariate data [67]; it has the capacity to work better than CoPlot with datasets containing outliers since it is not affected by their presence.

Both CoPlot and Robust CoPlot are able to reduce multidimensional data into a two-dimensional structure, by superimposing two graphs [68–70], simultaneously evaluating associations between variables and between observations. The first map uses a nonmetric version of MDS to spatially represent the distances between observations (in our case, the observations are the DMUs, that is, the bike-sharing stations in Malmöbybike): similar observations are located close to one another, and the goodness-of-fit of this representation is summarized by a single parameter, the Kruskal stress value, σ [71]. The second map, which is conditional on the first, generates vectors that display the relationships among the variables (which, in our case, are inputs and outputs, Section 3.3). Each variable has its vector: if two variables are highly correlated, the vectors describing them are close together, and if their correlation is negative, the vectors describing them go in opposite directions. In this case, we have a goodness-of-fit for each variable, which expresses the goodness of the regression with respect to the observations, and is visualized by the length (magnitude) of the vector (for more details, see [62,67]).

The procedure to identify correlated variables and outliers consists of repeating the Robust Co-Plot several times, removing, before each repetition, respectively, some variables correlated to each other and outliers. DMUs identified by a specific input/output variable are positioned in the same direction of that input/output vector. Correlated variables are represented by vectors having the same directions in space, while DMUs outliers are represented by points positioned far from the center of gravity (the point where the vectors diverge) compared to the other points of the chart.

Figure 7 shows, for example, the Robust CoPlot obtained for the 20 inputs and five outputs described in Section 3.3 in the first repetition.

The DMUs (bike-sharing stations) are graphically represented by red dots: as explained above, similarities between the stations in the dataset are transformed into distances on the map such that similar stations are closer together than less similar stations. The Kruskal stress value σ is 9.18%, showing a goodness-of-fit between good and fair [71].

The inputs and outputs are each represented by a black vector (labeled, with notation and magnitude). Those vectors having the same directions in space are highly correlated, hence we decided to not consider some of them and repeat the procedure, so to apply the DEA only considering the most significant variables.

Note that the analysis to remove the highly correlated inputs and outputs has to be done separately for inputs and outputs. Looking at the outputs (Figure 7), we can see that O2 and O3 are almost overlapping, and O4 and O5 have a similar direction. Hence, we selected O1, O3. and O5 since they seem to be the less correlated outputs and more significant for this dataset. Similar reasoning was applied to the 20 inputs, also taking into account those more meaningful in the Malmö context. The procedure was repeated three times, progressively removing those vectors with higher correlation, obtaining at the end the configuration shown by Figure 8, with 11 inputs and three outputs (the rule of Dyson et al. [27] is satisfied). When removing a variable, there is a rearrangement of the remaining ones in the Robust CoPlot map, depicting the associations in the new configuration.

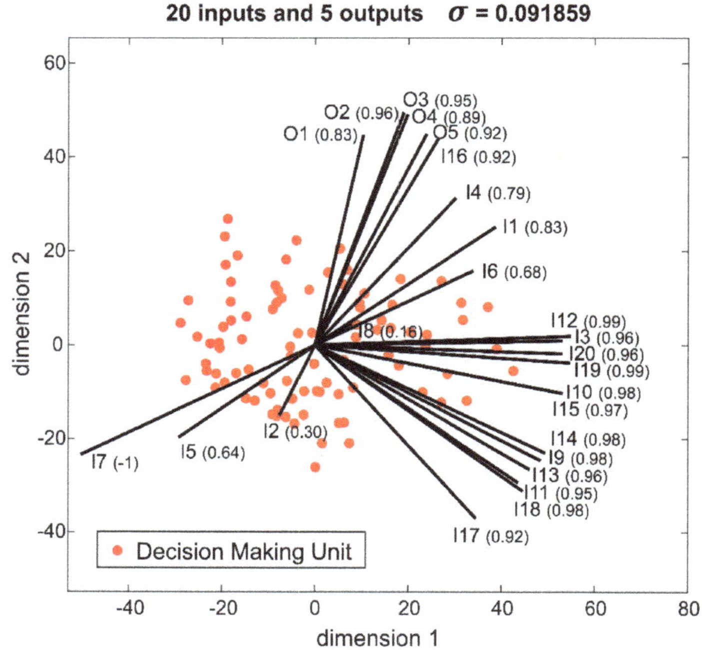

Figure 7. Robust CoPlot map of 25 variables (20 inputs and five outputs) describing the bike-sharing stations of the Malmöbybike system.

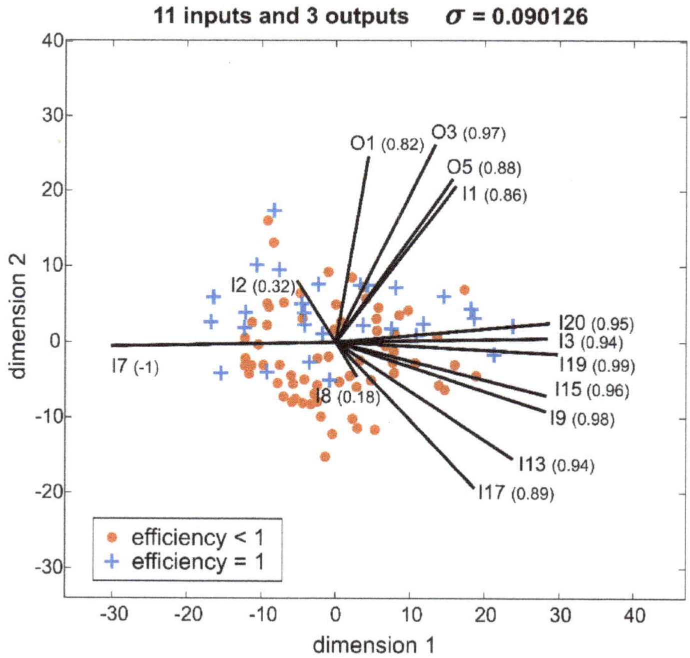

Figure 8. Final Robust CoPlot map of 14 selected variables (11 inputs and three outputs) describing the bike-sharing stations of the Malmöbybike system.

Looking at Figure 8, the efficient DMUs (bike-sharing stations) are represented with a blue cross (28 in total), while the less efficient are represented with a red dot. By eliminating variables with low correlations, the goodness-of-fit is slightly improved and the Kruskal stress value σ results equal to 9.01%. We did not remove any DMU since we did not notice any significant cluster/variable positioned too far from the center of gravity.

The estimated efficiency scores for the remaining DMUs as well as the inputs and outputs are presented and further discussed in the next section.

4. Results and Discussion

Figure 9 presents the efficiency scores yielded by DEA. It shows an overall pattern of the relative efficiency for the BSS stations included in the analysis based on the data from June 2020. As represented by the ramp color (dark green to light yellow), stations exhibit clear differences regarding their efficiency levels. Mapping the efficiency scores across space is helpful for both identifying the most/least efficient stations and comparing a subset of the stations to one another or to the contextual conditions. The variation in the relative efficiency scores demonstrate a meaningful pattern concerning the contextual factors and highlights three categories of stations according to their level of efficiency: (1) the efficient BSS stations (having efficiency = 1); (2) the medium efficient BSS stations; (3) the least efficient BSS stations. Each efficiency category is further addressed and discussed in the following subsections.

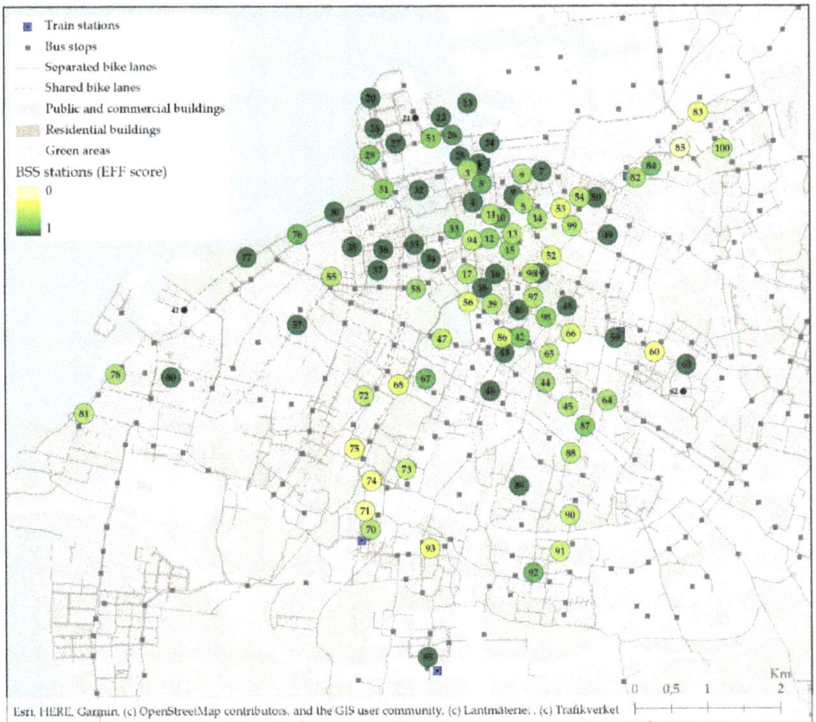

Figure 9. Monthly stations efficiency map for the Malmöbybike system using DEA, June 2020.

4.1. The Efficient BSS Stations

The stations visualized in the darkest green color represent efficient stations, that is, those having efficiency equal to one (for instance, stations no. 30, 18, or 63). Located in different areas of the city, the efficiencies of these stations may be attributed to varying land use contexts. However, the availability of separated cycling lanes indicates that the catchment areas for these stations contain a high level of bicycle infrastructure. This pattern reflects the results found in previous studies [35,72,73]. Consistent with the literature [4,15,20], another common property of this category is the proximity to a green area or an activity center such as commercial buildings, public facilities, and job centers. Considering the spatial properties and the urban context of the station locations, three groups can be identified.

The first group includes stations located in the northern part of the city with good access to nature, e.g., green areas and the waterfront. Trips originated from or ending at these stations are likely made by cyclists visiting the area for outdoor activities. Therefore, the presence of natural resources seems to positively contribute to the efficiency of these stations. This result is similar to the finding reported in the study by Kim et al. [74].

The weather or the seasonal conditions may be considered another external factor contributing to a larger number of trips connected to this area [18]. The last week in June coincides with the start of summer vacations in Sweden, hence the increased usage of shared bikes in areas with a larger share of recreational activities. In general, a combination of the mentioned contextual factors is likely to improve the DEA based evaluated efficiency for these stations.

The second group of efficient stations is located in those areas with a high level of access to public transport (no. 18, 16, 1, 24, 25 next to railway stations), and close to the city center. In this case, the shared bicycles users are likely the passengers who are travelling by public transport, using bikes as first/last-mile feeder mode. Such trips can be both commuting and noncommuting trips, meaning the efficiency of these stations may be less affected by the weather or seasonal conditions in June. Hence, good access to public transport may be a major contributor to the higher efficiency of these stations. This result confirms the findings of previous research suggesting that successful BSSs complement existing transport infrastructure such as public transport [16,75].

The third group includes those stations located in areas further from the city center (if compared with the first two groups), but still in the urban area, e.g., stations no. 46, 57, 89. Most of them are newly added stations that have a station age of less than one year. They are located in areas with high population density, next to the buildings which are public facilities or commercial centers, with good bicycle infrastructure available, and close to bus stops. Previous studies have provided strong evidence that these factors contribute to increased use of BSS services [4,74,76]. In some cases, the density of BSS station within 1 km is rather lower than the average level (farther than 500 m to the next station, e.g., stations no. 57, 89) which could contribute to the efficiency of these stations. The pattern of this group may indicate that, for the less dense areas that are located further away from the city center, locations next to the public facilities and commercial centers where often the bus stops are planned are likely to be the optimal spots for planning efficient BSS stations. At the same time, a good quality cycling infrastructure should be provided.

4.2. The Medium Efficient BSS Stations

Those stations colored in mid-range green are categorized as medium efficient stations, such as stations no. 11, 14, 99. Most of these stations are located in the central area of the city with a higher concentration of public facilities and commercial buildings. The central area is often characterized by a high density in terms of population and jobs which, in turn, implies that it generates or attracts a larger number of trips and, due to the densely built environment, makes traveling by bikes or public transport more convenient than by cars [77]. Similarly, this context may create a higher demand for cycling compared to the peripheral areas, which often motivates the need for a medium/high level of BSS service provision in urban centers.

In the case of Malmö, although these stations did not fall into the efficient station group, many of them have obtained an efficiency score close to 1 (that is, the maximum efficiency score in DEA). Their slightly lower efficiency scores are probably due to the very high density of the BSS stations in the area. Most of the stations in this category have overlapping catchment areas and/or more than one BSS station may be present within their 300 m catchment area. Reducing the density by removing some stations would likely make the remaining ones more efficient. However, given the urban form context in the city central area, the level of the current efficiency of all the stations rather demonstrates the success of the BSS service in the area. In a similar urban context, previous studies have suggested the buffer to be between 200 and 400 m when planning for new stations [18,29,78]. In general, a smaller radius seems to contribute positively to the usage of the service.

The stations located further away from the city center (no. 64, 87, 90) are commonly placed within a maximum of 600 m distance from another. While this radius falls within the reasonable distance range noted in the previous studies, these stations seem to further benefit from proximity to bus stops or large public facilities/commercial buildings. Additionally, despite the lower population size in the peripheral areas, a higher residential density in the form of apartment housings, as opposed to single family houses areas, could be observed in the catchment area of these stations. In general, the observed pattern further confirms the results discussed in the earlier section and previous studies that for the noncentral urban area, the density of the BSS stations, proximity to bus stops, and large public buildings, as well as the high population density could contribute to the efficiency of the BSS stations.

4.3. The Least Efficient BSS Stations

The least efficient stations, visualized in the lightest shade of green/yellow, mostly include those added during 2019, meaning that their age is less than one year (e.g., stations no. 52, 53, 56, 60, 66, 68, 71, 74, 83, 85, 86, 93). Most of these stations are located further away from the city center and in areas with lower population density. While some of the stations (such as no. 71, 74, 75) are located in proximity to small scale public facilities and commercial buildings, the low population density in their catchment areas indicates a low travel demand [28]. Similarly, the cycling infrastructure connected to the stations is rather poor which can significantly impact cycling behavior [18]. Station no. 60 is an exception to this, most likely because it is located next to two other BSS stations (no. 59 and 63) which are, respectively, next to a train station (no. 59) and public facility buildings (no. 63), providing sufficient service demand in the area. In this single case, removing station no. 60 perhaps would make stations no. 59 and 63 more efficient, reducing the running cost in general. This shows how in the areas far away from city center, where the population density is relatively low, even though there is demand due to the connection to the public transport and access to the public facilities or commercial areas, a higher density of BSS stations may not be needed. This issue has been discussed in the previous studies which have suggested different buffers according to the distance between the location of the stations and the central area [11,18].

Station no. 85 is located in a villa house area. The low efficiency of the station may be due to a low population density around the station and to the socioeconomic features of the population living in the catchment area. More specifically, the residents in the area seem to be associated with larger household size and being part of a higher income group who is more likely to travel by car than bicycle [79]. However, we would like to argue that, although the station has low efficiency, from a behavior nudging perspective, it is still worth placing the BSS service here for promoting and normalizing cycling for the groups living in these contexts.

Based on the examination of the three efficiency categories in relation to the urban contexts, the relative efficiencies evaluated by the DEA method seem highly reasonable and well supported by the previous studies.

5. Conclusions

The study proposed and tested a method, the data envelopment analysis, for evaluating the relative efficiency of BSS stations. The method was tested by applying DEA to a Swedish case study, the BSS Malmöbybike in Malmö.

The efficiencies were evaluated starting from a pool of input and output variables supported by literature, reports, and BSS planning guides, with declinations which allow the same procedure to be applied potentially to any city. This method does not only evaluate the efficiency of each shared-bicycle station but also enables the possibility of considering the influence of external variables, thereby contributing to the literature as a methodology for analyzing the efficient operation of shared-bicycle stations and the management of shared-bicycle systems.

The results provided by the application to the Malmöbybike BSS are meaningful in relation to both the specificities of the urban context and the findings reported in previous studies. This seems to indicate that the suggested method can provide a reliable evaluation of the BSS efficiency and that it can be used by decision-makers and planners for developing operational strategies to plan BSS stations and networks.

One of the limitations of the proposed methodology is related to the identification of a specific timeframe under evaluation. If external factors change during the days/weeks/months after the analysis, the calculated efficiencies are no longer correct. Furthermore, the analyst should have a good knowledge of the urban context under examination to be sure to include the most suitable variables capable of representing it.

It is important to point out that the objective of the study is to propose and test the DEA methodology rather than carrying out a comprehensive evaluation for the BSS in Malmö. In future studies, broader spatial and temporal information should be included and compared to achieve a more complete evaluation of the Malmöbybike efficiency. The evaluation should be carried out during the seasons when cycling is the most and the least popular. The differences between and within days, weeks, and months should all be analyzed and compared to gain a good overview of the efficiency for supporting effective operational and planning strategies.

Some of the input variables may be difficult to be expressed in a quantitative way, such as the station visibility. This type of variable could be defined through fuzzy sets. A new formulation of the methodology proposed here which considers a fuzzy DEA approach [66] is currently being prepared.

A further line of research should possibly investigate the inclusion of the suggested methodology in bike-sharing network design models, to take into account the potential efficiency of BSS stations when planning or expanding such a system.

Author Contributions: Conceptualization, L.C. and R.C.; Methodology, L.C. and R.C.; Software, L.C.; Formal Analysis, L.C., Z.H. and R.C.; Writing—original draft preparation, R.C. and C.Z.; Writing—review and editing, R.C., Z.H., C.Z. and L.C.; Visualization, L.C. and Z.H.; Supervision, L.C. and R.C. All authors have read and agreed to the published version of the manuscript.

Funding: This research received no external funding.

Institutional Review Board Statement: Not applicable.

Informed Consent Statement: Not applicable.

Data Availability Statement: Not applicable.

Acknowledgments: The authors would like to thank Clearchannel for their assistance with the collection of the Malmöbybike BSS data.

Conflicts of Interest: The authors declare no conflict of interest.

References

1. Cardenas, I.D.; Dewulf, W.; Vanelslander, T.; Smet, C.; Beckers, J. The E-Commerce Parcel Delivery Market and the Implications of Home B2C Deliveries Vs Pick-Up Points. *Int. J. Transp. Econ.* **2017**, *44*. [CrossRef]
2. Hickman, R.; Vecia, G. Discourses, Travel Behaviour and the "Last Mile" in London. Available online: https://www.ingentaconnect.com/content/alex/benv/2016/00000042/00000004/art00003;jsessionid=ouwsa66uvkmd.x-ic-live-02 (accessed on 4 April 2019).
3. Pal, A.; Zhang, Y. Free-Floating Bike Sharing: Solving Real-Life Large-Scale Static Rebalancing Problems-ScienceDirect. Available online: https://www.sciencedirect.com/science/article/pii/S0968090X17300992 (accessed on 26 March 2019).

4. Kabak, M.; Erbaş, M.; Çetinkaya, C.; Özceylan, E. A GIS-Based MCDM Approach for the Evaluation of Bike-Share Stations. *J. Clean. Prod.* **2018**, *201*, 49–60. [CrossRef]
5. Nikitas, A.; Wallgren, P.; Rexfelt, O. The Paradox of Public Acceptance of Bike Sharing in Gothenburg. *Proc. Inst. Civ. Eng. Sustain.* **2015**, *169*, 101–113. [CrossRef]
6. Lånecyklar, S.; Lånecyklar, S. Available online: https://goteborg.se/wps/portal?uri=gbglnk%3agbg.page.1edf2db0-9d4c-4aa8-9fff-86cd711a0541 (accessed on 10 January 2021).
7. LinBike. Linköpings Elcykelpool. Available online: https://linbike.se (accessed on 10 January 2021).
8. Sveriges Första Regionala Hyrcykelsystem. Available online: /regional-utveckling/samhallsplanering-och-infrastruktur/hallbara-transporter/hyrcykelsystem (accessed on 10 January 2021).
9. Macioszek, E.; Świerk, P.; Kurek, A. The Bike-Sharing System as an Element of Enhancing Sustainable Mobility—A Case Study Based on a City in Poland. *Sustainability* **2020**, *12*, 3285. [CrossRef]
10. Song, M.; Wang, K.; Zhang, Y.; Li, M.; Qi, H. Impact Evaluation of Bike-Sharing on Bicycling Accessibility. *Sustainability* **2020**, *12*, 6124. [CrossRef]
11. Hyland, M.; Hong, Z.; de Farias Pinto, H.K.R.; Chen, Y. Hybrid Cluster-Regression Approach to Model Bikeshare Station Usage. *Transp. Res. Part A Policy Pract.* **2018**, *115*, 71–89. [CrossRef]
12. Li, A.; Zhao, P.; He, H.; Axhausen, K.W. *Understanding the Variations of Micro-Mobility Behavior before and during COVID-19 Pandemic Period*; IVT, ETH Zurich: Zurich, Switzerland, 2020; Volume 1547.
13. Nikiforiadis, A.; Ayfantopoulou, G.; Stamelou, A. Assessing the Impact of COVID-19 on Bike-Sharing Usage: The Case of Thessaloniki, Greece. *Sustainability* **2020**, *12*, 8215. [CrossRef]
14. Henriksson, M.; Wallsten, A. Succeeding without Success: Demonstrating a Residential Bicycle Sharing System in Sweden. *Transp. Res. Interdiscip. Perspect.* **2020**, *8*, 100271. [CrossRef]
15. Nikitas, A. Understanding Bike-Sharing Acceptability and Expected Usage Patterns in the Context of a Small City Novel to the Concept: A Story of 'Greek Drama'. *Transp. Res. Part F Traffic Psychol. Behav.* **2018**, *56*, 306–321. [CrossRef]
16. Nikitas, A. How to Save Bike-Sharing: An Evidence-Based Survival Toolkit for Policy-Makers and Mobility Providers. *Sustainability* **2019**, *11*, 3206. [CrossRef]
17. Liu, J.; Li, Q.; Qu, M.; Chen, W.; Yang, J.; Xiong, H.; Zhong, H.; Fu, Y. Station Site Optimization in Bike Sharing Systems. In Proceedings of the 2015 IEEE International Conference on Data Mining, Atlantic City, NJ, USA, 14–17 November 2015; pp. 883–888.
18. Eren, E.; Uz, V.E. A Review on Bike-Sharing: The Factors Affecting Bike-Sharing Demand. *Sustain. Cities Soc.* **2020**, *54*, 101882. [CrossRef]
19. Nogal, M.; Jiménez, P. Attractiveness of Bike-Sharing Stations from a Multi-Modal Perspective: The Role of Objective and Subjective Features. *Sustainability* **2020**, *12*, 9062. [CrossRef]
20. Chen, E.; Ye, Z. Identifying the Nonlinear Relationship between Free-Floating Bike Sharing Usage and Built Environment. *J. Clean. Prod.* **2021**, *280*, 124281. [CrossRef]
21. Hong, J.; Tamakloe, R.; Tak, J.; Park, D. Two-Stage Double Bootstrap Data Envelopment Analysis for Efficiency Evaluation of Shared-Bicycle Stations in Urban Cities. *Transp. Res. Rec.* **2020**, *2674*, 211–224. [CrossRef]
22. Charnes, A.; Cooper, W.; Lewin, A.Y.; Seiford, L.M. Data Envelopment Analysis Theory, Methodology and Applications. *J. Oper. Res. Soc.* **1997**, *48*, 332–333. [CrossRef]
23. Mariz, F.B.; Almeida, M.R.; Aloise, D. A Review of Dynamic Data Envelopment Analysis: State of the Art and Applications. *Int. Trans. Oper. Res.* **2018**, *25*, 469–505. [CrossRef]
24. Markovits-Somogyi, R. Measuring Efficiency in Transport: The State of the Art of Applying Data Envelopment Analysis. *Transport* **2011**, *26*, 11–19. [CrossRef]
25. Cavaignac, L.; Petiot, R. A Quarter Century of Data Envelopment Analysis Applied to the Transport Sector: A Bibliometric Analysis. *Socio Econ. Plan. Sci.* **2017**, *57*, 84–96. [CrossRef]
26. Cheng, M.; Wei, W. An AHP-DEA Approach of the Bike-Sharing Spots Selection Problem in the Free-Floating Bike-Sharing System. *Discret. Dyn. Nat. Soc.* **2020**, *2020*. [CrossRef]
27. Dyson, R.G.; Allen, R.; Camanho, A.S.; Podinovski, V.V.; Sarrico, C.S.; Shale, E.A. Pitfalls and Protocols in DEA. *Eur. J. Oper. Res.* **2001**, *132*, 245–259. [CrossRef]
28. Ewing, R.; Cervero, R. Travel and the Built Environment. *J. Am. Plan. Assoc.* **2010**, *76*, 265–294. [CrossRef]
29. Croci, E.; Rossi, D. *Optimizing the Position of Bike Sharing Stations. The Milan Case*; Social Science Research Network: Rochester, NY, USA, 2014.
30. NACTO Urban Street Design Guide. Available online: https://nacto.org/publication/urban-street-design-guide (accessed on 11 January 2021).
31. The Bike Share Planning Guide. Available online: https://www.itdp.org/who-we-are/for-the-press/the-bike-share-planning-guide (accessed on 11 January 2021).
32. Bachand-Marleau, J.; Larsen, J.; El-Geneidy, A. Much-Anticipated Marriage of Cycling and Transit: How Will It Work? *Transp. Res. Rec. J. Transp. Res. Board* **2011**, 109–117. [CrossRef]
33. NACTO Walkable Station Spacing Is Key to Successful, Equitable Bike Share. Available online: https://nacto.org/2015/04/28/walkable-station-spacing-is-key-to-successful-equitable-bike-share (accessed on 11 January 2021).

34. Schoner, J.E.; Levinson, D.M. The Missing Link: Bicycle Infrastructure Networks and Ridership in 74 US Cities. *Transportation* **2014**, *41*, 1187–1204. [CrossRef]
35. Dill, J.; Voros, K. Factors Affecting Bicycling Demand: Initial Survey Findings from the Portland, Oregon, Region. *Transp. Res. Rec. J. Transp. Res. Board* **2007**, 9–17. [CrossRef]
36. Hamidi, Z.; Zhao, C. Shaping Sustainable Travel Behaviour: Attitude, Skills, and Access All Matter. *Transp. Res. Part D Transp. Environ.* **2020**, *88*, 102566. [CrossRef]
37. Yang, M.; Zacharias, J. Potential for Revival of the Bicycle in Beijing. *Int. J. Sustain. Transp.* **2016**, *10*, 517–527. [CrossRef]
38. Shaheen, S.; Cohen, A.; Martin, E. Public Bikesharing in North America. *Transp. Res. Rec. J. Transp. Res. Board* **2013**, *2387*, 83–92. [CrossRef]
39. Caggiani, L.; Colovic, A.; Ottomanelli, M. An Equality-Based Model for Bike-Sharing Stations Location in Bicycle-Public Transport Multimodal Mobility. *Transp. Res. Part A Policy Pract.* **2020**, *140*, 251–265. [CrossRef]
40. Fishman, E.; Washington, S.; Haworth, N. Barriers and Facilitators to Public Bicycle Scheme Use: A Qualitative Approach. *Transp. Res. Part F Traffic Psychol. Behav.* **2012**, *15*, 686–698. [CrossRef]
41. Zhao, C.; Nielsen, T.A.S.; Olafsson, A.S.; Carstensen, T.A.; Meng, X. Urban Form, Demographic and Socio-Economic Correlates of Walking, Cycling, and e-Biking: Evidence from Eight Neighborhoods in Beijing. *Transp. Policy* **2018**. [CrossRef]
42. Lu, W.; Scott, D.M.; Dalumpines, R. Understanding Bike Share Cyclist Route Choice Using GPS Data: Comparing Dominant Routes and Shortest Paths. *J. Transp. Geogr.* **2018**, *71*, 172–181. [CrossRef]
43. Maas, S.; Nikolaou, P.; Attard, M.; Dimitriou, L. Examining Spatio-Temporal Trip Patterns of Bicycle Sharing Systems in Southern European Island Cities. *Res. Transp. Econ.* **2020**, 100992. [CrossRef]
44. Cerna, M.; Harvey, A.F. *The Fundamentals of FFT-Based Signal Analysis and Measurement*; National Instruments: Batu Maung, Malaysia, 2000.
45. Statistics Sweden Population by Region, Marital Status, Age and Sex. Year 1968–2019. Available online: https://www.statistikdatabasen.scb.se/pxweb/en/ssd (accessed on 31 December 2020).
46. Malmöstad. Cykelbanor och Bilvägar. Available online: http://miljobarometern.malmo.se/trafik/cykling/cykelbanor-och-bilvagar (accessed on 6 January 2021).
47. Region Skåne Resevaneundersökning 2018 Interaktiv Webbplats. Available online: https://utveckling.skane.se/publikationer/rapporter-analyser-och-prognoser/resvaneundersokning-i-skane (accessed on 31 December 2020).
48. Lantmäteriet GSD-Property Map (SWEREF99TM). Available online: https://www.lantmateriet.se/globalassets/kartor-och-geografisk-information/kartor/e_fastshmi.pdf (accessed on 30 December 2020).
49. Trafikverket NVDB GCM-Vägtyp. Available online: https://www.trafikverket.se/TrvSeFiler/Dataproduktspecifikationer/V%C3%A4gdataprodukter/DPS_E-G/1054gcm-vagtyp.pdf (accessed on 30 December 2020).
50. Hamidi, Z.; Camporeale, R.; Caggiani, L. Inequalities in Access to Bike-and-Ride Opportunities: Findings for the City of Malmö. *Transp. Res. Part A Policy Pract.* **2019**, *130*, 673–688. [CrossRef]
51. Clearchannel Clearchannel.Se. Available online: https://www.clearchannel.se (accessed on 31 December 2020).
52. Average Weather in June in Malmö, Sweden-Weather Spark. Available online: https://weatherspark.com/m/76106/6/Average-Weather-in-June-in-Malm%C3%B6-Sweden (accessed on 31 December 2020).
53. Malmö by Bike. Available online: https://www.malmobybike.se/en/info/about-malmo-by-bike (accessed on 31 December 2020).
54. Kabra, A.; Belavina, E.; Girotra, K. Bike-Share Systems: Accessibility and Availability. *Manag. Sci.* **2020**, *66*, 3803–3824. [CrossRef]
55. Sarkis, J. Preparing your data for DEA. In *Modeling Data Irregularities and Structural Complexities in Data Envelopment Analysis*; Springer: Berlin/Heidelberg, Germany, 2007; pp. 305–320.
56. Trafikverket Geodata-Lantmäteriet, Dataproduktspecifikation-Hållplats. Available online: https://www.trafikverket.se/TrvSeFiler/Dataproduktspecifikationer/V%C3%A4gdataprodukter/DPS_H-K/1011Hallplats.pdf (accessed on 30 December 2020).
57. Statistics Sweden Population Statistics (SWEREF99TM). Available online: https://www.scb.se/en/finding-statistics/statistics-by-subject-area/population/population-composition/population-statistics/ (accessed on 30 December 2020).
58. Statistics Sweden Labour Market and Education Statistics (SWEREF99TM). Available online: http://www.scb.se/en/finding-statistics/statistics-by-subject-area/labour-market (accessed on 30 December 2020).
59. Statistics Sweden Income Statistics (SWEREF99TM). Available online: http://www.scb.se/en/finding-statistics/statistics-by-subject-area/household-finances/income-and-income-distribution/income-and-tax-statistics (accessed on 30 December 2020).
60. Boyd, T.; Docken, G.; Ruggiero, J. Outliers in data envelopment analysis. *J. Cent. Cathedra* **2016**, *9*, 168–183. [CrossRef]
61. Huang, H.; Liao, W. A Co-Plot-Based Efficiency Measurement to Commercial Banks. *JSW* **2012**, *7*, 2247–2251. [CrossRef]
62. Adler, N.; Raveh, A. Presenting DEA Graphically. *Omega* **2008**, *36*, 715–729. [CrossRef]
63. Nath, P.; Mukherjee, A.; Pal, M.N. Identification of Linkage between Strategic Group and Performance of Indian Commercial Banks: A Combined Approach Using DEA and Co-Plot. *Int. J. Digit. Account. Res.* **2001**, *1*, 125–152. [CrossRef]
64. Atilgan, Y.K. Robust Coplot Analysis. *Commun. Stat. Simul. Comput.* **2016**, *45*, 1763–1775. [CrossRef]
65. Atilgan, Y.K.; Atilgan, E.L. RobCoP: A Matlab Package for Robust CoPlot Analysis. *Open J. Stat.* **2017**, *7*, 23–35. [CrossRef]
66. Kuljanin, J.; Kalić, M.; Caggiani, L.; Ottomanelli, M. A Comparative Efficiency and Productivity Analysis: Implication to Airlines Located in Central and South-East Europe. *J. Air Transp. Manag.* **2019**, *78*, 152–163. [CrossRef]

67. Bravata, D.M.; Shojania, K.G.; Olkin, I.; Raveh, A. CoPlot: A Tool for Visualizing Multivariate Data in Medicine. *Stat. Med.* **2008**, *27*, 2234–2247. [CrossRef] [PubMed]
68. Krzanowski, W.J.; Dillon, W.R.; Goldstein, M. Multivariate Analysis-Methods and Applications. *Biometrics* **1986**, *42*, 222. [CrossRef]
69. Borg, I.; Groenen, P.J. *Modern Multidimensional Scaling: Theory and Applications*; Springer Science & Business Media: Berlin/Heidelberg, Germany, 2005.
70. Cox, M.A.A.; Cox, T.F. Multidimensional Scaling. In *Handbook of Data Visualization*; Chen, C., Härdle, W., Unwin, A., Eds.; Springer: Berlin/Heidelberg, Germany, 2008; pp. 315–347.
71. Kruskal, J.B. Multidimensional Scaling by Optimizing Goodness of Fit to a Nonmetric Hypothesis. *Psychometrika* **1964**, *29*, 1–27. [CrossRef]
72. Caulfield, B.; Brick, E.; McCarthy, O.T. Determining Bicycle Infrastructure Preferences-A Case Study of Dublin. *Transp. Res. Part D Transp. Environ.* **2012**, *17*, 413–417. [CrossRef]
73. Mateo-Babiano, I.; Bean, R.; Corcoran, J.; Pojani, D. How Does Our Natural and Built Environment Affect the Use of Bicycle Sharing? *Transp. Res. Part A Policy Pract.* **2016**, *94*, 295–307. [CrossRef]
74. Kim, D.; Shin, H.; Im, H.; Park, J. Factors Influencing Travel Behaviors in Bikesharing. Available online: https://nacto.org/wp-content/uploads/2012/02/Factors-Influencing-Travel-Behaviors-in-Bikesharing-Kim-et-al-12-1310.pdf (accessed on 30 December 2020).
75. McBain, C.; Caulfield, B. An Analysis of the Factors Influencing Journey Time Variation in the Cork Public Bike System. *Sustain. Cities Soc.* **2018**, *42*, 641–649. [CrossRef]
76. Kaltenbrunner, A.; Meza, R.; Grivolla, J.; Codina, J.; Banchs, R. Urban Cycles and Mobility Patterns: Exploring and Predicting Trends in a Bicycle-Based Public Transport System. *Pervasive Mob. Comput.* **2010**, *6*, 455–466. [CrossRef]
77. Ewing, R.; Cervero, R. Travel and the Built Environment: A Synthesis. *Transp. Res. Rec. J. Transp. Res. Board* **2001**, *1780*, 87–114. [CrossRef]
78. El-Assi, W.; Salah Mahmoud, M.; Nurul Habib, K. Effects of Built Environment and Weather on Bike Sharing Demand: A Station Level Analysis of Commercial Bike Sharing in Toronto. *Transportation* **2017**, *44*, 589–613. [CrossRef]
79. Maurer, L.K. Feasibility Study for a Bicycle Sharing Program in Sacramento, California. Available online: https://nacto.org/wp-content/uploads/2012/02/Feasibility-Study-for-a-Bicycle-Sharing-Program-in-Sacramento-California-Maurer-12-4431.pdf (accessed on 30 December 2020).

sustainability

MDPI

Article

How to Save Bike-Sharing: An Evidence-Based Survival Toolkit for Policy-Makers and Mobility Providers

Alexandros Nikitas

Huddersfield Business School, University of Huddersfield, Huddersfield HD1 3DH, UK; a.nikitas@hud.ac.uk

Received: 26 April 2019; Accepted: 31 May 2019; Published: 9 June 2019

check for updates

Abstract: A new mobility ethos is needed for cities looking to overcome the problems that have been accumulated for decades by a transport paradigm that prioritises automobiles over people. Bike-sharing, a measure promoting voluntary travel behaviour change, could be part of a refined toolbox that will help in forging this new ethos. Despite a rapid emergence during the last handful of years, as evidenced by 1956 operational local schemes and approximately 15,254,400 self-service public use bicycles across the world, bike-sharing has been attracting negative attention lately. Tens of schemes have closed down, deemed as financial or operational failures, stigmatising bike-sharing's brand and putting the future of the concept itself in jeopardy. However, discounting bike-sharing as flawed may not be fair or accurate. This paper identifies a formula of success for bike-sharing operations based on a state-of-the-art case study analysis, which is supported by primary data evidence from two survey-based studies in Sweden and Greece. This paper suggests that residents in cities hosting or looking to host bike-sharing schemes are usually very supportive of them but not always likely to use them. More importantly, this paper delivers some key policy and business lessons that form a survival guide for effectively introducing and running public bicycle schemes. These lessons include, among others, the need for: tailoring the system design and expansion strategy according to the host city needs, city-operator and commercial partner synergies, more bike-friendly infrastructure and legislation, pro-active cultural engagement, anti-abuse measures, enhanced fleet management and realistic profit expectations.

Keywords: bike-sharing; public bicycles; shared use mobility; cycling; sustainable transport

1. Introduction

An excessively car-centric transportation system has been the cornerstone of urban development for decades now; a cornerstone associated with adverse effects on social, economic and environmental sustainability. These effects that characterise the century-long reign of the conventional, fossil-fuelled, human-led, privately-owned car include increased traffic congestion, climate change, local air and noise pollution, road injuries and casualties, obesity and chronic diseases, decline in physical activity and a loss of social engagement [1–3]. All of them will continue to increase without appropriate interventions [4–6]. A new mobility ethos is needed, therefore, for cities looking to effectively address these challenges, which, among other actions, will require policy-makers and mobility providers to promote voluntary travel behaviour change via powerful active and shared transport initiatives [7]. This will push forward the transition to a paradigm change [8] that will help cities plagued by conventional unsustainable thinking to transform into smart cities [9,10]. Bike-sharing, the greenest form among shared use mobility interventions, can potentially be a key to this transition. Bike-sharing is defined as *a system referring to the provision of affordable short-term access to locally branded bicycles on an 'as-needed' basis that could extend the reach of public transit services to final destinations and be a door-opener for increased bicycle usage* [11].

Bike-sharing, typically framed as an ideal and innovative first- and last-mile travel solution for congested metropolitan environments [12,13], and less often tested for the context of smaller and medium cities [14], is now more relevant than ever before [15]. The recent game-changing introduction of stationless and free-floating schemes where bicycles are unleashed from docking stations, allowing users an extra layer of mobility freedom [16], expanded the traditional dock-based bike-sharing provision and transformed the concept into one with a door-to-door travel potential. Bike-sharing has, from that point onwards, emerged rapidly across the globe, with 1956 operational schemes and approximately 15,254,400 self-service public bicycles and pedelecs in use [17], as an easily implementable modal shift mechanism that is iconic for the host city's commitment to sustainable transport operations [18,19].

However, at the same time bike-sharing, especially with a growing number of reckless and under-designed implementations of dockless systems, is attracting negative attention, with numerous schemes shutting down as huge financial or operational failures. This has forced usually smaller bike-sharing companies to go bankrupt and even put some iconic ones in danger. The withdrawal of Mobike from Manchester and Ofo from London in the UK, reflecting the retreat of the two strongest and most recognisable international bike-sharing operators, and the closure of Bluegogo, China's third-biggest bike-sharing company with 20 M users and £226 M in deposits at its zenith, underline a burning need. Research needs to revisit what makes bike-sharing services efficient at an operational level and how local schemes can transform into valuable smart city ingredients that flourish in the long term.

This paper aims to re-invent the formula of long-term success for bike-sharing operations by developing policy and business lessons that will help policy-makers and transport providers in establishing and managing these schemes more effectively, and, to a degree, other micromobility systems. This paper provides support to the argument that the concept of bike-sharing is not a failing one that will be dismissed as a whole in the near future, but a resourceful one that is being harshly "tested" at present by substandard decision-making, planning and management, on many occasions.

Hereafter, the paper provides a literature review, a description of the chosen methodology, a concise analysis of the key results, a discussion that will identify lessons learnt and suitable policy and industrial recommendations, and a conclusion section that will epitomise the genuine contribution of this work to academia, policy and practice.

2. Literature Review

Cycling is being promoted as a travel mode with the capacity to increase sustainable transportation, alleviate environmental problems and support healthier lifestyles [20]. It is viewed as an integral part of any urban mobility policy intervention package looking to increase the quality of life in modern cities [21]. One natural promoter of cycling is bike-sharing [22], a sustainable mobility initiative that dates back to 1968, with the famous "White Bicycles" system in Amsterdam [23]. Bike-sharing has boomed over the last decade, especially with the introduction of dockless fleets, because of its potential to complement public transport services and provide affordable and green first- and last-mile travel solutions. As a matter of fact, bike-sharing has, according to recent research [24], experienced the fastest growth of any mode of transport in the history of the planet.

The possible benefits of bike-sharing include, among others, flexible mobility, emission reductions, physical activity benefits, reduced congestion and fuel use, individual financial savings and support for multimodal transport connections [25]. These benefits are difficult to measure per se because bike-sharing schemes are parts of complex multimodal transport systems and cannot be easily examined in isolation. Lately however, there is more and more evidence about some quantifiable bike-sharing gains. For example, recent research concludes that bike-sharing programmes have significant positive externalities in terms of economy, energy use, the environment and public health [26]. A panel dataset analysis of 96 urban areas in the US that have introduced local schemes argues that these schemes can reduce peak-hour congestion, at least in the context of larger cities [27]. The promotion of bike-sharing scheme use among car drivers can significantly increase health benefits [28]. Bike-sharing has great potential to reduce energy consumption and emissions based on its rapid development; in 2016 alone, bike-sharing in Shanghai saved 8358 tonnes of petrol and decreased CO_2 and NOX emissions by 25,240 and 64 tonnes, respectively [29]. A review of the existing evidence suggests that bike-sharing can increase cycling levels but, in principle, needs complementary pro-cycling measures and wider support for sustainable urban mobility in order to thrive [30].

In contrast, according to Nikitas (2019) [11], problems typically associated with bike-sharing include: (i) systematic underuse, (ii) vandalism and theft, (iii) lethargic and complicated planning procedures, (iv) sluggish or over-ambitious scheme expansion usually referring to station-based and station-less systems, respectively, (v) a one-bike-fits-all business model which may not be ideal for all populations and environments, (vi) strict cycling regulations including compulsory helmet use for some countries (e.g., Australia), which make schemes impractical or at least reliant upon a supporting rent-a-helmet mechanism, (vii) political friction if local authorities (or residents) are unwilling to forsake street parking space for bike stations, (viii) road traffic safety concerns generated by the co-existence of bicycles with other modes in a heavily car-dominated environment, but also the pedestrians versus bicycles narrative in mixed usage situations and (ix) lack of adequate cycling infrastructure (e.g., bike lanes, cycle paths, parking racks) that could complement and promote a bike-sharing scheme.

There is an extensive and still growing body of research about the traditional dock-based bike-sharing [31], but the sustainability performance of new-generation dockless schemes has not been thoroughly examined as of yet [32]. A Dutch study found that station-based business models are well institutionalised but harder to scale up, while the dockless model has the greatest scaling potential if institutional adaptations and geo-fencing technologies are successfully implemented [33]. This study also reported that peer-to-peer sharing is likely to remain a niche with special purpose bikes. In dock-based schemes, satisfaction is flawed by a set of factors such as the mechanics of the bikes, the picking and dropping system, and the apps used to organise the service [34], while dockless shared mobility models are potentially useful in generating participation but face substantial technical, analytical, and communication barriers [35]. The rapid expansion of dockless bike-sharing may be the reason behind a large-scale renaissance of the very concept since 2016, however, this is coincident with the serious oversupply of bikes [36] in many programmes. In these lesser used schemes, the presence of unusable bicycles increases the level of user dissatisfaction [37], while overuse is one of the most important issues faced by bike-sharing systems operating in China [38]. Bike-sharing usage is a key determinant of scheme success and has been studied by the literature. Some researchers [39] suggest that weather-related variables, land-use and built environment characteristics have significant effects on the overall bike-sharing usage. Larger bike fleets are associated with higher usage but also with diminishing marginal impact (i.e., each new bike may induce fewer new trips), while high land-use mixtures, easy access to public transportation, more supportive cycling facilities, and free-ride promotions positively impact the usage of dockless bikes [40]. Increasing bike-sharing fleet size does not necessarily increase performance according to other studies (e.g., [41]). Bike-sharing systems should also be carefully developed to appreciate the quality and timely interplay between the physical design of the system and the provision of services being offered [42].

Despite some progress, there is a need for more research as a means of evaluating the dynamic concept of bike-sharing as it spreads across the globe. On the one hand, there is still relatively limited evidence on existing schemes as to whether they achieved their objectives [30] and secured passenger satisfaction, since the vast majority of implemented schemes has not been closely examined as of yet; the reported findings in the literature are generated from less than 10% of the existing schemes. On the other hand, there is a paucity of research with large numbers of people who are not bike-share users, notwithstanding that these studies are critical to bike-sharing user growth [43]. The economic efficiency of public bikes is also being heavily questioned [44]. Theft, vandalism and uncivilized usage of public bicycles is a massive barrier, especially for dockless fleets, that can single-handedly endanger some schemes' economic viability [45,46]. A better understanding of the planning and management of bike-sharing as a key ingredient of smart cities, in conjunction with citizen's perception, is yet to be attained [47]. This can be partly achieved by generating a comprehensive evaluation framework and indicator system for researchers and operators to improve the sustainability performance of bike-sharing in practice [48].

The present study aims to fill in this important research gap and provide a comprehensive policy, business and academic guide that will indicate how bike-sharing schemes can be effectively safe-guarded from future failures. This work underlines that bike-sharing can survive its recent failures and be a tool supporting efforts, to some modest degree at least, aiming to restructure the still car-dominated mobility paradigm.

3. Methodology

This work adopts a two-stage methodology for generating evidence-based findings that could be widely representative and generalisable. It is a research effort analysing primary, secondary and news feed data that allows the author to develop an empirical and theoretical understanding of how bike-sharing can address current flaws and inefficiencies and re-establish its reputation as a pragmatic long-term transport alternative.

The first part of the work is based on quantitative evidence from two survey-based studies held in Gothenburg, Sweden (mid-sized city, ~520,000 residents) and in Drama, Greece (small city, ~60,000 residents), looking at acceptability and usage determinants. The two surveys, conducted and analysed by the author, captured road user attitudes for Gothenburg's 1000-bike station-based scheme Styr & Ställ that was established in 2010, and for a hypothetical small-scale forthcoming city-centre scheme in Drama. The combined sample refers to 1175 respondents that could be users, future users or non-users of the local schemes; 535 respondents in Sweden and 640 respondents in Greece. Both samples allowed the generation of statistically significant correlations for the two survey studies and thus sample sizes are deemed satisfactory. The choice of the cities aimed to: (i) address the severely understudied context of bike-sharing acceptance for non-metropolitan urban environments and (ii) understand differences between cities like Gothenburg with an established pro-cycling culture that are familiar with shared transport initiatives versus cities novel to shared use mobility that only recently started to develop pro-cycling initiatives, like Drama. Another factor that makes this comparison interesting is the contrasting contexts of Northern Europe (higher salaries, higher GDP, lower temperatures, more rainfall) and Southern Europe (lower salaries, lower GDP, higher temperatures, less rainfall). Some aspects of this work, including detailed statistical analysis, have been covered extensively in [1] and [11] for Gothenburg and Drama, respectively, but they have not been compared systematically with each other and contrasted with the latest bike-sharing developments that have aspired a sense that public bicycle schemes are losing their edge.

The two surveys have almost identical formats and thus contain questions and sections easily comparable with each other. The survey in Drama has 19 main questions organised in four parts, referring to: the respondents' general travel behaviour choices; their views on bicycles and cycling; their attitudes towards public bicycles and their suitability for the city of Drama; and their socio-demographic characteristics. The survey in Gothenburg had similar questions organised in the same order as the questionnaire in Drama, with the addition of an extra thematic part regarding the respondents' actual public bicycle experience. Five-point Likert-scales were used to record responses, varying from strongly agree to strongly disagree with a neutral mid-point. The time needed for completing the survey in both cases was approximately 10 to 12 minutes. Since pre-notification [49] and financial incentives [50] enhance response rates, both were adopted. The incentive was an entry into a prize-draw, whilst several reminders through social media alerted the volunteers to fill in the questionnaire. Also, the introduction of each survey discussed why its completion could be a meaningful and timely task for the respondents. The surveys were both available to the public in an online form; hard copies were also used in Drama to make sure that groups not having access to computers or the internet (which is far more typical in Greece than Sweden) are represented. The response rate for the hard copy part of the study in Drama was 13%.

Perhaps more importantly though, this paper identifies, lists and discusses for the first time the key findings of an extensive state-of-the-art analysis of the latest developments in bike-sharing operations, using exemplary bike-sharing case studies from across the world. Some of these case studies refer to successful implementation and service delivery stories, and others to systems that failed to establish commercially viable operations capable of adding value to the image of the city hosts. The originality and value of this case study approach is highly significant since the academic literature of local success and failure stories is extremely limited to date; this work analyses in depth, using for the first time academic lenses, these anecdotes as reported by local media and technical literature.

This unique mixed method approach allows the delivery of some important theoretical and empirical lessons, valuable to transport academics, urban policy-makers and mobility providers. These could be applicable to a broader context, at least for similar urban environments and bike-sharing schemes to the ones negotiated in the present work.

4. Results

4.1. Primary Data Analysis

This part of the article provides evidence that if bike-sharing is introduced, operated and promoted adequately, it can be acceptable in small and medium sized cities (including those that have yet to establish a pro-cycling culture), even for people not intending to ever use them. Usage or intended usage rates were, in contrast, significantly smaller than acceptance rates. The two common key usage barriers refer to road safety concerns and the lack of adequate cycling infrastructure. It is safe to argue that bike-sharing is not for everyone to use and not a relevant or feasible choice for many trips. Acceptability, however, is argued to be equally, if not more, important than strict usage-related criteria for measuring success; people's willingness to "allow" their cities to invest more in bike-sharing was very high (and was relatively unrelated to actual or potential usage of their local scheme) due to bike-sharing's potential to add to their city's capacity to be sustainable. This overwhelming local community acceptance should therefore underpin a drive to create win–win synergies between bike-sharing providers and city hosts, since the underusage of schemes should signify that these might not be commercially successful when only supported by subscriptions and short-term rentals, and need the addition of other streams of income to prosper.

These are valuable lessons providing robust evidence that bike-sharing is not a fundamentally flawed concept that somehow emerged because of a rush to establish relatively affordable sustainability-centric mechanisms; it is a sound shared use mobility initiative that in some cases suffers from mismanagement, a lack of long-term vision and ability to create channels with the city hosts and local communities, and unrealistic profit expectations.

4.1.1. Studying Bike-Sharing in Gothenburg

Gothenburg's dock-based Styr & Ställ is assessed as being a worthwhile institution by more than 90% of the city's residents, despite usage rates reflecting some level of participation to the scheme by only a quarter of the study's sample [1,51].

More specifically, the vast majority of the respondents believed that Styr & Ställ is an affordable travel mode with the capacity to promote healthy living, improve road traffic conditions, make cycling more popular, complement the city's other public transport services and help the city becoming more liveable. Also, 85.3% of them recognised bike-sharing's potential to make people's travel behaviour less car-dependent. The most important finding of the survey however was that the participants acknowledged the significance of bike-sharing for their city; 92.4% agreed or strongly agreed that the scheme is good for Gothenburg and 93.5% disagreed or strongly disagreed with the notion that

Styr & Ställ is a sub-standard transport initiative. Even the respondents that had never used it before or self-reported no (or little) intention to bike-share in the future were positive towards the scheme. Many respondents were also supportive of the scheme's further expansion through more bike-sharing investments by the local authorities, and considered that the scheme is a viable public service for the city; 86.5% and 96.1% agreed or strongly agreed with these notions, respectively. Also, only 6.5% of the people that had actually used the scheme considered that there is something fundamentally wrong with it. It was almost a unanimous decision for Gothenburg respondents that their local scheme is a good addition to their city's transport system that needs to continue its operations and expand further to more destinations.

Despite these high acceptability rates, the majority of the respondents rarely used the scheme even as a secondary travel option; 76.8% of the respondents stated that they never use the scheme while only 2.8% use it as their main mode choice. Nonetheless, longitudinal data collected for four full operating years of Styr & Ställ (based on respondents' self-reporting capacity) indicate a small but distinct annual increase in the number of participants that used the scheme, especially for those using it as a main or a secondary travel alternative to their typical modal choice. This underuse indicates that there is, in theory, a massive untapped potential for utilising, in real usage terms, the scheme's wide acceptance [1].

The reasons for which the respondents in Gothenburg do not bike-share frequently or at all have also been captured (each respondent was limited to give only two reasons when filling in the survey). A considerable 14.9% of the respondents answered that limited road safety could act as an obstacle for them in using bike-sharing, while 30.9% felt that a lack of good bike-sharing-related infrastructure was an issue. Improving safety and enhancing bike-sharing and pro-cycling infrastructure, therefore, are of vital importance for the attractiveness of a local scheme. Another 16.9% of the respondents thought that cycling was not ideal for the city, and 11.5% that bike-sharing was not convenient for all purposes. Nevertheless, the most popular response (i.e., 41.1%) was that respondents had no need to bike-share because they had their own bicycle. Figure 1 presents the full set of these results. A detailed statistical analysis solely focused on this study is presented in the author's recent work [1,51].

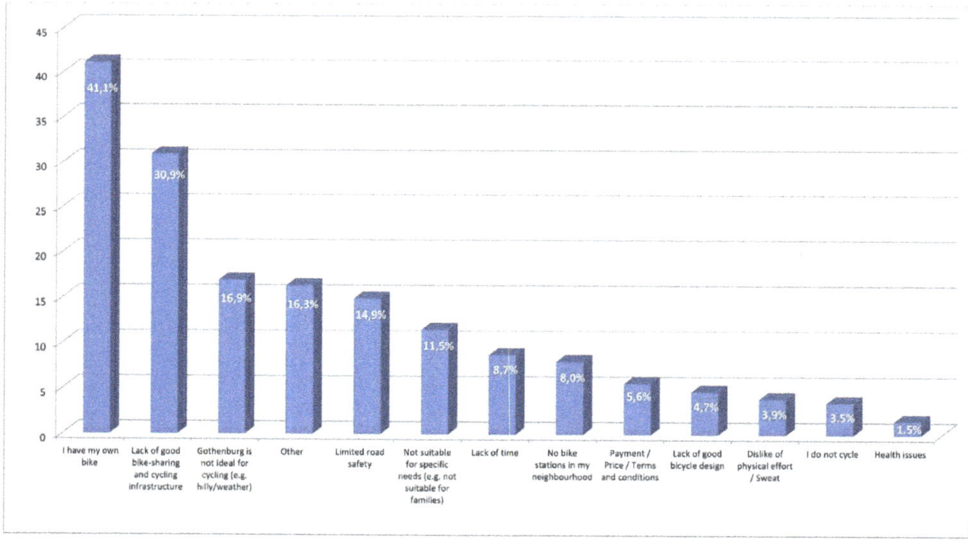

Figure 1. Reasons for not using bike-sharing frequently or at all (Gothenburg, Sweden).

4.1.2. Studying Bike-Sharing in Drama

The survey results from Drama were in many ways very similar to Gothenburg's key findings. Close to 90% of Drama's respondents thought that bike-sharing could be a good scheme for their city, despite lower usage intention rates that could have been still significantly exaggerated due to unintended optimism bias [11].

Overall, Drama's respondents recognised that bike-sharing could be beneficial for their city. The vast majority of them considered that bike-sharing is a sustainable modal option that could improve road traffic conditions, complement other means of public transport, offer an inexpensive transport option for the society, promote wellbeing, make cycling a more popular travel choice and reduce people's reliance on automobiles. However, perhaps the strongest finding, directly referring to the public acceptance of an eventual scheme, was people's disagreement to the notion that "public bicycles constitute an investment that they would not like to see being materialised". In absolute numbers, 86.5% of the respondents disagreed or strongly disagreed with this notion; only 5.1% agreed or strongly agreed and 8.4% were neutral respectively. The respondents self-reported relatively high levels of potential usage with expected usage on a somewhat frequent basis being 46.9%, and "rarely or never" being 31.5%. Still, these rates, which may be too optimistic, were considerably smaller than the reported acceptance rates.

The main reasons according to the respondents of Drama for being reluctant to cycle and potentially use public bicycles are principally associated with the lack of bike-friendly urban infrastructure and the feeling that currently there is only limited road safety for cyclists. Almost one in every two respondents made the case for each of these two specific answers, making clear that physical and cognitive barriers associated with the way a cyclist is hosted in one's respective urban environment constitute the key in giving up the ideas of cycling and bike-sharing. Well-designed bike-sharing infrastructure therefore and the establishment of a safer pro-cycling urban environment would be critical for making a local scheme a more attractive choice. It is true that the construction of bike lanes, bike roads and bicycle racks has only recently been initiated in the city of Drama; thus, these specific attitudes were well justified [11]. Dislike of physical effort was chosen by 12.5% of the respondents, lack of time by 13.8% and cycling and bike-sharing not being suitable for the city by 21.9%. Figure 2 presents the full set of these results. A detailed statistical analysis solely focused on this study is presented in the author's recent work [11].

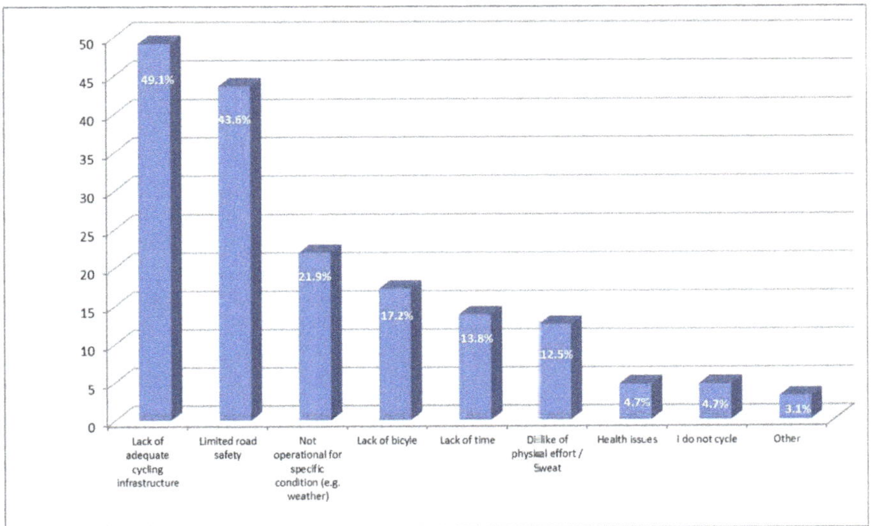

Figure 2. Reasons for not cycling and being reluctant to eventually bike-share (Drama, Greece).

4.2. Case Study Analysis

This part of the analysis concentrates on the examination of successful and unsuccessful bike-sharing applications as a means of identifying best and worst practice examples. This will allow the author to generate a set of policy and industrial recommendations that would inform the people responsible for local bike-sharing systems about the things that they need to do and need to avoid when establishing and running them.

Perhaps the most critical problem for conventional bike-sharing schemes is that station-based operations do not allow door-to-door convenience; traditional dock-based bike-sharing assumes that users can rent a bike from one of the existing bike-sharing stations and return it either to the original station or another station in a different location after using them. Schemes like Seattle's (U.S.) Pronto paid the price (among other reasons, including local politics, hilly topography and mismanagement) for its poorly and sparsely placed docking stations, and the lack of a systematic and incremental expansion strategy, by ceasing its operations [52]. Research found that subscribers used Pronto bikes more at stations that had more scheduled bus trips nearby; thus, bus–bike integration helps in promoting bike-sharing, but at the same time some users may shift to using buses during peak hours and rainy weather [53].

Over the last five years, many operators are actively trying to solve this door-to-door problem by providing station-less, maintenance-free, intelligent bicycles that lock and unlock through the use of mobile applications (i.e., dockless bike-sharing). This transition is led mainly by a few Chinese bike-sharing start-ups including Mobike and Ofo, a pair of operators with a combined valuation likely to have exceeded the £3bn threshold in 2018, since they have been supported by two of the biggest internet giants, Tencent and Alibaba.

Mobike has bike-sharing operations in 200 cities and 16 countries around the world and is responsible for as many as 30 million daily trips, with a fleet of 8 million bikes including electric options [54]. As of July 2018, Ofo claimed to have around 15 million bikes in operation in more than 300 cities across 22 countries, as well as 250 million global users, however, since then it has considerably scaled down its operations outside China [55].

Introducing, with an unprecedented speed, hundreds of schemes of this new breed of door-to-door bike-sharing systems, in China, wider Asia and finally over the last two to three years in North America, Australia and Europe, the homeland of conventional public bicycle programmes, has the power, in theory, to transform the world of cycle hire. This transformation can replicate to a degree the one that saw Uber and other similar ride-sourcing initiatives changing the car-sharing landscape by replacing many of the trips associated with carpooling and taxi-related services. Nevertheless, this monumental, and in many cases somewhat rushed, "embracement" is neither unproblematic nor without a fair share of early fiascos and overwhelming question-marks that have recently disrupted the rise and somewhat stigmatised the public image of bike-sharing.

4.2.1. Stories of Failure

In November 2017, China's third biggest bike-sharing company and the first dockless bike-share system to launch in the U.S, Bluegogo, went bankrupt, creating for the first time a dark cloud over the future of dockless bikes. The bankruptcy of a company that grew at an incredible pace to compete with Mobike and Ofo, peaking at 350,000 bicycles in China alone and poised to conquer San Francisco, raised concerns that there are simply too many bikes at very low prices on offer and insufficient demand. Bluegogo collapsed leaving vast bike-share graveyards that challenged the "sustainability" and "eco-friendly" value of the dockless bike-sharing concept, after falling $30 million in debt and struggling to repay customer deposits, with the surviving bikes being sold for as low as $5 per piece [56]. Bluegogo's chief executive apologised for this collapse saying that he had been "filled with arrogance" [57], which practically meant that the company had unrealistic revenue expectations, expanded too soon on an unsustainably big scale and was mismanaged.

Wukong Bicycle, a minor Chinese start-up of 1200 bikes in the notoriously hilly Chinese city of Chongqing, went out of business after only six months of operations in July 2017, since 90% of its bikes were lost, presumed either missing or stolen. This was the direct consequence of the operators' fatal mistake to not install GPS devices in their fleet [58]. Beijing-based bike-sharing firm 3Vbike also went bankrupt in June 2017 after losing more than 1000 of its bikes in just four months of operations; not having its own mobile app to track the bikes and having to depend on a cycle location tracking function on its WeChat page was the key reason for this downfall. This scheme also failed to gain traction due to limited fundraising; the owner had to purchase the bikes himself in the absence of other investors [58].

The operations of the Mobike Manchester scheme, the first of its kind in the UK, launched in June 2017, is another story of misfortune for the concept of dockless bike-sharing. In a span of 15 months, Mobike had to cease its operations in Manchester because of "unsustainable" losses from theft and vandalism, making this the first time the Chinese operator has abandoned a city because of anti-social behaviour. Manchester Mobikes have been found dumped in canals and bins, vandalised, and others have been stolen, making the company representatives suggest that the system has been "misunderstood" [58] and that "the learning is already being put to good use for creating a more suitable scheme in the not too distant future" [59].

Ofo, in an even more dramatic fashion, was forced to withdraw from most of its UK market including the cities of Norwich, Sheffield and Oxford, and as of January 2019 from London which was home to 3,000 Ofo bicycles. The company admitted that its UK business was loss-making and needed to move to a different direction [60]. Perhaps this retreat proves again that supply should not exceed demand and that international bike-sharing concepts should adapt to the city host for local applications. In the case of London, there was also direct competition with a well-established and popular dock-based system that has strong links with the city.

Similarly, Hong Kong-based bike-sharing operator Gobee shut down its operations in France in February 2018 after suffering what the company called a "mass destruction" of its fleet. Gobee, which had 2000 bikes in Paris alone and claimed around 150,000 users across the country, reported that 3400 of the company's bikes have been damaged and more than 1000 have been stolen [61]. Gobee pulled out of Belgium for similar reasons meaning that this was not an isolated country-specific issue.

OBike, another start-up with international reach, also exited from some of its key markets, with the most important ones being in Singapore and Melbourne, Australia, during 2018. These cities having a long track record of supporting environmentally friendly urban growth are now trying to ensure responsible bike-sharing use, introducing stricter licensing regulations that prevent visual pollution and unsustainable public space intrusion. Complying with these regulations, and in other cases with the helmet rules that govern bicycle use in Australia, can be expensive as it requires investments in technology, security, and management. Operators like OBike may sometimes decide that this extra investment is excessive, so they concentrate on other cities with softer regulations.

Another obvious case of mismanagement is the so-called Velib-gate, where a city project with a new operator intending to enhance the very popular local scheme in Paris, France went seriously wrong, creating huge technical and political issues. A notable dysfunction was the cyber-attack that Bycyklen suffered in Copenhagen, Denmark, which saw all its smart bikes being hacked. This caused serious operation disruption, since the only way to make the bikes usable again was to reboot all 1850 of them individually.

These fiascos have generated concerns about the long-term viability of bike-sharing and mainstream allegations that there is now a body of evidence proving that bike-sharing has been opportunistic or even failing as a concept. Recent research [62] concludes that most schemes typically benefit the privileged, help little to increase mass cycling transport and are used as easily deployable technological (false) solutions to contemporary problems, while advancing unjust tendencies to privatise public space and services.

4.2.2. Stories of Success

Nonetheless, as the primary data research has indicated there is much more than failure, misfortune and negativity associated with bike-sharing operations; there is also hope and genuine potential for improvement that could help to restore the image of bike-sharing. There are many examples indicating that bike-sharing can still be a viable transport option inspiring modal shift and making bicycle usage more popular and mainstream.

The government of the city of Hangzhou in China launched Hangzhou Public Bicycle in January 2008 starting with 2800 bicycles, 30 fixed stations, and 30 mobile stations (stations which can be moved to meet demand). This scheme went on to become the world's largest bike-sharing system with 100,000 bicycles and 4100 stations, as of December 2018. This scheme has been successful in acting as a complement to existing public transit and as a tool for modal shift; members exhibited a higher rate of auto ownership than non-members meaning that bike-sharing was attractive to car owners [63]. Some of its key success factors are: the low subscription fee, the availability of bicycles throughout the city, the subsidies from the local authorities, the fact that this was an initiative that the local transport agency created and is still a not-for-profit scheme (i.e., riding is free for the first 60 min), its complete integration with the other public transport services, its high-quality real-time information system, the upgrade of the bike hardware and the existence of a green corridor that promotes cycling in general. Hangzhou, a city with a registered population exceeding 9 million, widely considered as an emerging technology hub (home to the e-commerce giant Alibaba) also hosts some dockless bike-sharing fleets from Mobike, Ofo, Hello Bike and Qibei. Researchers [64] argue that Hangzhou Public Bicycle has already become rooted in the city as one of the public transportation modes, and because of its stable performance and the city's features fit for cycling, the habit of riding bikes has been awakened in Hangzhou. This has, in turn, provided an ideal environment for free-floating bikes to come into use.

Introduced in 2011, originally as Barclays Cycle Hire, London's bike-sharing scheme is a station-based system that can be accessed by anyone with a credit or debit card, with daily usage charged at £2 for unlimited journeys of up to 30 min. Santander Cycles offer the option of annual membership charged at £90 (~25p per day). The scheme, after incremental strategic expansions, now spreads across 100 km^2 of London and is the largest cycle hire scheme in Europe, with 11,500 bicycles available across 750 docking stations [65]. It has approximately 240,000 active members making over 10 million annual bicycle hires, and its continued expansion is viewed as a central component of the Mayor's policy to transform London's transport system into one which is based on sustainable modes [66]. Therefore, the scheme is subsidised and promoted heavily by Transport for London and is well-linked with all the public transport modes of the city. Santander Cycles have demonstrated the capacity to normalise the practice of cycling in city life; its users are not solely representatives of particular social cohorts such as sporty people [67]. Also, the scheme has become more equitable over time; it encourages women to use it, and with the eastern extension increased the share of trips made by residents in poorer areas, features that have been partly offset by increased prices [68]. As a whole, this is an award-winning intervention recognised not just for its impact on transport in England's capital city and its sustainability value, but also for its innovative design, the public relations exercise and the accurate delivery timescales.

Dublinbikes is a public bicycle rental scheme, which has operated in Dublin, Ireland since June 2009. At its launch, the scheme, which is sponsored by JCDecaux in exchange for 72 free advertising spaces around Dublin, used 450 bicycles organised in 40 stations. Now, the scheme has 114 stations and 42,000 active annual subscribers, and is one of the cheapest schemes in Europe with a €25 annual fee. Dublinbikes is widely considered as one of the most successful local applications in the world, as reported regularly in media [14,69,70], however occasionally there were reports suggesting that progress has stalled. Synergies with industry have been notable for the Dublin scheme. Coca-Cola Zero was a commercial partner with Dublinbikes for three years (June 2014 to June 2017), with Just Eat taking over in July 2017 for the next three years with plans to invest €2.25 million in the scheme over its tenure; in both cases the name of the scheme was rebranded accordingly.

Bicing, the local scheme of Barcelona, Spain, inaugurated in March 2007, is another system that is broadly regarded as successful. With 424 stations situated every 300 to 400 metres across the city, and more than 6,000 bicycles and electric bicycles, the initiative managed and maintained by the City Council and Clear Channel is an inexpensive option (annual fees range from €35 to €50). One of its unique features is that it is not a scheme for tourist use; it is open only to local subscribers and is marketed as an ideal complement to the traditional public transport of the city, intended to cover the small daily journeys that take place in Barcelona. A sign of early success was the fact that the usage and expansion targets were accomplished surprisingly fast. Bicing's initial foresight referred to reaching 15,000 subscribers at the end of the first year of operations, and 400 stations after a 10-year cycle, but the reality was that almost 100,000 subscribers subscribed in less than 12 months from the scheme's launch while the 10-year expansion plan happened in only 1.5 years [44]. The scheme is financed by subscriptions and by the local on-street parking control system profits. The system has been viewed as one that allows, through its digital footprints, the possibility to gain an understanding of human behaviour and city dynamics [71], and as a system with greater benefits than risks to health that reduces carbon dioxide emissions [72].

BIXI Montréal is a bike-sharing scheme serving Montreal, Canada. It launched its operations in May 2009, originally managed by Public Bike System Company (PBSC), to become North America's first large-scale bike-sharing scheme and an award-winning innovation. However, PBSC had to file for bankruptcy at the beginning of 2014 after problems including program mismanagement, breach of contract litigation and the surmounting of debt. BIXI Montréal from that point forward became a non-profit organisation owned and managed by the city of Montreal with 7250 bikes and 600 stations. The annual subscription costs $94. BIXI members can get a $15 discount on the local car-sharing service; packaging bike-sharing with car-sharing services is used as a promotion tool. BIXI attracts a substantial fraction of the population, accounting for more than one million trips annually [73] and is more likely to attract younger and more educated people who currently use cycling as a primary transportation mode [74]. The implementation of BIXI was associated with a shift toward active transportation, even if modal shift was complex and not simply the result of a discrete transition from one mode to another [75]. Also, research found that the accessibility of the bike-sharing docking stations in neighbourhoods was high, despite awareness inequalities that have decreased over time [76], something that is a key for increased usage. The city of Montreal announced that the service would be expanding in 2019 with 60 new stations offering 2625 docking points and 1000 additional bicycles in new areas.

Some of the schemes that have been labeled as good practice examples include among many: the French schemes Vélo'v in Lyon and Bicloo in Nantes with some of the highest annual growth rates in rentals and subscribers; many of the China-based operations of Mobike and Ofo; New York's (US) CitiBike system that averages 8.3 trips per bike and 42.7 trips per 1000 residents; Mexico City's (Mexico) Ecobici bike-share system and JUMP electric bikes and scooters that have spread in 19 cities in US and four in Europe.

5. Lessons to Be Learnt

The primary data analysis provides strong statistical evidence that bike-sharing is still a timely and meaningful proposition for urban policy-makers, and that it is widely accepted even from those citizens not expected to be scheme subscribers or occasional users. The survey respondents coming from two very different urban environments, in terms of size, socio-economic characteristics and pro-cycling culture, support in an almost identical degree (at around 90% of the sample) the notion that their respective cities should invest (i.e., Drama) or continue to invest (i.e., Gothenburg) in local bike-sharing schemes. This clearly reflects the acceptance of the bike-sharing concept and the general public willingness to see local city-specific schemes supported by their city hosts. Usage (or intended usage) rates are not mirroring acceptability rates closely; they are significantly lower. If there is a critical mass of scheme subscribers, however, this work makes the case that usage should not always be considered as the sole success parameter of a scheme; acceptability is equally important. The author also identified some of the key reasons that may make people reluctant to bike-share with traffic safety and insufficient bike-friendly infrastructure concerns being the two most critical.

The case study analysis identifies that inflexible standardised business models and operation strategies lacking the ability to tailor their offering to different areas, largely adopted from China, are not suitable for all urban environments. Dock-based systems, despite their inability to provide door-to-door services, seem to do better than dockless schemes for now, although the two can complement each other and work together, as seen in Hangzhou. Expansion tactics that have been over-aggressive, defied established competition, did not actively seek the collaboration and support of the hosting cities and were not adequately justified by travel demand data have failed. Schemes that were tailored to the city host needs, were integrated with or complemented the public transport provision, had a clear incremental expansion strategy, kept relatively inexpensive prices, secured the support of the local authorities, forged synergies with commercial partners and embraced technology have been successful and set the bar for the industry.

Some key recommendations that this work can provide, primarily to operators but also to cities looking to host sustainable and long-lived bike-sharing schemes, that will add value to their image suggest that they should:

- Make unique city-specific plans for delivering each scheme. The "one business model fits all" approach is fundamentally flawed. The ability to tailor, to some degree, an offering to reflect the character of a city and the norms of its citizens is critical.
- Prioritise the scheme's long-term success over easy profit and unrealistic revenue return expectations; bike-sharing should be user-centric and not profit-centric if it is to succeed.
- Realise that that the bulk of benefits that a bike-sharing scheme can deliver refers to avoiding the negative externalities of excessive car usage. This is a profit worth paying so operators should actively seek the support of the cities when possible in subsidies and supporting infrastructure. Profits from subscriptions and rentals may not be enough for a scheme's long-term commercial viability so there is a need to establish other streams of funding.
- Work together (operators and cities) so that the latter will be incentivised to support their local schemes. Private–public partnerships can work.
- Seek strategic commercial collaborations like the London and Dublin schemes. Extra financial support and a brand adding value might be a key in sustaining business.
- Acknowledge that an oversaturated bike-sharing market can be lethal. Travel supply should mirror travel demand. This is critical, especially for smaller providers, although bigger operators could also face problems as evidenced by the examples of Bluegogo primarily and Ofo to a smaller extent. When there is established competition, a new scheme needs to offer a different "twist" to what is already available (e.g., electric bicycles) in order to have success potential.
- Invest more efforts in regulating the responsible usage of bikes. Anti-social behavior, theft and vandalism have plagued many schemes and led to their closure. Protection mechanisms and penalties for vandalism and theft should be in place from day one. Cities should support these efforts with better policing.
- Initiate market research and education campaigns to understand better users and non-users and promote bike-sharing culture to encourage people to adopt a positive attitude towards these bikes. Cultural engagement from the outset is a prerequisite for success.
- Focus on providing fair and affordable fares, member subscriptions, and "ways out".
- Finance better business planning and bike management; companies need to constantly innovate to stay on top of this very dynamic market.
- Acknowledge that dock-based systems, for now at least, have been more successful in securing long-term viability than most dockless schemes. The option of docking stations, and enough

people on the ground to ensure that schemes are reliable and serve their purpose, should be provided when possible.

- Appreciate that technology is not always a panacea; it is only one of the several tools in the toolbox of successful bike-sharing provision [77]. However, mobile apps, rental machines, GPS tracking, and locking systems are of vital importance for scheme success and should be modern and user-friendly.
- Concentrate on providing better bike-sharing infrastructure; bikes (and stations when applicable) should be well-designed, attractive, safe and prevent anti-social behaviour.
- Diversify the provision of bikes so that some of them are more suitable for usually underserved populations. More specifically, explore ways to increase female participation in schemes by offering some bikes with more feminine designs. Also, introduce electric bicycles, pedal assist systems, tricycles and other inclusive vehicles, and make bike-sharing technologies more easily accessible (e.g., easier pick up and drop off services) as another way [11] to enable some individuals, and especially older people who can be open to a transport intervention if they feel that it is pro-social [78], to engage.
- Collaborate with city and even national authorities to resolve concerns referring to limited traffic safety for cyclists and inadequate pro-cycling infrastructure provision. Complementary to the schemes, bike-friendly road infrastructure (e.g., bike lanes, bike racks, bike prioritisation) and wider bicycling investments are very important for supporting bike-sharing and alleviating traffic safety concerns.
- Manage the distribution of bikes more effectively and responsibly. Operators need to be more accountable about visual pollution. In exchange, cities could be more flexible with their regulations (e.g., easing helmet use regulations or supplying shared use helmet schemes).
- Help and push the relevant decision-makers to establish more pro-cycling national and local legislation and governance.

6. Conclusions

This work provides an evidence-based roadmap, generated by blending two "twin" survey studies and a best-practice versus worst-practice international case study comparison. This mixed method approach intends to help a still growing, but somewhat jeopardised and stigmatised, mobility innovation to avoid a path that leads to an unsustainable and short-lived future. This paper disengages "scheme success" from a strict usage rate perspective, and informs policy-makers and scheme suppliers that citizens want to see systems that deliver sustainability benefits being supported by the local authorities. This result, combined with field evidence that bike-sharing schemes could be underused and thus may not be commercially profitable when income depends solely on subscription and short rental rates, signifies the need for establishing strong links and commercial partnerships with the city hosts and private industries interested to associate their brand (and pay for it) with bike-sharing, respectively. The city support can therefore take the form not only of complementary cycling infrastructure investment and more bike-friendly legislation, but also of direct city funding under the precondition that schemes meet local authority and general public expectations.

More importantly though, this study generates policy and business lessons "reading carefully through" the current public bicycle practice, by identifying successful and unsuccessful bike-sharing implementation cases. These lessons include, among others, the need for: tailored, according to the explicit host requirements, system design and expansion strategy; city-operator and commercial partner synergies; more bike-friendly infrastructure and legislation; pro-active cultural engagement of the local communities; anti-theft and anti-vandalism measures; easy to understand and fair usage terms; enhanced fleet management; and realistic profit expectations.

All in all, this paper's intended contribution is to function as a policy-minded survival guide for establishing, running and securing the future financial and operational viability of bike-sharing schemes, something that could be of significant value to academics, mobility providers and policy-makers.

Funding: No direct external funding is associated with this work.

Acknowledgments: The city of Drama is motivated to introduce bike-sharing based on the results of the relevant study presented in Section 4.1.2.

Conflicts of Interest: No conflicts of interest to report.

References

1. Nikitas, A.; Wallgren, P.; Rexfelt, O. The paradox of public acceptance of bike sharing in Gothenburg. *Proc. Inst. Civ. Eng.-Eng. Sustain.* **2016**, *169*, 101–113. [CrossRef]
2. Thomopoulos, N.; Nikitas, A. Editorial: Smart urban mobility futures. *Int. J. Automot. Technol. Manag.* **2019**, *19*, 1–9.
3. Mueller, N.; Rojas-Rueda, D.; Cole-Hunter, T.; de Nazelle, A.; Dons, E.; Gerike, R.; Goetschi, T.; Panis, L.I.; Kahlmeier, S.; Nieuwenhuijsen, M. Health impact assessment of active transportation: A systematic review. *Prev. Med.* **2015**, *76*, 103–114. [CrossRef] [PubMed]
4. Şimşekoğlu, Ö.; Nordfjærn, T.; Rundmo, T. Predictors of car use habit strength in an urban Norwegian public. *Transportation* **2017**, *44*, 575–588. [CrossRef]
5. Buehler, R.; Pucher, J.; Gerike, R.; Götschi, T. Reducing car dependence in the heart of Europe: lessons from Germany, Austria, and Switzerland. *Transp. Rev.* **2017**, *37*, 4–28. [CrossRef]
6. Gärling, T.; Gärling, A.; Loukopoulos, P. Forecasting psychological consequences of car use reduction: a challenge to an environmental psychology of transportation. *Appl. Psychol.* **2002**, *51*, 90–106. [CrossRef]
7. Nikitas, A.; Wang, J.Y.; Knamiller, C. Exploring parental perceptions about school travel and walking school buses: A thematic analysis approach. *Transp. Res. Part A Policy Pract.* **2019**, *124*, 468–487. [CrossRef]
8. Banister, D. The sustainable mobility paradigm. *Transp. Policy* **2008**, *15*, 73–80. [CrossRef]
9. Russo, F.; Rindone, C.; Panuccio, P. European plans for the smart city: From theories and rules to logistics test case. *Eur. Plan. Stud.* **2016**, *24*, 1709–1726. [CrossRef]
10. Russo, F.; Rindone, C.; Panuccio, P. The process of smart city definition at an EU level. WIT Transact. *Ecol. Environ.* **2014**, *191*, 979–989.
11. Nikitas, A. Understanding bike-sharing acceptability and expected usage patterns in the context of a small city novel to the concept: A story of Greek Drama. *Transp. Res. Part F Traffic Psychol. Behav.* **2018**, *56*, 306–321. [CrossRef]
12. DeMaio, P. Bike-sharing: History, impacts, models of provision, and future. *J. Public Transp.* **2009**, *12*, 41–56. [CrossRef]
13. Shaheen, S.; Chan, N. Mobility and the sharing economy: Potential to facilitate the first-and last-mile public transit connections. *Built Environ.* **2016**, *42*, 573–588. [CrossRef]
14. Caulfield, B.; O'Mahony, M.; Brazil, W.; Weldon, P. Examining usage patterns of a bike-sharing scheme in a medium sized city. *Transp. Res. Part A Policy Pract.* **2017**, *100*, 152–161. [CrossRef]
15. Gu, T.; Kim, I.; Currie, G. To be or not to be dockless: Empirical analysis of dockless bikeshare development in China. *Transp. Res. Part A Policy Pract.* **2019**, *119*, 122–147. [CrossRef]
16. Xu, Y.; Chen, D.; Zhang, X.; Tu, W.; Chen, Y.; Shen, Y.; Ratti, C. Unravel the landscape and pulses of cycling activities from a dockless bike-sharing system. *Comput. Environ. Urban Syst.* **2019**, *75*, 184–203. [CrossRef]
17. Meddin, R.; DeMaio, P. The Bike-Sharing World Map. Available online: http://www.bikesharingworld.com (accessed on 22 May 2019).
18. Nikitas, A.; Rahe, U.; Karjalainen, T.M. Encouraging sustainable urban access: An exploratory student approach to design of product service systems. In Proceedings of the 15th International Conference on Engineering and Product Design Education "Growing our Future", Dublin, Ireland, 5–6 September 2013.
19. Nikitas, A.; Wallgren, P.; Rahe, U. Public bicycles: How the concept of human-oriented "mobility sharing" technology can influence travel behaviour norms and reshape design education. In Proceedings of the 16th International Conference on Engineering and Product Design Education "Design Education & Human Technology Relations", Enschede, The Netherlands, 4–5 September 2014.
20. Shokoohi, R.; Nikitas, A. Urban growth, and transportation in Kuala Lumpur: Can cycling be incorporated into Kuala Lumpur's transportation system? *Case Stud. Transp. Policy* **2017**, *5*, 615–626. [CrossRef]

21. Cirianni, F.; Monterosso, C.; Panuccio, P.; Rindone, C. A Review Methodology of Sustainable Urban Mobility Plans: Objectives and Actions to Promote Cycling and Pedestrian Mobility. In Proceedings of the International Conference on Smart and Sustainable Planning for Cities and Regions, Bozen, Italy, 22–24 March 2017; Springer: Cham, Switzerland, 2018; pp. 685–697.
22. Handy, S.; Van Wee, B.; Kroesen, M. Promoting cycling for transport: research needs and challenges. *Transp. Rev.* **2014**, *34*, 4–24. [CrossRef]
23. García-Palomares, J.C.; Gutiérrez, J.; Latorre, M. Optimizing the location of stations in bike-sharing programs: A GIS approach. *Appl. Geogr.* **2012**, *35*, 235–246. [CrossRef]
24. Bakogiannis, E.; Vassi, A.; Siti, M.; Christodoulopoulou, G. Developing a sustainable mobility plan in Piraeus with special emphasis on cycling. *J. Traffic Transp. Eng.* **2016**, *4*, 61–74.
25. Shaheen, S.A.; Guzman, S.; Zhang, H. Bikesharing in Europe, the Americas, and Asia: Past, present, and future. *Transp. Res. Rec.* **2010**, *2143*, 159–167. [CrossRef]
26. Qiu, L.Y.; He, L.Y. Bike sharing and the economy, the environment, and health-related externalities. *Sustainability* **2018**, *10*, 1145. [CrossRef]
27. Wang, M.; Zhou, X. Bike-sharing systems and congestion: Evidence from US cities. *J. Transp. Geogr.* **2017**, *65*, 147–154. [CrossRef]
28. Otero, I.; Nieuwenhuijsen, M.J.; Rojas-Rueda, D. Health impacts of bike sharing systems in Europe. *Environ. Int.* **2018**, *115*, 387–394. [CrossRef] [PubMed]
29. Zhang, Y.; Mi, Z. Environmental benefits of bike sharing: A big data-based analysis. *Appl. Energy* **2018**, *220*, 296–301. [CrossRef]
30. Ricci, M. Bike sharing: A review of evidence on impacts and processes of implementation and operation. *Res. Transp. Bus. Manag.* **2015**, *15*, 28–38. [CrossRef]
31. Li, Y.; Zhu, Z.; Guo, X. Operating Characteristics of Dockless Bike-Sharing Systems near Metro Stations: Case Study in Nanjing City, China. *Sustainability* **2019**, *11*, 2256. [CrossRef]
32. Shi, J.G.; Si, H.; Wu, G.; Su, Y.; Lan, J. Critical factors to achieve dockless bike-sharing sustainability in China: A stakeholder-oriented network perspective. *Sustainability* **2018**, *10*, 2090. [CrossRef]
33. van Waes, A.; Farla, J.; Frenken, K.; de Jong, J.P.; Raven, R. Business model innovation and socio-technical transitions. A new prospective framework with an application to bike sharing. *J. Clean. Prod.* **2018**, *195*, 1300–1312. [CrossRef]
34. Manzi, G.; Saibene, G. Are they telling the truth? Revealing hidden traits of satisfaction with a public bike-sharing service. *Int. J. Sust. Transp.* **2018**, *12*, 253–270. [CrossRef]
35. Rahim Taleqani, A.; Hough, J.; Nygard, K.E. Public opinion on dockless bike sharing: A machine learning approach. *Transp. Res. Rec.* **2019**, 0361198119838982. [CrossRef]
36. Tu, Y.; Chen, P.; Gao, X.; Yang, J.; Chen, X. How to make dockless bikeshare good for cities: Curbing oversupplied bikes. *Transp. Res. Rec.* **2019**, 0361198119837963. [CrossRef]
37. Kaspi, M.; Raviv, T.; Tzur, M. Bike-sharing systems: User dissatisfaction in the presence of unusable bicycles. *IISE Transact.* **2017**, *49*, 144–158. [CrossRef]
38. Wang, Z.; Zheng, L.; Zhao, T.; Tian, J. Mitigation strategies for overuse of Chinese bikesharing systems based on game theory analyses of three generations worldwide. *J. Clean. Prod.* **2019**, *227*, 447–456. [CrossRef]
39. Wang, K.; Akar, G.; Chen, Y.J. Bike sharing differences among millennials, Gen Xers, and baby boomers: Lessons learnt from New York City's bike share. *Transp. Res. Part A Policy Pract.* **2018**, *116*, 1–14. [CrossRef]
40. Shen, Y.; Zhang, X.; Zhao, J. Understanding the usage of dockless bike sharing in Singapore. *Int. J. Sustain. Transp.* **2018**, *12*, 686–700. [CrossRef]
41. de Chardon, C.M.; Caruso, G.; Thomas, I. Bicycle sharing system 'success' determinants. *Transp. Res. Part A Policy Pract.* **2017**, *100*, 202–214. [CrossRef]
42. Zhang, L.; Zhang, J.; Duan, Z.Y.; Bryde, D. Sustainable bike-sharing systems: characteristics and commonalities across cases in urban China. *J. Clean. Prod.* **2015**, *97*, 124–133. [CrossRef]
43. Fishman, E. Bikeshare: A review of recent literature. *Transp. Rev.* **2016**, *36*, 92–113. [CrossRef]
44. Anaya, E. Are public bikes systems good for cycling? The view from Barcelona. *Work* **2010**, *56*, 7.
45. Jia, L.; Liu, X.; Liu, Y. Impact of different stakeholders of bike-sharing industry on users' intention of civilized use of bike-sharing. *Sustainability* **2018**, *10*, 1437. [CrossRef]
46. Castillo-Manzano, J.I.; Sánchez-Braza, A. Can anyone hate the bicycle? The hunt for an optimal local transportation policy to encourage bicycle usage. *Environ. Politics* **2013**, *22*, 1010–1028. [CrossRef]

47. Cerutti, P.S.; Martins, R.D.; Macke, J.; Sarate, J.A.R. "Green, but not as green as that": An analysis of a Brazilian bike-sharing system. *J. Clean. Prod.* **2019**, *217*, 185–193. [CrossRef]
48. Si, H.; Shi, J.G.; Wu, G.; Chen, J.; Zhao, X. Mapping the bike sharing research published from 2010 to 2018: A scientometric review. *J. Clean. Prod.* **2018**, *213*, 415–427. [CrossRef]
49. Shiono, P.H.; Klebanoff, M.A. The effect of two mailing strategies on the response to a survey of physicians. *Am. J. Epidem.* **1991**, *134*, 539–542. [CrossRef] [PubMed]
50. Gilbart, E.; Kreiger, N. Improvement in cumulative response rates following implementation of a financial incentive. *Am. J. Epidemiol.* **1998**, *48*, 97–99. [CrossRef] [PubMed]
51. Nikitas, A.; Sochor, J. Analysing the Acceptability and Usage Patterns of Bike-sharing in a City with an Established Pro-cycling Culture. In Proceedings of the 15th International Conference on Travel Behavior Research (IATBR 2018), Santa Barbara, CA, USA, 15–20 July 2018.
52. Small, A. City Lab Article: The Four Horsemen of the Bike Share Apocalypse. Available online: https://www.citylab.com/transportation/2017/01/seattle-bike-share-pronto-goes-under/513575/ (accessed on 8 April 2019).
53. Sun, F.; Chen, P.; Jiao, J. Promoting public bike-sharing: A lesson from the unsuccessful Pronto system. *Transp. Res. Part D Transp. Environ.* **2018**, *63*, 533–547. [CrossRef]
54. Mobike. Press Release: World's Largest Smart Bicycle Sharing Company Rides into Pune, India. Available online: https://mobike.com/global/public/MobikePuneIndia.pdf (accessed on 21 May 2019).
55. Griswold, A. Quarz Article: Bike-sharing Company Ofo Is Dramatically Scaling back in North America. Available online: https://qz.com/1331368/bike-sharing-company-ofo-is-dramatically-scaling-back-in-north-america/ (accessed on 21 May 2019).
56. Gillin, A. GJEL Accident Attorneys Article: Bankrupt Bluegogo Auctions Leftover Bikes for as Little as $1 Each. Available online: https://www.gjel.com/blog/bicycling-information/bankrupt-bluegogo-auctions-leftover-bikes-little-1.html (accessed on 8 April 2019).
57. Haas, B. The Guardian Article: Chinese Bike Share Graveyard a Monument to Industry's 'Arrogance'. Available online: https://www.theguardian.com/uk-news/2017/nov/25/chinas-bike-share-graveyard-a-monument-to-industrys-arrogance?page=with%3Aimg-3 (accessed on 9 April 2019).
58. Nikitas, A. The Conversation Article: Bike-Sharing Fiascoes and How to Avoid Them—An Expert's Guide. Available online: https://theconversation.com/bike-sharing-fiascoes-and-how-to-avoid-them-an-experts-guide-84926 (accessed on 8 April 2019).
59. Pidd, H. The Guardian Article: Mobike Pulls Out of Manchester Citing Thefts and Vandalism. Available online: https://www.theguardian.com/uk-news/2018/sep/05/theft-and-vandalism-drive-mobike-out-of-manchester (accessed on 20 May 2019).
60. Kollewe, J.; McIntyre, N. The Guardian Article: Ofo Cycle Hire Firm Pulls Out of London. Available online: https://www.theguardian.com/uk-news/2019/jan/10/ofo-cycle-hire-firm-pulls-out-of-london (accessed on 22 May 2019).
61. Agence France-Presse. The Guardian Article: Gobee Bike Pulls Out of France Due to 'Mass Destruction' of Its Dockless Bike Fleet. Available online: https://www.theguardian.com/world/2018/feb/25/gobeebike-france-mass-destruction-dockless-bikes (accessed on 8 April 2019).
62. de Chardon, C.M. The contradictions of bike-share benefits, purposes and outcomes. *Transp. Res. Part A Policy Pract.* **2019**, *121*, 401–419. [CrossRef]
63. Shaheen, S.A.; Zhang, H.; Martin, E.; Guzman, S. China's Hangzhou public bicycle: Understanding early adoption and behavioral response to bikesharing. *Transp. Res. Rec.* **2011**, *2247*, 33–41. [CrossRef]
64. Chen, M.; Wang, D.; Sun, Y.; Waygood, E.O.D.; Yang, W. A comparison of users' characteristics between station-based bikesharing system and free-floating bikesharing system: Case study in Hangzhou, China. *Transportation* **2018**, *45*, 1–16. [CrossRef]
65. Transport for London. Cycling: Santander Cycles. Available online: https://tfl.gov.uk/modes/cycling/santander-cycles/find-a-docking-station?intcmp=2321 (accessed on 8 April 2019).
66. Morton, C. Appraising the market for bicycle sharing schemes: Perceived service quality, satisfaction, and behavioural intention in London. *Case Stud. Transp. Policy* **2018**, *6*, 102–111. [CrossRef]
67. Goodman, A.; Green, J.; Woodcock, J. The role of bicycle sharing systems in normalising the image of cycling: An observational study of London cyclists. *J. Transp. Health* **2014**, *1*, 5–8. [CrossRef] [PubMed]

68. Goodman, A.; Cheshire, J. Inequalities in the London bicycle sharing system revisited: impacts of extending the scheme to poorer areas but then doubling prices. *J. Transp. Geogr.* **2014**, *41*, 272–279. [CrossRef]
69. Daly, M. The Guardian Article: What's the Secret of the Dublin Bike Hire Scheme's Success? Available online: https://www.theguardian.com/money/2017/feb/25/best-and-worst-city-cycle-schemes-bike-sharin-london (accessed on 10 April 2019).
70. Collinson, P. The Guardian Article: On Your Bike: The Best and the Worst of City Cycle Schemes? Available online: https://www.theguardian.com/environment/bike-blog/2011/aug/04/dublin-bike-hire-scheme (accessed on 10 April 2019).
71. Froehlich, J.; Neumann, J.; Oliver, N. Measuring the pulse of the city through shared bicycle programs. In Proceedings of the International Workshop on Urban, Community, and Social Applications of Networked Sensing System (UrbanSense08), Raleigh, NC, USA, 4 November 2008; pp. 16–20.
72. Rojas-Rueda, D.; de Nazelle, A.; Tainio, M.; Nieuwenhuijsen, M.J. The health risks and benefits of cycling in urban environments compared with car use: Health impact assessment study. *BMJ* **2011**, *343*, d4521. [CrossRef] [PubMed]
73. Faghih-Imani, A.; Eluru, N.; El-Geneidy, A.M.; Rabbat, M.; Haq, U. How land-use and urban form impact bicycle flows: Evidence from the bicycle-sharing system (BIXI) in Montreal. *J. Transp. Geogr.* **2014**, *41*, 306–314. [CrossRef]
74. Fuller, D.; Gauvin, L.; Kestens, Y.; Daniel, M.; Fournier, M.; Morency, P.; Drouin, L. Use of a new public bicycle share program in Montreal, Canada. *Am. J. Prev. Med.* **2011**, *41*, 80–83. [CrossRef] [PubMed]
75. Fuller, D.; Gauvin, L.; Kestens, Y.; Morency, P.; Drouin, L. The potential modal shift and health benefits of implementing a public bicycle share program in Montreal, Canada. *Int. J. Behav. Nutr. Phys. Act.* **2013**, *10*, 66. [CrossRef] [PubMed]
76. Bernatchez, A.C.; Gauvin, L.; Fuller, D.; Dubé, A.S.; Drouin, L. Knowing about a public bicycle share program in Montreal, Canada: Are diffusion of innovation and proximity enough for equitable awareness? *J. Transp. Health* **2015**, *2*, 360–368. [CrossRef]
77. Sochor, J.; Nikitas, A. Vulnerable users' perceptions of transport technologies. *Proc. Inst. Civ. Eng. Urban Des. Plan.* **2016**, *169*, 154–162. [CrossRef]
78. Nikitas, A.; Avineri, E.; Parkhurst, G. Understanding the public acceptability of road pricing and the roles of older age, social norms, pro-social values and trust for urban policy-making: The case of Bristol. *Cities* **2018**, *79*, 78–91. [CrossRef]

MDPI
St. Alban-Anlage 66
4052 Basel
Switzerland
Tel. +41 61 683 77 34
Fax +41 61 302 89 18
www.mdpi.com

Sustainability Editorial Office
E-mail: sustainability@mdpi.com
www.mdpi.com/journal/sustainability

www.ingramcontent.com/pod-product-compliance
Lightning Source LLC
Chambersburg PA
CBHW040808250926
51850CB00007BA/1117
9783036517063